# Miktoarm Star Polymers
From Basics of Branched Architecture to Synthesis, Self-assembly and Applications

**Polymer Chemistry Series**

*Editor-in-chief:*
Ben Zhong Tang, *The Hong Kong University of Science and Technology, Hong Kong, China*

*Series editors:*
Alaa S. Abd-El-Aziz, *University of Prince Edward Island, Canada*
Stephen L. Craig, *Duke University, USA*
Jianhua Dong, *National Natural Science Foundation of China, China*
Toshio Masuda, *Shanghai University, China*
Christoph Weder, *University of Fribourg, Switzerland*

*Titles in the series:*
1: Renewable Resources for Functional Polymers and Biomaterials
2: Molecular Design and Applications of Photofunctional Polymers and Materials
3: Functional Polymers for Nanomedicine
4: Fundamentals of Controlled/Living Radical Polymerization
5: Healable Polymer Systems
6: Thiol-X Chemistries in Polymer and Materials Science
7: Natural Rubber Materials: Volume 1: Blends and IPNs
8: Natural Rubber Materials: Volume 2: Composites and Nanocomposites
9: Conjugated Polymers: A Practical Guide to Synthesis
10: Polymeric Materials with Antimicrobial Activity: From Synthesis to Applications
11: Phosphorus-Based Polymers: From Synthesis to Applications
12: Poly(lactic acid) Science and Technology: Processing, Properties, Additives and Applications
13: Cationic Polymers in Regenerative Medicine
14: Electrospinning: Principles, Practice and Possibilities
15: Glycopolymer Code: Synthesis of Glycopolymers and their Applications
16: Hyperbranched Polymers: Macromolecules in-between Deterministic Linear Chains and Dendrimer Structures
17: Polymer Photovoltaics: Materials, Physics, and Device Engineering
18: Electrical Memory Materials and Devices
19: Nitroxide Mediated Polymerization: From Fundamentals to Applications in Materials Science
20: Polymers for Personal Care Products and Cosmetics
21: Semiconducting Polymers: Controlled Synthesis and Microstructure
22: Bio-inspired Polymers

23: Fluorinated Polymers: Volume 1: Synthesis, Properties, Processing and Simulation
24: Fluorinated Polymers: Volume 2: Applications
25: Miktoarm Star Polymers: From Basics of Branched Architecture to Synthesis, Self-assembly and Applications

*How to obtain future titles on publication:*
A standing order plan is available for this series. A standing order will bring delivery of each new volume immediately on publication.

*For further information please contact:*
Book Sales Department, Royal Society of Chemistry, Thomas Graham House, Science Park, Milton Road, Cambridge, CB4 0WF, UK
Telephone: +44 (0)1223 420066, Fax: +44 (0)1223 420247
Email: booksales@rsc.org
Visit our website at www.rsc.org/books

# *Miktoarm Star Polymers*

## *From Basics of Branched Architecture to Synthesis, Self-assembly and Applications*

Edited by

**Ashok Kakkar**
*McGill University, Montreal, Canada*
*Email: ashok.kakkar@mcgill.ca*

ROYAL SOCIETY OF CHEMISTRY

THE QUEEN'S AWARDS FOR ENTERPRISE: INTERNATIONAL TRADE 2013

Polymer Chemistry Series No. 25

Print ISBN: 978-1-78262-575-9
PDF eISBN: 978-1-78801-042-9
EPUB eISBN: 978-1-78801-125-9
ISSN: 2044-0790

A catalogue record for this book is available from the British Library



The Royal Society of Chemistry is a charity, registered in England and Wales, Number 207890, and a company incorporated in England by Royal Charter (Registered No. RC000524), registered office: Burlington House, Piccadilly, London W1J 0BA, UK, Telephone: +44 (0) 207 4378 6556.

For further information see our web site at www.rsc.org

Printed in the United Kingdom by CPI Group (UK) Ltd, Croydon, CR0 4YY, UK

# Preface

Miktoarm polymers constitute an intriguing and important class of macromolecules, in which the diversity in the key elements of the polymeric architecture *i.e.*, overall shape, well-defined molecular weight, and composition, offers a platform to develop materials for a wide variety of applications. Our ability to fine-tune the branched and multifunctional structure can provide insight into structure–property relationships toward the design of macromolecules with pre-determined properties. It is also of significance in the tailored and unique self-assembly behavior of these macromolecules in different media, which is of particular interest in constructing nanocarriers for biological applications.

Since the inception of the term 'miktoarm' in the early 1990s for asymmetric branched macromolecules, there has been a tremendous effort devoted to developing and simplifying the synthetic methodologies, as well as exploring their self-assembly aspects, and exploiting their unique properties for a variety of applications. This book brings together some of the eminent players in the field, and makes an attempt to provide a coherent and collective look at the state-of-the-art in miktoarm polymer research. Chapter 1 gives a historical perspective first, and then walks the reader through the synthetic developments in miktoarm polymers and miktoarm polymer-based complex architectures. Chapter 2 introduces the reader to a new iterative methodology using living anionic polymerization to synthesize multi-component miktoarm star polymers, including the evolution of the methodology into its second generation using the diphenylethylene anion. Chapter 3 explores the synthesis of multicomponent star copolymers using controlled polymerization and 'click chemistry' methods. The adoption of 'click chemistry' in the synthesis of miktoarm star polymers is then detailed in Chapter 4. In Chapter 5, the self-assembly behavior of miktoarm star polymers is compared to that

Polymer Chemistry Series No. 25
Miktoarm Star Polymers: From Basics of Branched Architecture to Synthesis, Self-assembly and Applications
Edited by Ashok Kakkar

Published by the Royal Society of Chemistry, www.rsc.org

of block-copolymers with particular emphasis on micellar and emulsion-assisted drug delivery. Chapter 6 provides a look at how the construction of miktoarm polymers can be tailored for applications in biology, especially for drug delivery. Finally, in Chapter 7, the reader is introduced to the concept of supramolecular chemistry; it then provides a detailed look at how this concept is applied to miktoarm star polymers in terms of their self-assembly aspects.

Miktoarm polymers continue to be a topical area of research. This class of macromolecules is still relatively young, but some success has been achieved and demonstrated in terms of their synthesis, self-assembly, and applications in biology in particular. This book brings forward the versatility and uniqueness of these branched architectures, and will be of great benefit to young budding chemists, as well as researchers exploring this vibrant class of macromolecules. It is hoped that this book will give the reader a foundation to work upon and, in addition, stimulate them to seek unexpected and not yet explored domains in their synthesis, self-assembly, and applications.

Ashok Kakkar

# Contents



Polymer Chemistry Series No. 25
Miktoarm Star Polymers: From Basics of Branched Architecture to Synthesis, Self-assembly and Applications
Edited by Ashok Kakkar

Published by the Royal Society of Chemistry, www.rsc.org

CHAPTER 1

# *Miktoarm Star (μ-Star) Polymers: A Successful Story*

HERMIS IATROU[a], APOSTOLOS AVGEROPOULOS[†b,c], GEORGIOS SAKELLARIOU[a], MARINOS PITSIKALIS[a] AND NIKOS HADJICHRISTIDIS[‡*a,b]

[a]University of Athens, Department of Chemistry, Industrial Chemistry Laboratory, Panepistimiopolis Zografou, 15771, Athens, Greece; [b]King Abdullah University of Science and Technology (KAUST), Physical Sciences and Engineering Division, KAUST Catalysis Center, Polymer Synthesis Laboratory, Thuwal, 23955, Saudi Arabia; [c]Department of Materials Science Engineering, University of Ioannina, University Campus-Dourouti, Ioannina 45110, Greece
*E-mail: nikolaos.hadjichristidis@kaust.edu.sa

## 1.1 The Genesis of Miktoarm (μ-Star) Star Polymers

Well-defined polymers with narrow molecular, structural, and compositional (in the case of copolymers) dispersity are essential for establishing structure or composition–property relationships and indispensable to accomplish one of the ultimate goals of polymer chemistry: designing macromolecules with predetermined properties/applications.

[†]On Sabbatical to KAUST from 9/2015 until 3/2016.
[‡]Professor Emeritus at the University of Athens from 10/2010.

Polymer Chemistry Series No. 25
Miktoarm Star Polymers: From Basics of Branched Architecture to Synthesis, Self-assembly and Applications
Edited by Ashok Kakkar

Published by the Royal Society of Chemistry, www.rsc.org

Among branched polymers, regular or symmetric stars consisting of several identical linear chains linked together at one chain-end initially attracted the attention of scientists since the star structure has the simplest form of branching. The earliest attempt to synthesize star polymers was that by Schaefgen and Flory in 1948,[1] who synthesized the first 4- and 8-arm star homopolymers (polyamides) by polymerizing ε-caprolactam in the presence of either cyclohexanone-tetrapropionic or dicyclohexanone-octacarboxylic acid.

Fourteen years later, Morton and coworkers,[2] taking advantage of the living character of anionic polymerization, succeeded to synthesize 4-arm star homopolystyrenes (PS) by 'terminating' living polystyryllithium with tetrachlorosilane (linking agent). Although the produced materials were mixtures of 3- and 4-arm PS, this work eventually led to the preparation of star polymers with up to 128 arms.

In 1963, Orofino and Wenger[3] were the first to use tri(chloromethyl) benzene in combination with anionic polymerization as a linking agent to prepare 3-arm star PS. Mayer[4] used 1,2,4,5-tetra-(chloromethyl)benzene to prepare 4-arm star di- and triblock copolymers of styrene and isoprene. It was difficult to extend the functionality ($f$) of stars beyond $f = 6$ with chloromethylbenzene derivatives due to the unavailability of chloromethylbenzene-based linking agents.[5] Other compounds used as linking agents, such as the cyclic trimer of phosphonitrilic chloride,[6] 2,4,6-tri(allyloxy)triazine,[7] 1,1,4,4-tetraphenyl-1,4-bis (diallyloxytriazine)butane,[8] tin tetrachloride,[9] and phosphorus trichloride,[10] suffer the same disadvantage. Decker and Rempp[11] demonstrated for the first time the validity of divinylbenzene (DVB) as a linking agent by preparing and properly characterizing PS stars with 6 to 15 arms. The DVB method was apparently first alluded to by Milkovich[12] but, unfortunately, in his patent there was no clear indication that star-branched polymers had been prepared. It should be noted that the DVB method does not allow the accurate control of the number of star arms since the polymerization of DVB (difunctional monomer) with the living chains is not well controlled.[13] Considering the disadvantages of the aforementioned compounds as linking agents, multifunctional chlorosilane compounds became the reagents of choice for the preparation of well-defined stars. Table 1.1 summarizes the evolution of the synthesis of symmetric star polymers with chlorosilane linking agents.

In 1989, Roovers and collaborators,[24] using a multifunctional linking agent designed/prepared by hydrosilylation of a low molecular weight linear or star 1,2-polybutadiene, succeeded to synthesize star polybutadienes (PB) with 200 and 270 arms. The exhaustive studies of the properties of these well-defined stars led to important conclusions concerning the influence of the star architecture on their properties in solution and bulk.[25] In addition, these model polymers were used to test the existing related theories.[26]

Many other interesting linking systems using cationic, group transfer, or living ring-opening metathesis polymerization were later developed, leading to symmetric star vinyl ethers,[27–31] isobutylenes,[32–36] methacrylates,[37] and norbornenes.[38,39]

**Table 1.1** Evolution of the synthesis of symmetric star polymers with various functionalities (*f*) from anionic polymerization and chlorosilane linking agents.

| Year | Author | Monomer | Functionality (*f*) |
|---|---|---|---|
| 1962 | Morton, Helminiak, Gadkary, Bueche[2] | Styrene | Mixt. of 3/4 |
| 1965 | Zelinski, Wofford[14] | Butadiene | 3, 4 |
| 1972 | Roovers, Bywater[15] | Styrene | 4 |
| 1974 | Roovers, Bywater[16] | Styrene | 6 |
| 1974 | Hadjichristidis, Roovers[17] | Isoprene | 4, 6 |
| 1978 | Hadjichristidis, Guyot, Fetters[18] | Isoprene | 8, 12 |
| 1980 | Hadjichristidis, Fetters[19] | Isoprene | 18 |
| 1983 | Roovers, Hadjichristidis, Fetters[20] | Styrene | 12, 18 |
| 1986 | Nguyen, Hadjichristidis, Fetters[21] | Styrene-*b*-isoprene | 18 |
| 1992 | Zhou, Hadjichristidis, Toporowski, Roovers[22] | Butadiene | 32 |
| 1993 | Roovers, Zhou, Toporowski, van der Zwan, Iatrou, Hadjichristidis[23] | Butadiene | 64, 128 |

Later on, the synthesis of stars with different arms either in molecular weight (molecular weight asymmetry; asymmetric stars) or chemistry (chemical asymmetry; miktoarm stars) was achieved (Scheme 1.1a and 1.1b). The term miktoarm stars (coming from the Greek word μικτός meaning *mixed*) was adopted by our group for stars with chemical asymmetry. The term heteroarm stars (hetero coming from the Greek word έτερος, meaning *the other*) is not appropriate for this class of polymers, since it does not convey the concept of a star composed of dissimilar arms. Later, the term was expanded to stars with molecular weight asymmetry, which can be considered miktoarm homopolymers. Stars having arms of similar chemical nature but different end-functional groups also belong to this category. Finally, topologically asymmetric stars are also μ-stars. They consist of diblock copolymer arms attached by different chain-ends to the star center. π-Shaped, H-shaped, super-H or pom-pom copolymers can be considered as double miktoarm stars (Scheme 1.1c).

Pennisi and Fetters[40] were the first to report the synthesis of 3-arm asymmetric star homopolymers of PB and PS. Mays[41] and our group,[42] based on the work of Fetters, were the first to synthesize a 3μ-star copolymer consisting of two polyisoprene (PI) and one PS arm[41] and a 3μ-star terpolymer of PS, PI, and polybutadiene (PBd or PB),[42] respectively. Later, different miktoarm stars were synthesized by anionic polymerization and selective chlorosilane-based linking chemistry, as reported mainly by our group,[43–61] representative examples of which are given in Scheme 1.1. The evolution of the synthesis of different miktoarm structures are summarized in Table 1.2.

Synthetic efforts based on anionic polymerization leading to miktoarm stars (μ-stars) are reviewed in this chapter. In addition, a few examples of the striking influence of the star structure on the morphology of block co- and terpolymers are given. The structures synthesized by anionic polymerization guided scientists working with other types of polymerization techniques, such as controlled/living, ring opening, catalytic, ring opening metathesis

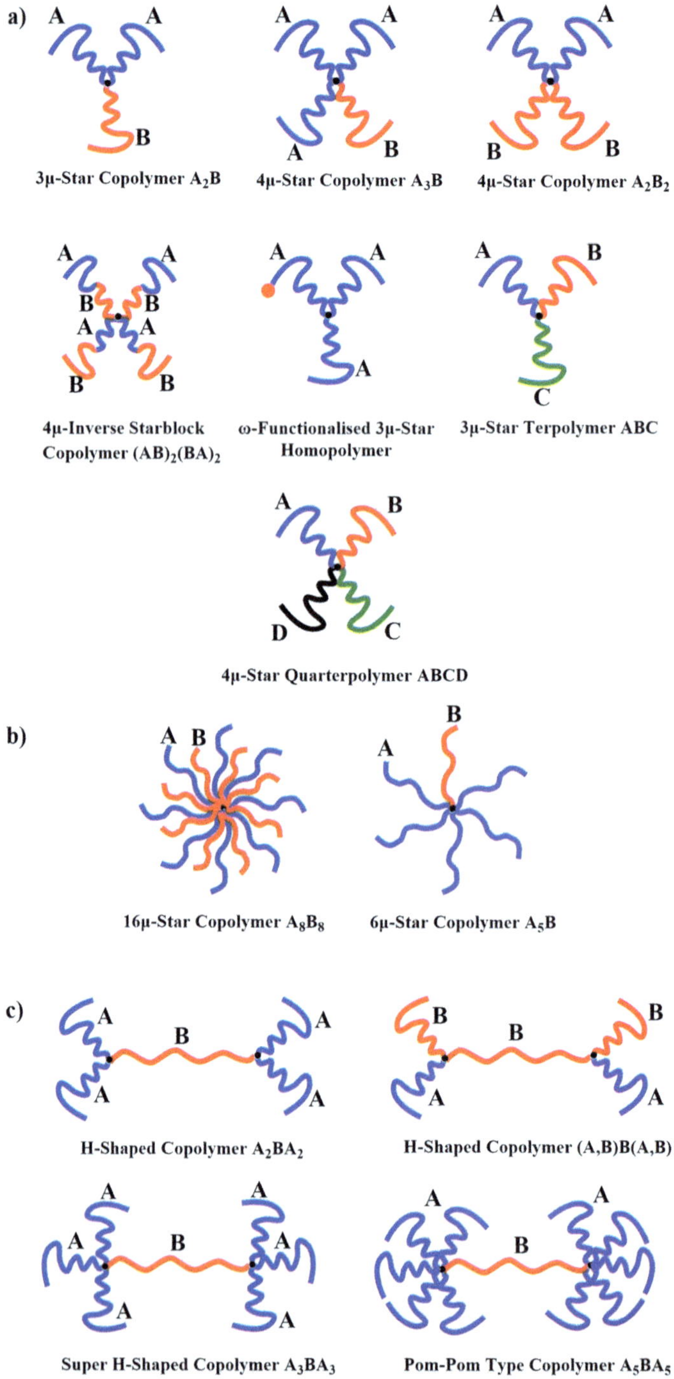


**Scheme 1.1** a) Miktoarm star (μ-star) polymers with 3 and 4-arms, (b) miktoarm star (μ-star) copolymers with more than 4 arms, and (c) double miktoarm star copolymers.

**Table 1.2** Evolution of the synthesis of miktoarm star polymers with various structures from anionic polymerization and chlorosilane linking agents.

| Year | Author | Monomers | Functionality/Architecture |
|---|---|---|---|
| 1988 | Pennisi, Fetters[40] | Styrene, butadiene | $3/(PS)_2PS'$, $(PBd)_2PBd'$ |
| 1990 | Mays[41] | Styrene, isoprene | $3/PS(PI)_2$ (one sample) |
| 1992 | Iatrou, Hadjichristidis[42] | Styrene, isoprene, butadiene | 3/(PS)(PI)(PBd) |
| 1993 | Iatrou, Hadjichristidis[43] | Styrene, isoprene, butadiene, 4-methylstyrene | $4/(PS)_2(PBd)_2$ and 4/(PS)(PI)(PBd) (P4MeS) |
| 1994 | Iatrou, Avgeropoulos, Hadjichristidis[44] | Styrene, isoprene | $6/(PI)_3PS(PI)_3$ |
| 1995 | Iatrou, Siakali-Kioulafa, Hadjichristidis, Roovers, Mays[45] | Styrene, isoprene, butadiene | $3/PS(PI)_2$, $PI(PS)_2$, $PS(PBd)_2$, and $PBd(PS)_2$ |
| 1995 | Pitsikalis, Hadjichristidis[46] | Butadiene | $3/(PBd)_2(PBd_F)$ |
| 1996 | Tselikas, Iatrou, Hadjichristidis, Liang, Mohanty, Lohse[47] | Styrene, isoprene | $3/PS(PI)_2$ and $4/PS(PI)_3$ |
| 1996 | Allgaier, Young, Efstratiadis, Hadjichristidis[48] | Isoprene, butadiene | $4/(PI)_2(PBd)_2$ |
| 1996 | Tselikas, Hadjichristidis, Lescanec, Honeker, Wohlgemuth, Thomas[49] | Styrene, isoprene | $4/(PI\text{-}b\text{-}PS)_2(PS\text{-}b\text{-}PI)_2$ (inverse μ-star-block) |
| 1996 | Avgeropoulos, Poulos, Hadjichristidis, Roovers[50] | Styrene, isoprene | $16/(PS)_8(PI)_8$ |
| 1997 | Avgeropoulos, Hadjichristidis[51] | Styrene, isoprene | $3/PS(PI\text{-}b\text{-}PS)_2$, $4/PS(PI\text{-}b\text{-}PS)_3$, $6/(PS\text{-}b\text{-}PI)_3PS(PI\text{-}b\text{-}PS)_3$ |
| 1997 | Sioula, Tselikas, Hadjichristidis[52] | Styrene, isoprene, methyl methacrylate (MMA) | 3/(PS)(PI)(PMMA) |
| 1999 | Velis, Hadjichristidis[53] | Styrene, isoprene | $6/PS(PI)_5$ |
| 2000 | Zioga, Sioula, Hadjichristidis[54] | Styrene, isoprene, 2-vinyl pyridine (2VP) | 3/(PS)(PI)(P2VP) |
| 2000 | Bellas, Iatrou, Hadjichristidis[55] | Styrene, isoprene, hexamethylcyclotrisiloxane | $3/(PS)_2PDMS$ and 3/(PS)(PI)(PDMS) |
| 2002 | Tsoukatos, Hadjichristidis[56] | 1,3-Cyclohexadiene, butadiene | $3/(PCHD)(PBd)_2$ and $4/(PCHD)(PBd)_3$ |
| 2003 | Mavroudis, Avgeropoulos, Hadjichristidis, Thomas, Lohse[57] | Styrene, 2-methyl-1,3-pentadiene (2MP) | $3/PS(P2MP)_2$ and $4/PS(P2MP)_3$ |
| 2006 | Mavroudis, Avgeropoulos, Hadjichristidis, Thomas, Lohse[58] | Styrene, 2-methyl-1,3-pentadiene (2MP) | $6/PS(P2MP)_5$ |
| 2006 | Fragouli, Iatrou, Hadjichristidis, Sakura, Hirao[59] | Hexamethylcyclotrisiloxane, 2-vinyl pyridine (2VP) | $3/(PDMS)_2(P2VP)$ |
| 2006 | Fragouli, Iatrou, Hadjichristidis, Sakura, Machunaga, Hirao[60] | Styrene, isoprene, hexamethylcyclotrisiloxane | $4/(PDMS)_2(PS)_2$, $4/(PDMS)_2(PS)(PI)$ and $5/(PDMS)_2(PS)_2(PI)$ |
| 2011 | Krkensgaard, Fragouli, Hadjichristidis, Mortensen[61] | Styrene, isoprene, hexamethylcyclotrisiloxane, 2-vinyl pyridine (2VP) | $3/(PS\text{-}b\text{-}PI\text{-}b\text{-}PDMS)_2P2VP$ |

polymerization reactions, *etc.*, towards the synthesis of star architectures. The tremendous influence of miktoarm stars on polymer science is evidenced by: (i) the need for a book on miktoarm stars, (ii) the high number of references miktoarm stars have produced in the last years (h-index = 70, references = 21 275, from our first paper to August 2016, source: ISI web of science), and (iii) the significant amount of novel nanostructures reported by this class of materials.

## 1.2 Synthesis of Miktoarm Star (μ-Star) Polymers

Two general strategies have been developed for the synthesis of miktoarm stars *via* living anionic polymerization. The first one is based on divinyl compounds, either homopolymerizables (*e.g.*, divinylbenzene) or non-homopolymerizables (*e.g.*, double diphenylethylene, DDPE), and the second one on multifunctional linking agents. Several linking agents have been used for the synthesis of star polymers.[62] The most commonly used ones are chlorosilanes,[2] adopted mostly for non-polar chains, and chloro-/bromomethyl benzenes,[3,4] adopted for polar ones. A few more complex linking agents have also been used in the synthesis of star-like copolymers.[63–65]

### 1.2.1 Divinylbenzene (Homopolymerizable Linking Agent)

The use of DVB for the synthesis of miktoarm stars was first recognized by Eschwey and Burchard[66] and developed mainly by Rempp and colleagues.[67–70] The general route is given in Scheme 1.2.

The living macroinitiator/precursor ALi polymerizes a small amount of DVB,[71] leading to the formation of a star molecule bearing a number of active sites within its core (microgel nodule of DVB), which is theoretically equal to the number of incorporated A arms of the star polymer. Subsequent addition of another monomer, B, or the same monomer A, yields a μ-star copolymer or asymmetric homopolymer star, respectively. The growing B arms have anionic sites at their outer ends, thus providing the possibility of reacting with electrophilic compounds or other monomers towards the preparation of end-functionalized stars or star-block copolymers. Because of its simplicity,

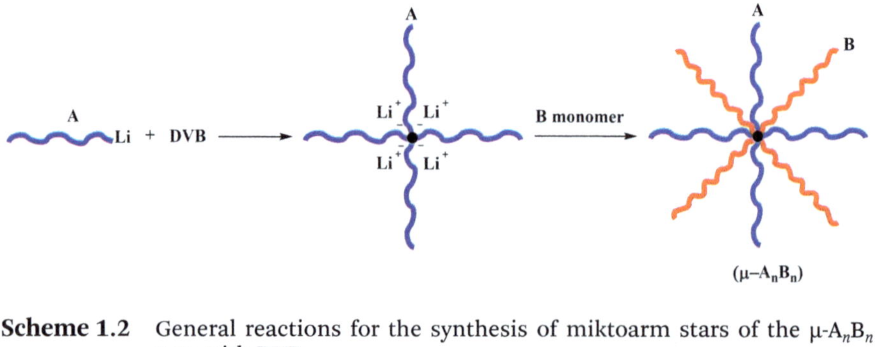


**Scheme 1.2** General reactions for the synthesis of miktoarm stars of the μ-$A_nB_n$ type with DVB.

this method can be carried out under inert atmosphere, thereby avoiding the use of highly demanding and time-consuming vacuum techniques. The DVB method has been applied in the synthesis of μ-star copolymers of the $A_nB_n$ type, with the A arm being polystyrene and the B arm poly(*tert*-butyl methacrylate),[68] poly(*tert*-butyl acrylate),[69–71] poly(ethylene oxide),[72] or poly(2-vinyl pyridine).[73] Usually, $n$ varies between 6 and 20. PS μ-star homopolymers of the type $A_nA'_n$ have also been prepared by this method.[74]

The DVB method is characterized by several disadvantages, the foremost being architectural limitations. Only stars with equal number of arms different in chemical nature or molecular weight, $A_nB_n$, can be prepared. In fact, since the polymerization of DVB by living chains is not controllable, $n$ is an average value influenced by several parameters. Specifically, $n$ increases with the decreasing molecular weight of the precursor A and the molar ratio of DVB to living chains. Another disadvantage is that the B arms cannot be isolated and characterized independently. Finally, reaction of the living ends with the remaining double bond of the DVB nodule can lead to the formation of loops (intramolecular reaction) or networks (intermolecular reaction).

### 1.2.2 Double Diphenylethylenes (Non-Homopolymerizable Linking Agents)

Hocker and Latterman[75] recognized in 1976 the usefulness of the addition of living chains to non-homopolymerizable divinyl compounds. They were the first to propose that 4μ-stars could be obtained by adding living chains to bis(1-phenylvinyl)benzenes, followed by the subsequent addition/polymerization of another monomer. In 1978, Szwarc and coworkers[76] studied the kinetics of addition of PSLi to several divinyl compounds in benzene by UV spectroscopy. They found that, in the case of the *para*-double diphenyl ethylene (PDDPE) 1,4-bis(1-phenylethenyl)benzene, the ratio of the rate constants of the first and second additions was equal to 13. In 1983, Leitz and Hocker[77] reported that the reaction of two moles of *sec*-BuLi with the *meta*-double diphenylethylene (MDDPE) 1,3-bis(1-phenylethenyl)benzene proceeds rapidly and efficiently to produce a dilithium initiator. The ratio of the rate constants of the first and second additions is almost identical in toluene.

Quirk and coworkers[78] have further developed this 'living linking method' for the synthesis of 3μ- and 4μ-stars. The general reactions are given in Schemes 1.3 and 1.4, respectively. PDDPE is usually employed for the synthesis of 3μ-$A_2B$ and 3μ-ABC, whereas MDDPE is used for $A_2B_2$ 4μ-stars. More recently, Quirk's group extended this method to 6μ-stars by using a triple diphenylethylene, 1,3,5-tris(1-phenylethyl)benzene.[79]

The key to the living linking procedure is the control of the stoichiometry of the reaction between the living A chains and the DDPE; otherwise, a mixture of star and linear polymers is produced. A major problem is that the rate constants of initiation of the two new active sites differ, resulting in a bimodal distribution. To overcome this problem, polar compounds have to be added. It is well known that they dramatically affect the microstructure of the polydienes formed in later stages. However, addition of lithium

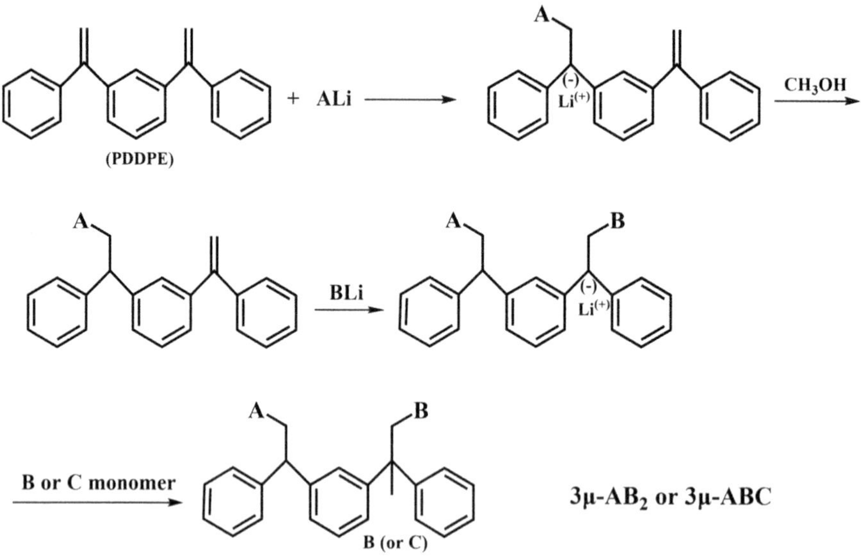


**Scheme 1.3** Synthesis of 3-miktoarm star co/terpolymers by living linking chemistry.

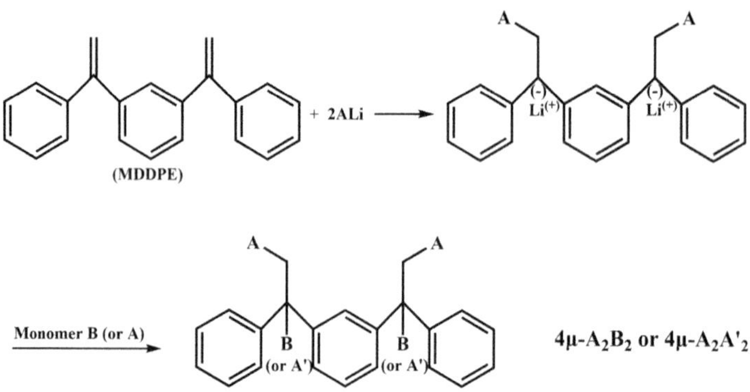


**Scheme 1.4** Synthesis of 4-miktoarm star copolymers by living linking chemistry.

*sec*-butoxide to the living DDPE derivative prior to the addition of the diene monomer was found to produce monomodal well-defined μ-stars with high 1,4 content. Again, the B arms cannot be isolated from the reaction mixture and characterized separately. Nevertheless, Quirk's method is valuable for the synthesis of ω-functionalized μ-stars.

## 1.2.3 Chlorosilanes

Regular homopolymer and block copolymer stars can be synthesized by reaction of an excess of living chains, prepared by anionic polymerization, with the appropriate chlorosilane. An excess of the living polymer is needed to force the linking reaction to completion. For the synthesis of μ-stars, each chlorine atom should be replaced stepwise by a different chain. To achieve this goal, the different reactivity of the living chain ends towards the SiCl group must be taken into consideration. The reactivity of the living chain end decreases with the charge delocalization and by increasing the steric hindrance as follows: butadienyl lithium (BdLi) > isoprenyl lithium (IsLi) > styryl lithium (SLi) > diphenyl ethylenyl lithium (DPELi). The reactivity of the living ends is also affected by the chain molecular weight (the lower the molecular weight, the lower the steric hindrance and, consequently, the higher the reactivity), the polarity of the environment (the higher the polarity, the lower the association of the living chains and, consequently, the higher the reactivity), and the temperature (same as the environment).

Chlorosilanes cannot be combined with macroanions of polar monomers, such as (meth)acrylates and 2-vinyl pyridine, since the linking reaction either leads to unstable products [poly(meth)acrylates] or it does not occur at all [poly(2-vinyl pyridine), P2VP]. Instead, linking agents including chloro-/bromomethyl benzene derivatives[3,4] are used although, unfortunately, they display significant drawbacks such as lithium–halogen exchange, leading to linking agents with higher functionalities and, consequently, to a mixture of stars with different functionalities.[62] To overcome this problem, potassium (instead of lithium) counter ions and polar solvents at low temperature (−78 °C) should be used.

### 1.2.3.1 *Trichloromethylsilane and Tetrachlorosilane*

Trichloromethylsilane ($CH_3SiCl_3$) and tetrachlorosilane ($SiCl_4$) are appropriate linking agents for the synthesis of 3μ- and 4μ-stars. The replacement of only one Cl by one chain can be achieved by very fast addition of the living polymer into a large excess of the chlorosilane. Before the addition of the second living chain, the unreacted chlorosilane is removed to avoid contamination of the μ-stars with the homostars with three ($CH_3SiCl_3$) or four ($SiCl_4$) arms. The above chlorosilanes can be easily removed on a high vacuum line owing to their low boiling point ($CH_3SiCl_3$, b.p.: 66 °C and $SiCl_4$, 57.6 °C). This method was developed by Pennisi and Fetters,[40] and was used for the synthesis of 3μ-star homopolymers of styrene and butadiene with arm molecular weight asymmetry. The general reactions for the synthesis of 3μ-star homopolymers are given in Scheme 1.5.

The excess A'Li needed for completion of the linking reaction is removed by fractionation, as in the case of regular stars. In the case of asymmetric PS

$$\text{ALi} + \text{excess } CH_3SiCl_3 \longrightarrow \text{ASi}(CH_3)Cl_2 + \text{unreacted } CH_3SiCl_3\uparrow + \text{LiCl}$$

$$\text{ASi}(CH_3)Cl_2 + \text{excess A'Li} \longrightarrow 3\mu\text{-AA'}_2 \text{ (homopolymer)} + \text{LiCl} + \text{excess A'Li}$$

**Scheme 1.5** Synthesis of 3-miktoarm stars with trichloromethylsilane as a linking agent.

stars, to ensure complete reaction of PSLi with $ASi(CH_3)Cl_2$, the living PS is end-capped with a few units of butadiene to increase its reactivity towards the chlorosilane.

Mays prepared[41] one sample of 3μ-star $PS(PI)_2$ by this method. The method was further developed by our group to all possible combinations of $A_2B$ μ-stars, where A and B were PS, PI, or PB.[45] Furthermore, by using $SiCl_4$, we prepared $PS(PI)_3$ 4μ-stars.[46] A more sophisticated high vacuum technique was used to ensure the synthesis of well-defined μ-stars.[42] A high degree of molecular and compositional homogeneity was identified by size exclusion chromatography (SEC), with refractive index and UV-detectors, low-angle laser light-scattering (LALLS), membrane (MO) as well as vapor-pressure osmometry (VPO), and NMR spectroscopy. The chlorosilane approach was also adopted for the synthesis of μ-stars of the $PS(PI\text{-}b\text{-}PS)_{2\text{ or }3}$ type.[51]

When only one chlorine needs to be replaced by a highly reactive living chain, for example low molecular-weight PBLi or PI, the living end has to be transformed by end-capping into a less reactive carbanion. By decreasing the reactivity, the selectivity is increased and the replacement of only one chlorine is achieved. Using DPE, we prepared model mono- and difunctional 3μ-star PBs[46] and $(d\text{-}PB)_2(PI)$, $(d\text{-}PB)(PI)_2$, where d-PB is deuterated PB,[80] having arm molecular weights below 10 000 g $mol^{-1}$. The same goal could be achieved more easily (avoiding the necessity for end-capping) by linking the first arm or arms at a temperature low enough to create selectivity between the successive steps of replacement of chlorines.[48]

Using the chlorosilane method, 4μ-stars $(PS\text{-}b\text{-}PB)(PB)_3$ were synthesized by Tsiang.[81] Living polybutadiene PB chains were first reacted with $SiCl_4$ in a molar ratio 3 : 1, followed by addition of the living diblock (PS-*b*-PB)Li. The key step of this method was the successful synthesis of the $(PB)_3SiCl$ intermediate. The high reactivity of the PBLi chain end posed questions about the purity of this polymer, since several byproducts such as $(PB)_2SiCl_2$, $(PB)_4Si$, or $PBSiCl_3$ can be formed in the first step. SEC analysis was performed to monitor the reaction sequence. It is obvious that the method developed by Tsiang is very demanding with regard to the stoichiometry of the reagents. The byproducts are almost impossible to be separated from the main miktoarm star.

For the synthesis of 3μ-ABC terpolymers (Scheme 1.6), three chlorines have to be consecutively and selectively replaced. The first chlorine can be

$$\text{PILi} + \text{excess } CH_3SiCl_3 \longrightarrow \text{PISi}(CH_3)Cl_2 + \text{unreacted } CH_3SiCl_3\uparrow + \text{LiCl}$$

$$\text{PISi}(CH_3)Cl_2 + \text{PSLi} \xrightarrow{\text{titration}} \text{(PI)(PS)Si}(CH_3)\text{Cl} + \text{LiCl}$$

$$\text{(PI)(PS)Si}(CH_3)\text{Cl} + \text{excess PBdLi} \longrightarrow 3\mu\text{–ABC} + \text{LiCl} + \text{excess PBd}$$

**Scheme 1.6** Synthesis of 3-miktoarm star terpolymers by selective replacement of the three chlorines of trichloromethylsilane by three different chains.

replaced by PI using an excess of $CH_3SiCl_3$, the second is replaced by titration with PSLi, and the third by an excess of PBLi.[42] The order of linking of the different living chains to the chlorosilane plays an essential role. The most sterically hindered chain, PSLi, has to be added by titration in the second step, the less sterically hindered chain end, PBLi, has to be added at the last step of the synthesis. The whole procedure can be monitored by SEC, as shown in Figure 1.1.

Using the same route, asymmetric AA'B miktoarm stars were also prepared.[82] These are stars having two chemically identical A arms but of different molecular weights. The synthesis of the 4μ-star copolymer $(PS)_2(PI)_2$ was accomplished in a similar way.[43] PSLi was added in two separate steps (excess and titration) to achieve maximum control over the polymer architecture.

Using the selective replacement of the chlorines of $SiCl_4$, a miktoarm star quarterpolymer was prepared consisting of PS, poly(4-methyl styrene) (P4MeS), PI, and PB.[43] The reaction sequence for the preparation of this miktoarm star is presented in Scheme 1.7. The main feature of this method is

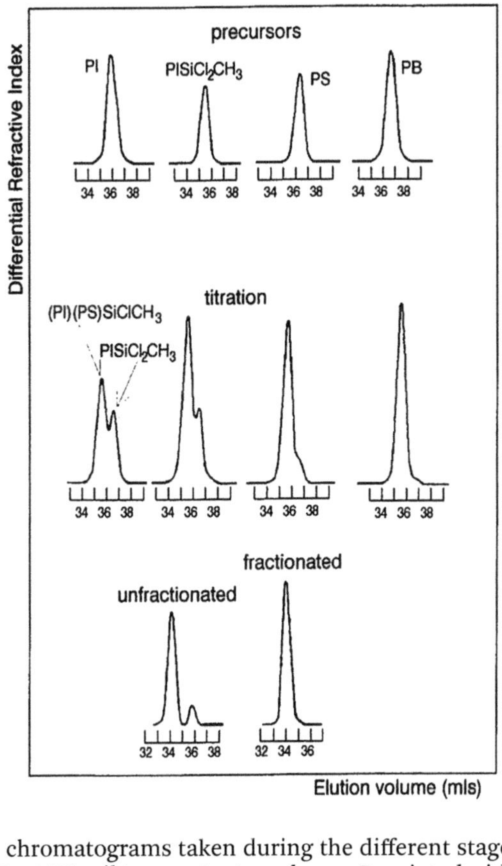


**Figure 1.1** SEC chromatograms taken during the different stages of the synthesis of a 3μ-ABC miktoarm star terpolymer. Reprinted with permission from ref. 42. Copyright 1992. American Chemical Society.

that two of the arms are incorporated by titration. PS was chosen to react first with an excess of $SiCl_4$, followed by titration with the more sterically hindered P4MeS after evaporation of the excess silane. The moderately hindered PILi anion was then added by titration. The least sterically hindered PBLi anion was used to complete the linking reaction.

Xie and Xia were the first to prepare a 4μ-star $A_2B_2$.[83] Living PS chains were reacted with $SiCl_4$ in a 2:1 molar ratio leading to the formation of a two-arm product. The two remaining SiCl bonds were used for a linking reaction with living poly(ethylene oxide) (PEO) chains. The process is facilitated by the increased steric hindrance of the living PS chain ends. It is, indeed, very difficult to prepare 3- or 4-arm PS stars. From this point of view, control of the stoichiometry is less important. $A_2B$ miktoarm stars have also been prepared using $CH_3SiCl_3$ instead of $SiCl_4$.

### 1.2.3.2 1,2-Bis(trichlorosilyl)ethane

1,2-Bis(trichlorosilyl)ethane ($SiCl_6$) has been used for the synthesis of $PS(PI)_5$ 6μ-stars.[53] Stoichiometric addition of PSLi to the $SiCl_6$ linking agent was chosen in this case, instead of an excess of $SiCl_6$, since the unreacted solid chlorosilane cannot be removed by distillation. The reaction sequence is outlined in Scheme 1.8.

Living PS chains were reacted with 1,2-bis(trichlorosilyl)ethane in a 5:1 ratio. Dropwise addition of the living polymer solution into the vigorously

$$\text{PSLi} + \text{excess SiCl}_4 \longrightarrow \text{PSSiCl}_3 + \text{unreacted SiCl}_4\uparrow + \text{LiCl}$$

$$\text{PSSiCl}_3 + \text{P4MeSLi} \xrightarrow{\text{titration}} \text{(PS)(P4MeS)SiCl}_2 + \text{LiCl}$$

$$\text{(PS)(P4MeS)SiCl}_2 + \text{PILi} \xrightarrow{\text{titration}} \text{(PS)(P4MeS)(PI)SiCl} + \text{LiCl}$$

$$\text{(PS)(P4MeS)(PI)SiCl} + \text{excess PBdLi} \xrightarrow{\text{titration}} 4\mu\text{–ABCD} + \text{LiCl} + \text{excess PBd}$$

**Scheme 1.7** Synthesis of 4-miktoarm star quartepolymers by selective replacement of the chlorines of tetrachlorosilane by four different chains.

$$\text{PSLi} + \text{SiCl}_6 \xrightarrow{\text{titration}} \text{PSSiCl}_5 + \text{LiCl}$$

$$\text{PSSiCl}_5 + \text{excess PILi} \longrightarrow \text{PS (PI)}_5 + \text{LiCl} + \text{excess PI}$$
$$(6\mu\text{–star copolymer, }6\mu\text{–A}_5\text{B})$$

**Scheme 1.8** Synthesis of 6-miktoarm star copolymers.

stirred solution of the linking agent was performed to minimize multiple coupling products. Under these conditions, 15% of the dimeric product was formed. The pentachlorosilane-capped PS was then reacted with an excess of PILi, followed by fractionation to isolate the desired polymer.

#### *1.2.3.3 Hexadecachlorosilane ($SiCl_{16}$) and Tetrahexicontachlorosilane ($SiCl_{64}$)*

Model 16μ-star copolymers with eight PS and eight PI arms, $PS_8PI_8$, were synthesized using a dendritic carbosilane $Si\{[CH_2CH_2Si(CH_3)][CH_2\text{-}CH_2Si(CH_3)Cl_2]_2\}_4$ or $SiCl_{16}$ with 16 SiCl bonds.[50] This compound was prepared using tetravinylsilane as the initial core molecule, methyldichlorosilane as the propagating units, and vinylmagnesium bromide for the transformation of silicon chloride to the silicon vinyl group. It was found that PS must be linked first, followed by addition of an excess of PI, in order to successfully synthesize these miktoarm stars due to two basic constrains: (a) the inability of the sterically hindered polystyryl anions to react with both chlorines at each peripheral silicon atom and (b) the ability of the less sterically hindered polyisoprenyl anions to react completely with the remaining chlorosilane bonds. Living PS chains were reacted with the linking agent in an 8 : 1 molar ratio for the preparation of the PS 8-arm star. Even a slight excess of PSLi (~5%) could be used without the incorporation of more than eight arms due to the steric hindrance of the already attached PS chains on each Si atom. A small excess of PILi to SiCl was finally added to prepare the final product.

The term *Vergina* star copolymers was coined by our group for $A_8B_8$ miktoarm stars due to their similarity with the 16-rayed star emblem discovered by the late Professor M. Andronikos of the University of Thessaloniki in the ancient tomb of King Philippos of Macedonia in Vergina, a place close to Thessaloniki (Figure 1.2).

By using the same methodology and a chlorosilane with 64 peripheral SiCl groups, Roovers[84] prepared a 64μ-star copolymer of the $A_{32}A'_{32}$ type, where A and A′ are PB arms with different molecular weights.

## 1.3 Miktoarm-Based Polymers with Complex Architectures

Double μ-stars have been synthesized by selective coupling bifunctional living chains instead of monofunctional ones. The structures obtained are shown in Scheme 1.1c. As an example, the synthetic procedure for the super H-shaped block copolymer[44] is outlined in Scheme 1.9.

More complex miktoarm stars (Figure 1.3) have been synthesized by our group.[85] Styrenic single and double star-tailed macromonomers were first obtained by reaction of living homo/miktoarm stars with the SiCl groups of 4-(chlorodimethylsilyl)- and 4-(dichloromethylsilyl)styrene, respectively. The *in situ* anionic homopolymerization of macromonomers with *sec*-BuLi or

**Figure 1.2** The Golden Larnax (chest) (40.9 × 34.1 × 17 cm) from the Great Tomb of the King Philippos discovered by the late Professor M. Andronicos in Vergina, a place in Northern Greece close to Thessaloniki. On the lid, the 16-rayed star emblem of the Macedonian Dynasty (Thessaloniki Archeological Museum). Reprinted with permission from ref. 116. Copyright 2008, Wiley.

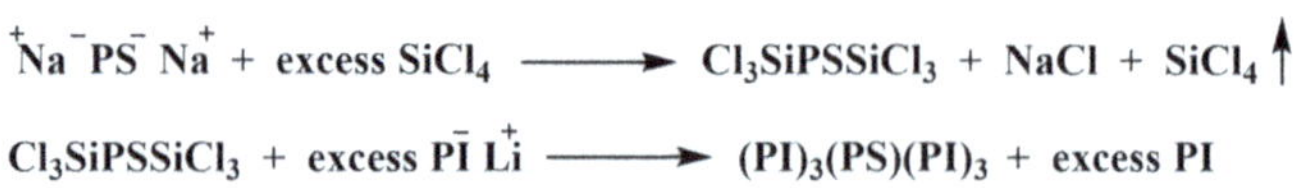


**Scheme 1.9** Synthetic procedure for the synthesis of super H-shaped block copolymers.

copolymerization with butadiene or styrene, led to single/double homo/miktoarm star-tailed molecular brushes and combs, as well as to a block copolymer consisting of a linear polystyrene chain and a double miktoarm (PBd/PS) star-tailed brush-like polymer.

Miktoarm stars involving diblock copolymers as arms have also been reported by our group.[49,51,86,87] Typical cases are (a) inverse 4-arm star copolymers, $(PS\text{-}b\text{-}PI)_2$-junction-$(PS\text{-}b\text{-}PI)_2$, where each arm is a block copolymer of styrene and isoprene but two of the arms are connected to the junction point with the PS block, whereas the other two are so with the PI block, and (b) $A(BA)_{n=2,3}$ and super-H shaped stars of the $(AB)_3A(BA)_3$ type, where A = PS and B = PI. Recently, similar but asymmetric non-linear copolymers $A(BA')_{n=2,3}$ (Scheme 1.10) have been reported.[88] These complex miktoarm-based copolymers have tremendous impact on the microdomain morphology of block copolymers.[49,86]

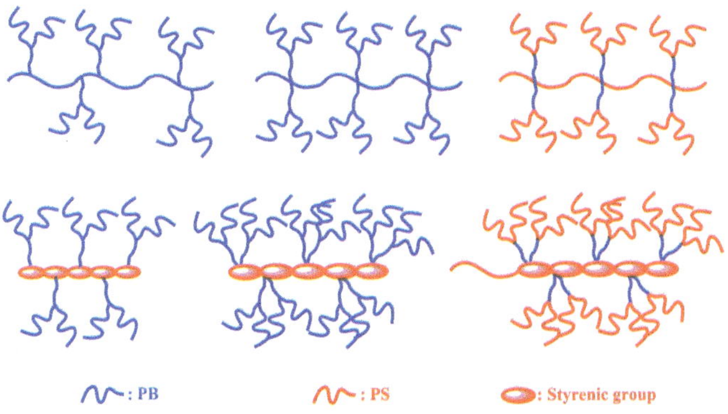


**Figure 1.3** Schematic presentation of miktoarm star-based complex macromolecular architectures prepared by homopolymerization or copolymerization of single/double homo/mikto star-tailed styrenic macromonomers with PB (blue arms) and/or PS (red arms). Reprinted with permission from ref. 85. Copyright 2008. Wiley.

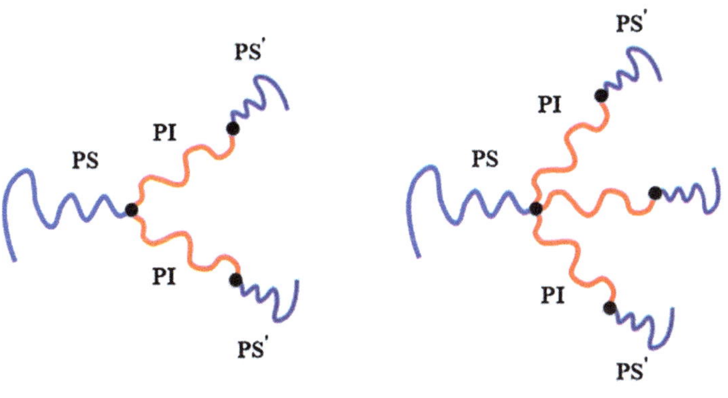


**Scheme 1.10** Asymmetric $A(BA')_{n=2,3}$ block copolymers.

## 1.4 Model Polyethylenes

By applying the chlorosilane approach, our group was able to prepare model 1,4-poly(butadienes) of the $A_2A$, $AA'A''$, $A_2A'A_2$, $A_3A'A_3$, and $A_5A'A_5$ types and, by hydrogenation, the corresponding branched polyethylenes.[89]

## 1.5 Individual Methods for the Synthesis of Miktoarm Stars

### 1.5.1 3μ-Star Copolymers of the $A_2B$ Type

$A_2B$ stars with two PS arms and one poly(2-vinyl pyridine) (P2VP) arm, $(PS)_2(P2VP)$, were prepared by Khan *et al.*[90] Living PS chains were linked to dichloromethylsilane, $CH_3SiCl_2H$, to produce the two arms of the star.

In another reactor, living P2VP was terminated with allyl bromide. The 3μ-star was produced through hydrosilylation addition of SiH to the allyl group. Due to incomplete hydrosilylation, the μ-stars were found to have high polydispersity (1.33 to 1.50).

Anionic polymerization techniques and naphthalene chemistry were used by Teyssie *et al.* to prepare $A_2B$ miktoarm stars, where A was PEO and B was PS, PI, poly(α-methyl styrene), or poly(*t*-butyl styrene).[91]

A special technique was employed by Naka *et al.* for the preparation of $A_2B$ stars, A being PEO and B polyoxazoline (POX), by forming Ru(III) complexes with bipyridyl terminated polymers.[92]

### 1.5.2 μ-Stars of the $A_nB$ Type

Star polymers with several PS arms and only one poly(2-vinyl naphthalene) (PVN) arm were prepared by Takano *et al.* by anionic polymerization techniques.[93] Sequential anionic block copolymerization of (4-vinylphenyl)dimethylvinylsilane (VS) and VN was employed. The double bonds attached to silicon have to remain unaffected during the polymerization of VS. This was accomplished with K counterions in THF and short polymerization times. The PVS block with the unreacted double bonds was used as a multifunctional linking agent. Subsequent addition of living PS chains produced miktoarm stars of the type $(PS)_n$PVN. Characterization studies revealed that $n = 13$.

Similar structures of the $AB_n$ type were prepared by Wang and Roovers, where A = PS and B = PB or P2VP.[94] Due to the much higher molecular weight of the PS arm, these μ-stars were called umbrella copolymers (Scheme 1.11). The reaction sequence for the preparation of the $PS(PB)_n$ copolymers is given in Scheme 1.11.

Butadiene was polymerized anionically in the presence of dipiperidinoethane (dipip), followed by addition of styrene. Hydrosilylation chemistry was employed to add the $Si(CH_3)Cl_2$ or $Si(CH_3)_2Cl$ groups to the 1,2-PBd double bonds. Subsequent addition of 1,4-PBdLi or P2VPK led to the formation of umbrella copolymers. The limited control exercised over the hydrosilylation reaction is the main reason why the number of arms cannot be accurately controlled.

Roovers and collaborators[95] also succeeded in preparing umbrella-star copolymers $[(PS\text{-}uPI)_n]_m$ (Scheme 1.12). The synthesis is based on the reaction of (1,2-PB-*b*-PS)Li with a chlorosilane having 32 peripheral SiCl bonds, followed by hydrosilylation of the PB double bonds and reaction with PILi.

### 1.5.3 μ-Star Copolymers of the $A_nB_m$ Type

Ishizu and Kuwahara[96] have developed a macromonomer technique for the synthesis of miktoarm stars of the $A_nB_m$ type. PS and PI macromonomers with vinyl end-groups were anionically copolymerized in benzene solutions with *n*-BuLi. The products can be considered miktoarm stars of the type $A_nB_m$, as evidenced by their solution and solid-state properties.

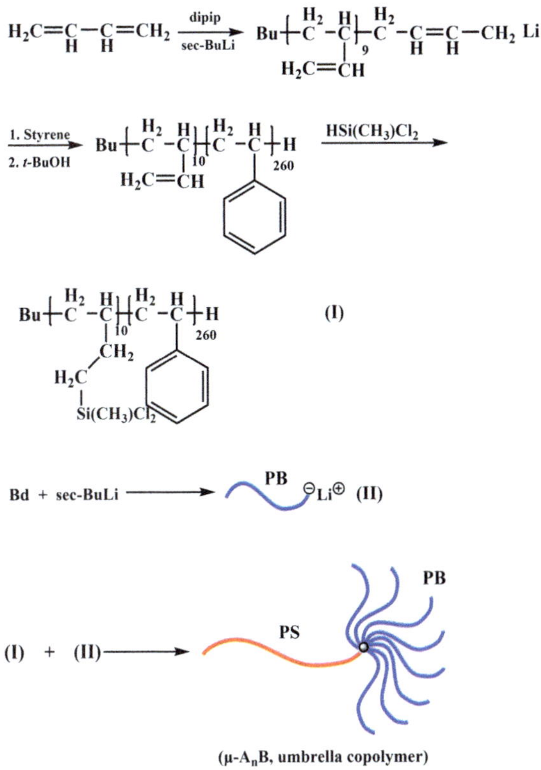


**Scheme 1.11** General reactions for the synthesis of umbrella copolymers.

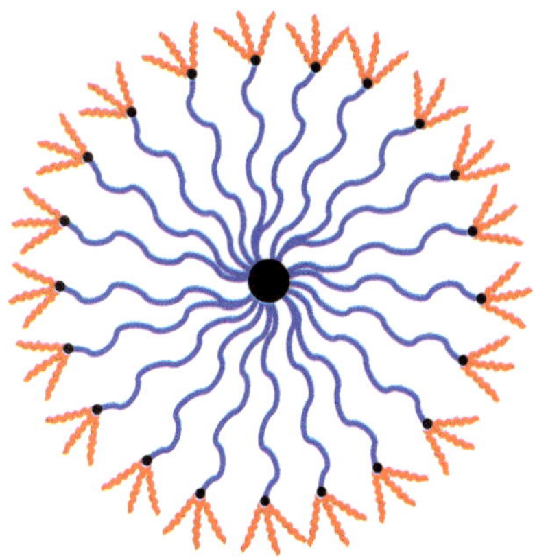

**Scheme 1.12** Umbrella-star copolymer.

Diblock macromonomers with central vinyl groups were used for the synthesis of $(PS)_n(PtBuMA)_n$ μ-stars.[97,98] The macromonomers were prepared by sequential anionic polymerization of styrene, 1,4-divinylbenzene (DVB), and *t*-butyl methacrylate (*t*BuMA). The DVB monomer was left to react with the living PS chains for short times (5 min) so that only a few DVB units were incorporated at the end of the PS chains, and the formation of PS stars was avoided. Free radical polymerization in solution and in bulk using AIBN as the initiator, tetramethylthiuram as the photosensitizer, and ethylene glycol dimethacrylate as the crosslinking agent was carried out for the synthesis of the μ-stars. A similar experiment was performed using PS-*b*-P2VP diblocks having central isoprene units.

Finally, a cyclophosphazene derivative was used as a linking agent to produce μ-stars consisting of PS and Nylon-6 arms.[99]

### 1.5.4 μ-Star Terpolymers of the ABC Type

Fujimoto *et al.* prepared (PS)(PDMS)(P*t*BuMA) μ-stars,[100] as described in Scheme 1.13. The lithium salt of *p*-(dimethylhydroxy)silyl-diphenylethylene was used as the initiator for the polymerization of hexamethylcyclotrisiloxane ($D_3$). Living PS chains were reacted with the end double bond of the diphenylethylene, followed by the anionic polymerization of *t*BuMA.

PDMS has a rather high molecular weight distribution ($I \sim 1.4$) and fractionation needs to be performed before the following synthetic steps. The polymethacrylate branch cannot be isolated and studied independently.

A similar synthetic route was adopted by Stadler *et al.* for the synthesis of (PS)(PB)(PMMA) stars,[101] as shown in Scheme 1.14.

Living PS chains were end-capped with *p*-bromomethyldiphenylethylene. The next step involved the linking of living PB chains, prepared in THF at −70 °C, to the end double bond. This reaction produces a new active center used to initiate the polymerization of MMA.

The chlorosilane method was also employed by our group for the synthesis of μ-stars with PS, PI, and PMMA arms.[52,102] The reaction sequence is presented in Scheme 1.15.

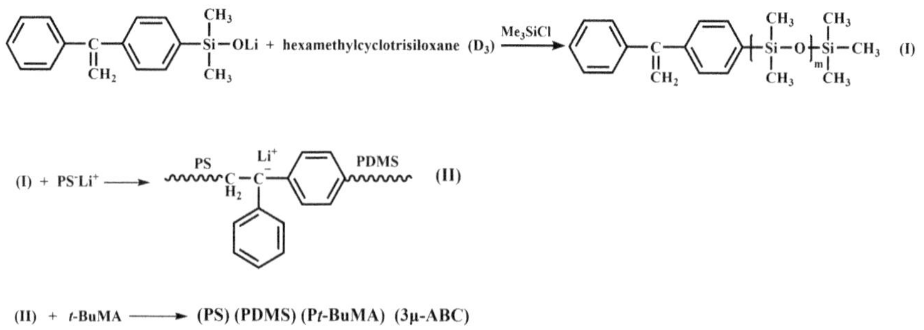


**Scheme 1.13** General reactions for the synthesis of 3-miktoarm star terpolymers of polystyrene, poly(dimethyl siloxane) and poly(*tert*-butyl methacrylate).

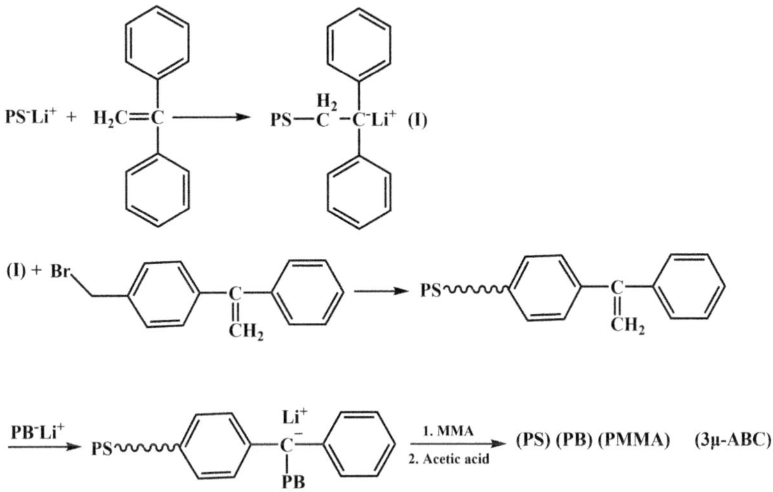


**Scheme 1.14** General reactions for the synthesis of 3-miktoarm star terpolymers of polystyrene, polybutadiene and poly(methyl methacrylate).

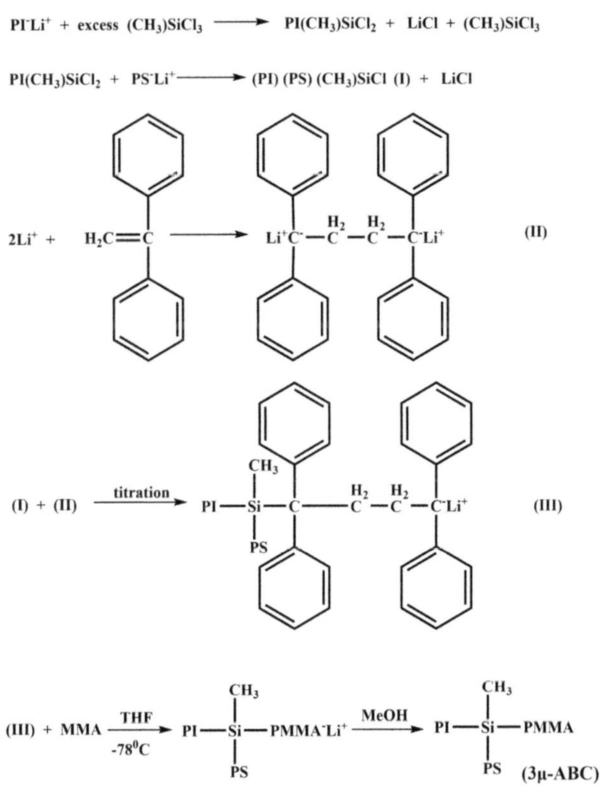


**Scheme 1.15** General reactions for the synthesis of 3-miktoarm star terpolymers of polystyrene, polyisoprene and poly(*methyl* methacrylate).

The monofunctional macromolecular linking agent (PS)(PI)($CH_3$)SiCl was prepared through a procedure similar to those described above. This linking agent was reacted with a stoichiometric amount of the dilithium initiator prepared from 1,1-diphenylethylene (DPE) and Li. The remaining active center was used to polymerize MMA in THF at −78 °C. In this procedure, the PMMA arm cannot be isolated and cannot be characterized independently.

A technique similar to Stadler's was employed for the synthesis of miktoarm stars with PS, PEO, poly(ε-caprolactone) (PCL), or PMMA arms.[103] A PS-*b*-PMMA diblock copolymer possessing a central DPE derivative bearing a protected hydroxyl function was prepared. After deprotection and transformation of the hydroxyl group into an alkoxide, the anionic ring-opening polymerization of the third monomer (EO or ε-CL) was initiated. Only limited characterization data were given.

### 1.5.5 6μ-Star Copolymers of the $A_2B_4$ Type

Recently, Hirao and coworkers[104] have proposed a promising methodology for the synthesis of 3-, 4-, 5-, and 6-arm regular and μ-stars. Anionic living chains react with specially designed terminators with a predetermined number of methoxy benzyl groups. The methoxy groups are transformed into a chloromethyl group, followed by reaction with another living chain. As an example of the concept, the synthesis of a 6μ-star $A_2B_4$ is given in Scheme 1.16.

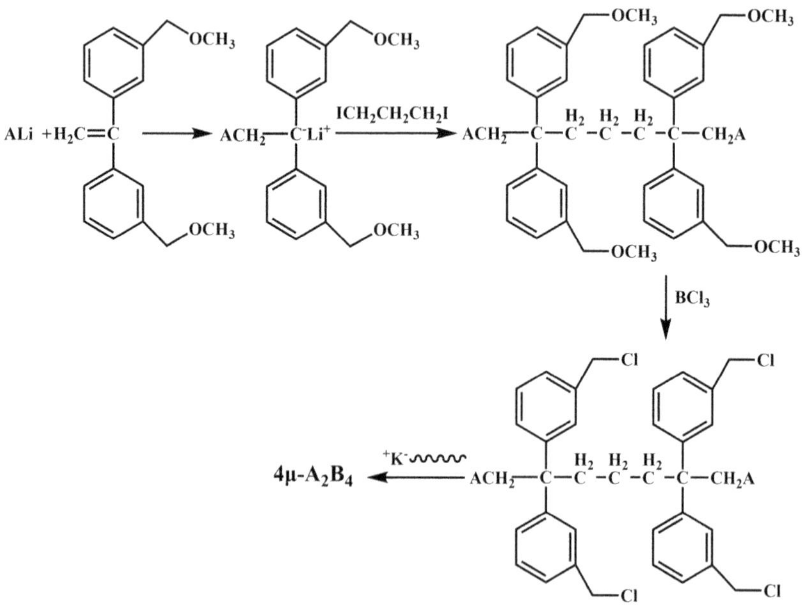


**Scheme 1.16** General reactions for the synthesis of 6-miktoarm star copolymers.

### 1.5.6 Miktoarm Macromolecular *Chimeras*

By exploiting the unique features of anionic polymerization to synthesize end- or in-chain amino-functionalized polymers, as well as the living nature of the ring opening polymerization of α-amino acid *N*-carboxyanhydrites (NCAs) under high vacuum conditions, our group in collaboration with Hirao's group succeeded to synthesize the following miktoarm star (μ-star) polymer/polypeptide hybrids (macromolecular *chimeras*): $(PS)_2$(PBLG or PBLL) and (PS)(PI)(PBLG or PBLL) (3μ-stars), and $(PS)_2$[P(α-MeS)](PBLG or PBLL) and $(PS)_2$(PBLG or $PBLL)_2$ (4μ-stars), where P(α-MeS) is poly(α-methylstyrene), PBLG is poly(γ-benzyl-L-glutamate), and PBLL is poly(ε-*tert*-butyloxycarbonyl-L-lysine).[105] DPE-functionalized polymers were first prepared and subsequently activated by reaction with a living chain or *sec*-BuLi, followed by reaction with 1-(3-bromopropyl)-2,2,5,5-tetramethyl-aza-2,5-disilacyclopentane. The silyl-protected group was cleaved under acidic conditions to afford amine-functionalized macroinitiators for the polymerization of BLG and BLL NCAs to afford the desired miktoarm stars. An example is given in Scheme 1.17. Detailed characterization revealed the efficiency of the synthetic strategy and the homogeneity of the synthesized miktoarm *chimeras*.

## 1.6 Microphase Separation of Miktoarm Stars

The morphology and consequently the properties of a given linear block copolymer are controlled mainly by the volume fraction of its components. Upon changing from linear to a non-linear architecture, the

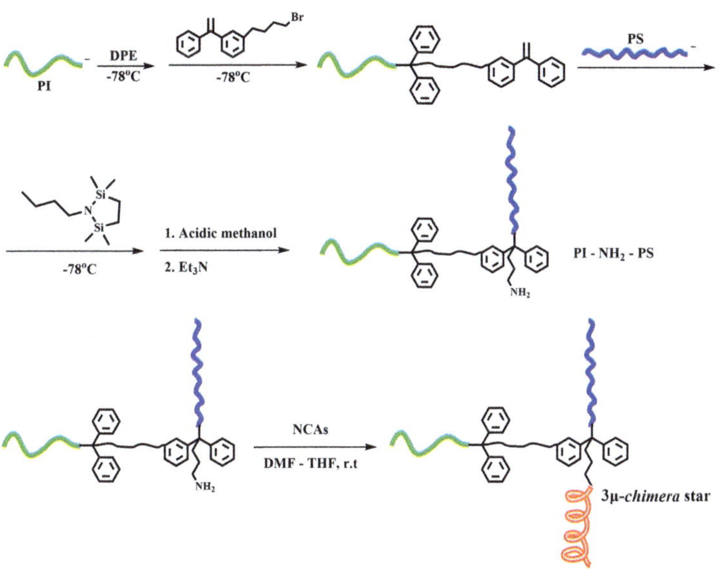


**Scheme 1.17** General reactions for the synthesis of 3-miktoarm star *chimeras*. Reprinted with permission from ref. 105. Copyright 2008. American Chemical Society.

study of miktoarm star copolymer microphase separation becomes very important, since these materials shift the boundaries between different morphologies compared to the corresponding linear copolymers or form new structures (μ-ABC).[106] As an example, the PS-*b*-PI (PS: 40% by volume) linear copolymer has a lamellar morphology, whereas the non-linear $(PI)_2PS$ and $(PI)_3PS$ block copolymers, with the same overall molecular weight and composition, form hexagonally packed cylinders and spheres in a body centered cubic (BCC) array, respectively.

In the first microphase separation study of $A_2B$ miktoarm stars synthesized by our group,[107] where A and B were PS or PI, respectively, a $PS(PI)_2$ sample with 37 vol% PS was found by transmission electron microscopy (TEM) to microphase separate, exhibiting PS cylinders in the PI matrix. This observation is in contrast to the alternating lamellar structure expected for a linear diblock copolymer with the same volume fraction. The subsequent studies[47,108,109] with a larger number of samples and covering a wider range of compositions revealed that differences exist in the phase diagram of miktoarm star copolymers in comparison to what is predicted and expected for the corresponding linear diblock copolymers (in composition and molecular characteristics). Analogous shifts were also observed in the case of $A_3B$ (A = PI and B = PS) miktoarm stars.[47] These findings are in qualitative agreement with the theoretical predictions reported by Milner,[109] which were made just after the first experimental findings.[107]

Gido's and our group[110] have demonstrated the discrepancies between theoretical predictions and experimental results for $A_nB$ miktoarm stars as the value of $n$ increases. More significant deflections from the theory for $(PI)_5PS$ miktoarm star were reported, even when eliminating the formation of spherical and cylindrical morphologies in such highly asymmetric miktoarm stars. Based on the initial phase diagram introduced by Milner,[110] Gido's and our group proposed the phase diagram shown in Figure 1.4, where the discrepancies with the theoretical predictions are indicated with dark shaded symbols.

An inverse 4-arm miktoarm star block copolymer of styrene and isoprene (PS: 50% by volume) synthesized by our group and studied morphologically by the Thomas group showed an ordered bicontinuous double-diamond morphology[49] never seen before for the strong segregation limit of a neat copolymer (Figure 1.5).

The self-assembly of three-component ABC miktoarm star terpolymers is governed by two independent composition variables ($\varphi_A$, $\varphi_B$, $\varphi_C = 1 - \varphi_A - \varphi_B$) and by three interaction parameters ($\chi_{AB}$, $\chi_{AC}$, and $\chi_{BC}$), whereas the block sequence is eliminated in such architectures (due to the existence of only one junction point through which all blocks are connected) in contrast to that of ABC linear triblock terpolymers. Furthermore, the star junction points are restricted to lie on periodically spaced parallel lines (not necessarily straight) defined by the mutual intersection of the different domains, as theory predicts.

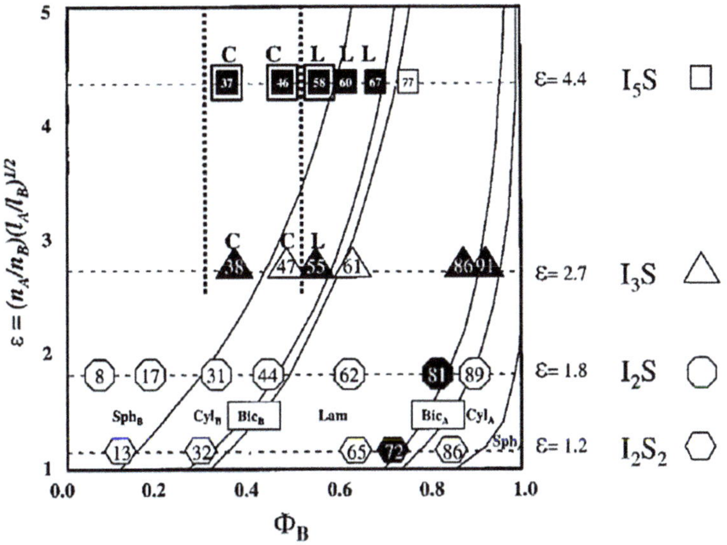


**Figure 1.4** Phase diagram of molecular asymmetry $\varepsilon$ *versus* the volume fraction of the B component $\Phi_B$. Shaded symbols indicate samples whose morphology disagrees with the theory. Reprinted with permission from ref. 111. Copyright 2001. American Chemical Society.

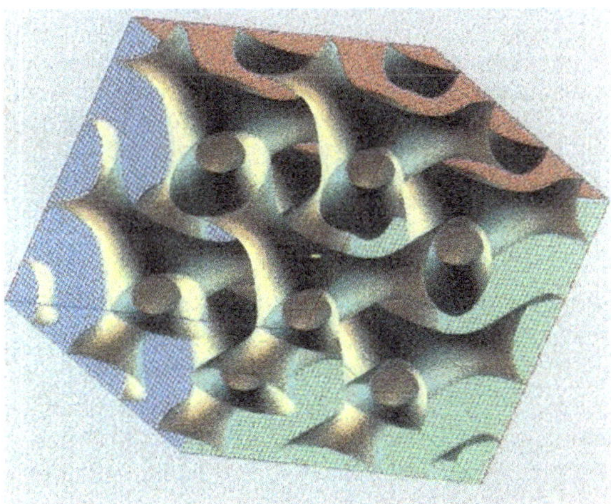

**Figure 1.5** Simulation of the unit cell of the Ordered Bicontinuous Double Diamond (OBDD) morphology observed at the inverse m-star copolymers. The unit cell of OBDD has dimensions on the scale of a few hundred Angstroms, cubic symmetry, and consists of eight tetrahedral elements that form two independent networks every four elements. These two networks are continuous, periodic in the three dimensions, do not intersect, and are formed by the phase of one of the components (PS), whereas the other component (PI) forms the matrix. Reprinted with permission from ref. 49. Copyright 1998. American Chemical Society.

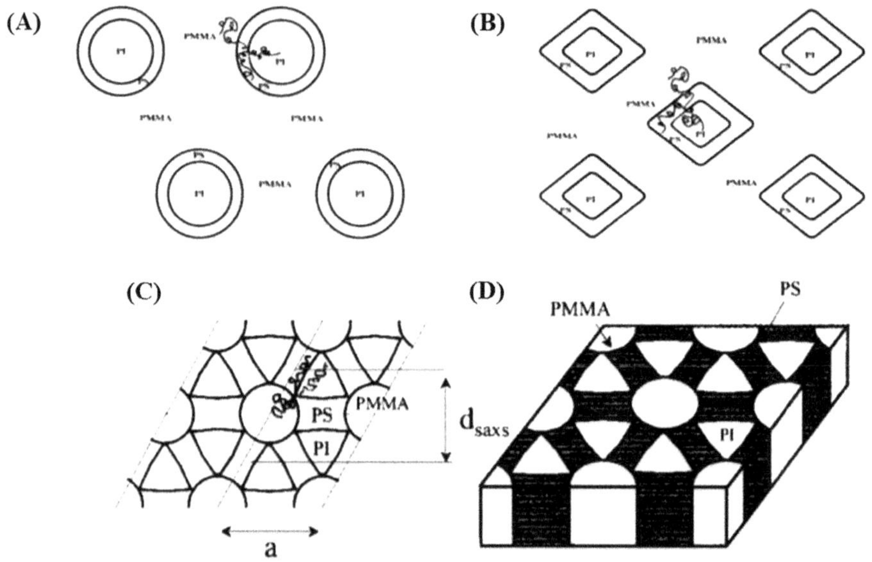


**Figure 1.6** (A) Schematic presentation of the two-dimensional microdomain morphology exhibiting hexagonally packed cylinders of PI with a concentric PS annulus in a matrix of PMMA. The PS and PMMA arms are partially mixed within the PS domain. (B) Two-dimensional microdomain morphology exhibiting the PI arm not forming hexagonally packed cylinders and the PI/PS and PS/PMMA surfaces with rhombohedral shape. (C) Representative chain conformation of the miktoarm star terpolymer, the three-dimensional microdomain morphology and the location of the junction point are indicated. (D) Schematic showing the junction points residing at the vertexes where the three types of microdomains intersect. Reprinted with permission from ref. 112 and 113. Copyright 1998. American Chemical Society.

In order to verify this prediction, the morphology of a series of 3-miktoarm star terpolymers comprised of PS, PI, and PMMA (SIM) were reported by Thomas' and our group.[112,113] These samples were synthesized *via* anionic polymerization with a specific method of alternating the functionality of (PS)(PI)Si($CH_3$)Cl in order to initiate polymerization of MMA,[52] as already described previously. In all cases, the materials exhibited distinct three-phase microdomain structures. In Figure 1.6, the resulting morphology schematics are given for these very interesting miktoarm star terpolymers of the (PS)(PI)(PMMA) type.

In the first case,[111] the PS and PMMA arms showed pronounced incompatibility towards PI, whereas the PS and PMMA arms were organized into a novel annulus-matrix structure. In the second case,[112] the self-assembly results indicated that, in these (PS)(PI)(PMMA) miktoarm star terpolymers, the composition between the three different domains is favorable towards a structure that allows the junction points to be confined

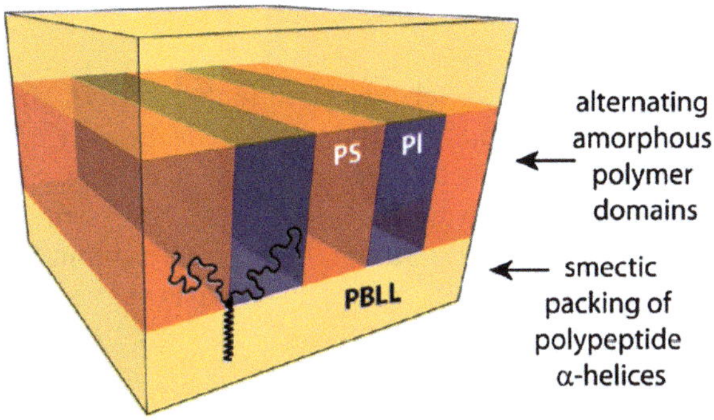


**Figure 1.7** Schematic illustration of the self-assembly in a miktoarm macromolecular *chimera* composed of two coil-like arms (PS and PI) and a mesogenic α-helical polypeptide arm (poly(ε-*tert*-butyloxycarbonyl L-lysine), PBLL). Reprinted with permission from ref. 114. Copyright 2010. American Chemical Society.

within lines where the three different types of microdomains intersect, verifying therefore the predictions by the constrains involved in the star-type architecture. Such an observation was the first one of its kind in the literature.[113]

Ikkala's and our group have reported the first hierarchical smectic self-assembly in a miktoarm macromolecular *chimera* composed of two coil-like arms (PS and PI) and a mesogenic α-helical polypeptide arm [poly(ε-*tert*-butyloxycarbonyl L-lysine), PBLL].[105] The PBLL α-helices are packed within lamellar nanodomains, which leads to an overall smectic alteration of the rod- and coil-containing layers typically observed in rod-coil block copolymers. Furthermore, the coil-containing lamellae have an inner structure composed of PS and PI rectangular cylinders, leading to what we call a hierarchical smectic phase (Figure 1.7).[114] Similar hierarchical self-assembly have been found in other miktoarm *chimeras* too.[115]

## 1.7 Concluding Remarks

Miktoarm (a term coined by our group in 1992) star polymers prepared by anionic polymerization high vacuum techniques play an important role in polymer science. In polymer chemistry, polymerization methodologies have evolved towards complex architectures, including miktostar structures and, in polymer physics, they have inspired scientists to elaborate/perform theories/experiments on the influence of the architecture on the self-assembly of block copolymers. In addition, miktoarm stars have led to the discovery of a significant number of novel nanostructures, in solution and bulk, as potential candidates for nanotechnology applications, such as nanomedicine, nanolithography, *etc.*

## Acknowledgements

N. H. wishes to thank his former Ph.D. students, without their enthusiasm and dedication the μ-star project would not have been possible: Prof. Hermis Iatrou, Prof. Marinos Pitsikalis, Dr Stergios Pispas, Dr Yiannis Tselikas, Dr Vasilis Efstratiadis, Professor Apostolos Avgeropoulos, Professor George Sakellariou, Dr Yiannis Poulos, Dr Stella Sioula, Dr Maria Xenidou, Dr Gabriel Velis, Dr Stella Paraskeva, Dr Panagiota Fragouli, Dr Paraskevi Driva. N.H. wishes also to thank his collaborators Dr G. Polymeropoulos and Dr K. Ntetsikas for the design of the structures comprised in the Schemes. All authors wish to thank the corresponding Universities of Athens, of Ioannina and King Abdullah University of Science and Technology (KAUST) for their continuous support.


## References


1. J. R. Schaefgen and P. J. Flory, *J. Am. Chem. Soc.*, 1948, **70**, 2709.
2. M. Morton, T. E. Helminiak, S. D. Gadkary and F. Bueche, *J. Polym. Sci.*, 1962, **57**, 471.
3. T. A. Orofino and F. Wenger, *J. Phys. Chem.*, 1963, **67**, 566.
4. R. Mayer, *Polymer*, 1974, **15**, 137.
5. J. C. Meunier and R. Van Leemput, *Makromol. Chem.*, 1971, **142**, 1.
6. J. A. Gervasi and A. B. Gosnell, *J. Polym. Sci., Part A: Polym. Chem.*, 1966, **4**, 1391.
7. J. E. Herz, M. Hert and C. Strazielle, *Makromol. Chem.*, 1972, **160**, 213.
8. C. Strazielle and J. E. Herz, *Eur. Polym. J.*, 1977, **13**, 223.
9. C. A. Uraneck and J. N. Short, *J. Appl. Polym. Sci.*, 1970, **14**, 1421.
10. H. L. Hsieh, *Rubber Chem. Technol.*, 1976, **49**, 1305.
11. D. Decker and P. Rempp, *C. R. Acad. Sci.*, 1965, **261**, 1977.
12. R. Milkovich, Can. Patent (to Shell Oil Co.), 716645, Aug. 24, 1965.
13. H. Eschwey, M. L. Hallensleben and W. Burchard, *Makromol. Chem.*, 1973, **173**, 235.
14. R. P. Zelinski and C. F. Wofford, *J. Polym. Sci., Part A: Gen. Pap.*, 1965, **3**, 93.
15. J. Roovers and S. Bywater, *Macromolecules*, 1972, **5**, 384.
16. J. Roovers and S. Bywater, *Macromolecules*, 1974, **7**, 443.
17. N. Hadjichristidis and J. Roovers, *J. Polym. Sci., Part B: Polym. Phys.*, 1974, **12**, 2521.
18. N. Hadjichristidis, A. Guyot and L. J. Fetters, *Macromolecules*, 1978, **11**, 668.
19. N. Hadjichristidis and L. J. Fetters, *Macromolecules*, 1980, **13**, 191.
20. J. Roovers, N. Hadjichristidis and L. J. Fetters, *Macromolecules*, 1983, **16**, 214.
21. A. B. Nguyen, N. Hadjichristidis and L. J. Fetters, *Macromolecules*, 1986, **19**, 768.
22. L.-L. Zhou, N. Hadjichristidis, P. M. Toporowski and J. Roovers, *Rubber Chem. Technol.*, 1992, **65**, 303.

23. J. Roovers, L.-L. Zhou, P. M. Toporowski, M. vander Zwan, H. Iatrou and N. Hadjichristidis, *Macromolecules*, 1993, **26**, 4324.
24. J. Roovers, P. M. Toporowski and J. Martin, *Macromolecules*, 1989, **22**, 1897.
25. G. Grest, L. J. Fetters, J. S. Huang and D. Richter, in *Advances Chemical Physics, XCIV*, ed. I. Prigogine and S. A. Rice, John Wiley & Sons, New York, 1996.
26. J. F. Douglas, J. Roovers and K. F. Freed, *Macromolecules*, 1990, **23**, 4168.
27. S. Kanaoka, M. Sawamoto and T. Higashimura, *Macromolecules*, 1991, **24**, 2309.
28. S. Kanaoka, M. Sawamoto and T. Higashimura, *Macromolecules*, 1991, **24**, 5741.
29. H. Shohi, M. Sawamoto and T. Higashimura, *Macromolecules*, 1991, **24**, 4926.
30. H. Shohi, M. Sawamoto and T. Higashimura, *Makromol. Chem.*, 1992, **123**, 2027.
31. E.-H. Fukui, M. Sawamoto and T. Higashimura, *Macromolecules*, 1994, **27**, 1297.
32. S. Jacob, I. Majoros and J. P. Kennedy, *Macromolecules*, 1996, **29**, 8631.
33. N. Omura and J. P. Kennedy, *Macromolecules*, 1997, **30**, 3204.
34. J. P. Kennedy, L. R. Ross, J. E. Lackey and O. Nuyken, *Polym. Bull.*, 1981, **4**, 67.
35. K. J. Huang, M. Zsuga and J. P. Kennedy, *Polym. Bull.*, 1988, **19**, 43.
36. T. M. Marsalka, I. Majoros and J. P. Kennedy, *Polym. Bull.*, 1993, **31**, 1993.
37. Z. Zhu, J. Rider, C. Y. Yang, M. E. Gilmartion and G. E. Wnek, *Macromolecules*, 1992, **25**, 7330.
38. G. C. Bazan and R. R. Schrock, *Macromolecules*, 1991, **24**, 817.
39. R. S. Saunders, R. E. Cohen, S. J. Wong and R. R. Schrock, *Macromolecules*, 1992, **25**, 2055.
40. R. W. Pennisi and L. J. Fetters, *Macromolecules*, 1988, **21**, 1094.
41. J. W. Mays, *Polym. Bull.*, 1990, **23**, 247.
42. H. Iatrou and N. Hadjichristidis, *Macromolecules*, 1992, **25**, 4649.
43. H. Iatrou and N. Hadjichristidis, *Macromolecules*, 1993, **26**, 2479.
44. H. Iatrou, A. Avgeropoulos and N. Hadjichristidis, *Macromolecules*, 1994, **27**, 6232.
45. H. Iatrou, E. Siakali-Kioulafa, N. Hadjichristidis, J. Roovers and J. W. Mays, *J. Polym. Sci., Part B: Polym. Phys.*, 1995, **33**, 1925.
46. M. Pitsikalis and N. Hadjichristidis, *Macromolecules*, 1995, **28**, 3904.
47. Y. Tselikas, H. Iatrou, N. Hadjichristidis, K. S. Liang, K. Mohanty and D. J. Lohse, *J. Chem. Phys.*, 1996, **105**, 2456.
48. J. Allgaier, R. N. Young, V. Efstratiadis and N. Hadjichristidis, *Macromolecules*, 1996, **29**, 1794.
49. Y. Tselikas, N. Hadjichristidis, R. L. Lescanec, C. C. Honeker, M. Wohlgemuth and E. L. Thomas, *Macromolecules*, 1996, **29**, 3390.
50. A. Avgeropoulos, Y. Poulos, N. Hadjichristidis and J. Roovers, *Macromolecules*, 1996, **29**, 6076.

51. A. Avgeropoulos and N. Hadjichristidis, *J. Polym. Sci., Part A: Polym. Chem.*, 1997, **35**, 813.
52. S. Sioula, Y. Tselikas and N. Hadjichristidis, *Macromolecules*, 1997, **30**, 1518.
53. G. Velis and N. Hadjichristidis, *Macromolecules*, 1999, **32**, 534.
54. A. Zioga, S. Sioula and N. Hadjichristidis, *Macromol. Symp.*, 2000, **157**, 239.
55. V. Bellas, H. Iatrou and N. Hadjichristidis, *Macromolecules*, 2000, **33**, 6993.
56. T. Tsoukatos and N. Hadjichristidis, *J. Polym. Sci., Part A: Polym. Chem.*, 2002, **40**, 2575.
57. A. Mavroudis, A. Avgeropoulos, N. Hadjichristidis, E. L. Thomas and D. J. Lohse, *Chem. Mater.*, 2003, **15**, 1976.
58. A. Mavroudis, A. Avgeropoulos, N. Hadjichristidis, E. L. Thomas and D. J. Lohse, *Chem. Mater.*, 2006, **18**, 2164.
59. P. G. Fragouli, H. Iatrou, N. Hadjichristidis, T. Sakura and A. Hirao, *J. Polym. Sci., Part A: Polym. Chem.*, 2006, **44**, 614.
60. P. Fragouli, H. Iatrou, N. Hadjichristidis, T. Sakura, Y. Machunaga and A. Hirao, *J. Polym. Sci., Part A: Polym. Chem.*, 2006, **44**, 6587.
61. J. J. K. Krkensgaard, P. Fragouli, N. Hadjichristidis and K. Mortensen, *Macromolecules*, 2011, **44**, 575.
62. H. L. Hsieh and R. P. Quirk, *Anionic Polymerization. Principles and Practical Applications*, Marcel Dekker, New York, NY, 1996.
63. S. Turner and R. Blevins, *Macromolecules*, 1990, **23**, 1856.
64. T. Otsu, A. Matsumoto and M. Yoshiaka, *Indian J. Technol.*, 1993, **31**, 172.
65. R. P. Quirk and B. Lee, *Polym. Int.*, 1992, **27**, 359.
66. H. Eschwey and W. Burchard, *Polymer*, 1975, **16**, 180.
67. C. Tsitsilianis, P. Chaumont and P. Rempp, *Makromol. Chem.*, 1990, **191**, 2319.
68. C. Tsitsilianis, S. Graff and P. Rempp, *Eur. Polym. J.*, 1990, **27**, 243.
69. C. Tsitsilianis, P.-J. Lutz, S. Graff, J.-P. Lamps and P. Rempp, *Macromolecules*, 1991, **24**, 5897.
70. D. Rein, P. Rempp and P.-J. Lutz, *Makromol. Chem., Macromol. Symp.*, 1993, **67**, 237.
71. The DVB used is usually a mixture of *para-*, *meta-*, and ethylvinylbenzenes. Kinetic studies on pure products showed that the *para*-isomer reacts slightly faster with the living chain than the *meta*-isomer. Nonetheless, the *meta*-isomer was found to be more efficient in linking than the *para* form; *i.e.*, at given ratio of divinylbenzene/dienyllithium the higher degree of branching was achieved with the *meta*-isomer.
72. C. Tsitsilianis, D. Papanagopoulos and P. Lutz, *Polymer*, 1995, **36**, 3745.
73. C. Tsitsilianis and D. Voulgaris, *Macromol. Chem. Phys.*, 1997, **198**, 997.
74. D. J. Frater, J. W. Mays and C. Jackson, *J. Polym. Sci., Part B: Polym. Phys.*, 1997, **35**, 141.
75. H. Hocker and G. J. Latterman, *J. Polym. Sci., Part C: Polym. Symp.*, 1976, **54**, 361.

76. A. Yanagishi, M. Szwarc, L. H. Tung and G. Y.-S. Lo, *Macromolecules*, 1978, **11**, 607.
77. E. Leitz and H. Hocker, *Macromol. Chem. Phys.*, 1983, **184**, 1893.
78. R. P. Quirk, T. Yoo and B. Lee, *J. Macromol. Sci., Pure Appl. Chem.*, 1994, **A31**(8), 911.
79. T. Yoo, Ph.D. Thesis, Akron University, 1994.
80. A. Ramzi, M. Prager, D. Richter, V. Efstratiadis, N. Hadjichristidis, R. N. Young and J. Allgaier, *Macromolecules*, 1997, **30**, 7171.
81. R. C. C. Tsiang, *Macromolecules*, 1994, **27**, 4399.
82. C. Lee, S. P. Gido, M. Pitsikalis, J. W. Mays, N. Beck Tan, S. F. Trevino and N. Hadjichristidis, *Macromolecules*, 1997, **30**, 3738.
83. H. Xie and J. Xia, *Makromol. Chem.*, 1987, **188**, 2543.
84. J. Roovers, *Macromolecules*, 1994, **27**, 5359.
85. P. Driva, D. J. Lohse and N. Hadjichristidis, *J. Polym. Sci., Part A: Polym. Chem.*, 2008, **46**, 1826.
86. K. Chrissopoulou, Y. Tselikas, S. H. Anastasiadis, G. Fytas, A. N. Semenov, G. Fleischer, N. Hadjichristidis and E. L. Thomas, *Macromolecules*, 1999, **32**, 5115.
87. A. Avgeropoulos, B. J. Dair, E. L. Thomas and N. Hadjichristidis, *Polymer*, 2002, **43**, 3257.
88. W. Shi, N. A. Lynd, D. Montarnal, Y. Luo, G. H. Fredrickson, E. J. Kramer, C. Ntaras, A. Avgeropoulos and A. Hexemer, *Macromolecules*, 2014, **47**, 2037.
89. N. Hadjichristidis, M. Xenidou, H. Iatrou, M. Pitsikalis, Y. Poulos, A. Avgeropoulos, S. Sioula, S. Paraskeva, G. Velis, D. J. Lohse, D. N. Schulz, L. J. Fetters, P. J. Wright, R. A. Mendelson, C. A. Garcia-Franco, T. Sun and C. J. Ruff, *Macromolecules*, 2000, **33**, 2424.
90. I. M. Khan, Z. Gao, K. Khougaz and A. Eisenberg, *Macromolecules*, 1992, **25**, 3002.
91. H. Ba-Gia, T. Jerome and P. Teyssie, *J. Polym. Sci., Part A: Polym. Chem.*, 1980, **18**, 3483.
92. A. Naka, K. Sada, Y. Chujo and T. Saegusa, *Polym. Prepr. Jpn.*, 1991, **40**, E1161.
93. A. Takano, M. Okada, T. Nose and T. Fujimoto, *Macromolecules*, 1992, **25**, 3596.
94. F. Wang, J. Roovers and P. M. Toporowski, *Macromol. Symp.*, 1995, **95**, 205.
95. F. Wang, J. Roovers and P. M. Toporowski, *Macromol. Rep.*, 1995, **A32**(suppls. 5, 6), 951.
96. K. Ishizu and K. Kuwahara, *Polymer*, 1994, **35**, 4907.
97. K. Ishizu, S. Yikimasa and R. Saito, *Polym. Commun.*, 1991, **32**, 386.
98. K. Ishizu, S. Yikimasa and R. Saito, *Polymer*, 1992, **33**, 1982.
99. K. Miyata, Y. Watanabe, T. Itaya, T. Tanigaki and K. Inoue, *Macromolecules*, 1996, **29**, 3694.
100. T. Fujimoto, H. Zhang, T. Kazama, Y. Isono, H. Hasegawa and T. Hashimoto, *Polymer*, 1992, **33**, 2208.

101. H. Huckstadt, V. Abetz and R. Stadler, *Macromol. Rapid Commun.*, 1996, **17**, 599.
102. S. Sioula, Y. Tselikas and N. Hadjichristidis, *Macromol. Symp.*, 1997, **117**, 167.
103. O. Lambert, S. Reutenauer, G. Hurtrez, G. Riess and P. Dumas, *Polym. Bull.*, 1998, **40**, 143.
104. M. Hayashi, K. Kojima, S. Nakahama and A. Hirao, *ACS Polym. Prepr.*, 1998, **39**, 478.
105. A. Karatzas, H. Iatrou, N. Hadjichristidis, K. Inoue, K. Sugiyama and A. Hirao, *Biomacromolecules*, 2008, **9**, 2072.
106. D. J. Lohse and N. Hadjichristidis, *Curr. Opin. Colloid Interface Sci.*, 1997, **2**, 171.
107. N. Hadjichristidis, H. Iatrou, S. K. Behal, J. J. Chludzinski, M. M. Disko, R. T. Garner, K. S. Liang, D. J. Lohse and S. T. Milner, *Macromolecules*, 1993, **26**, 5812.
108. D. J. Pochan, S. P. Gido, S. Pispas and J. W. Mays, *Macromolecules*, 1996, **29**, 5099.
109. D. J. Pochan, S. P. Gido, S. Pispas, J. W. Mays, A. J. Ryan, J. P. A. Fairclough, I. W. Hamley and N. J. Terrill, *Macromolecules*, 1996, **29**, 5091.
110. S. T. Milner, *Macromolecules*, 1994, **27**, 2333.
111. L. Yang, S. Hong, S. P. Gido, G. Velis and N. Hadjichristidis, *Macromolecules*, 2001, **34**, 9069.
112. S. Sioula, N. Hadjichristidis and E. L. Thomas, *Macromolecules*, 1998, **31**, 5272.
113. S. Sioula, N. Hadjichristidis and E. L. Thomas, *Macromolecules*, 1998, **31**, 8429.
114. S. Junnila, N. Houbenov, S. Hanski, H. Iatrou, A. Hirao, N. Hadjichristidis and O. Ikkala, *Macromolecules*, 2010, **43**, 9071.
115. S. Junnila, N. Houbenov, A. Karatzas, N. Hadjichristidis, A. Hirao, H. Iatrou and O. Ikkala, *Macromolecules*, 2012, **45**, 2850.
116. N. Hadjichristidis, *J. Polym. Sci., Part A: Polym. Chem.*, 1999, **37**, 867.

CHAPTER 2

# *Precise Synthesis of Multi-Component Miktoarm Star Polymers by a New Conceptual Iterative Methodology Using Living Anionic Polymerization*

SHOTARO ITO*[a], TOMOYA HIGASHIHARA[b], TAKASHI ISHIZONE[a] AND AKIRA HIRAO*[a,c,d]

[a]Tokyo Institute of Technology, Graduate School of Science and Engineering, Organic and Polymeric Materials Department, 2-12-1-S1-13, Ohokayama, Meguro-ku, Tokyo 152-8552, Japan; [b]Yamagata University, Graduate School of Science and Engineering, Department of Organic Device Engineering, 4-3-16 Jonan, Yonezawa, Yamagata 992-8510, Japan; [c]National Taiwan University, Department of Chemical Engineering, No.1, Sec. 4, Roosevelt Road, Taipei 10617, Taiwan; [d]National Chiao Tung University, Department of Applied Chemistry, 1001 University Road, Hsinchu 30010, Taiwan

*E-mail: sito@polymer.titech.ac.jp, ahirao@email.plala.or.jp

Polymer Chemistry Series No. 25
Miktoarm Star Polymers: From Basics of Branched Architecture to Synthesis, Self-assembly and Applications
Edited by Ashok Kakkar

Published by the Royal Society of Chemistry, www.rsc.org

## 2.1 Introduction

Miktoarm star polymers (μSPs), composed of chemically different arms, have attracted attention due to their unique self-assembling behavior in selective solvents or in bulk on the basis of their thermodynamically incompatible arms and branched structure.[1,2] However, the synthesis of μSPs is still challenging because selective multistep reactions corresponding to the introduction of different arms are inevitable.[3–6] The synthesis of μSPs generally requires the use of multifunctional core agents, to which each arm is step by step introduced by either a linking reaction between a polymer and a reaction site of the core or polymerization from an initiation reaction site of the core.[3–6] Therefore, each site must be selectively reacted in each step in order to introduce different arms at the different reaction stages. For this reason, most reported μSPs are composed of only two-component arm segments, and even three-, four-, or five-component arms are highly difficult to introduce in star polymers.[6–8] More complex multi-arm and multi-component μSPs have not yet been synthesized.

In this chapter, we demonstrate that an iterative methodology based on a novel concept is able to overcome this synthetic difficulty. The proposed iterative methodology is designed in such a way that the same reaction site is always reintroduced after the introduction of arm(s) in the reaction sequence, and this reaction sequence can be repeated. Therefore, a significant number of different arm polymers can be successively and, in principle, limitlessly introduced, resulting in the synthesis of multi-arm and multi-component μSPs. Very importantly, since each of the different arms is introduced in each reaction step, selective multistep reactions are not needed in this iterative methodology.

For the development of the iterative methodology, a living anionic polymerization system is always used based on the following advantageous viewpoints: (1) molecular weights (MWs) can be precisely controlled from $10^3$ g $mol^{-1}$ to even $10^6$ g $mol^{-1}$; (2) extremely narrow molecular weight distributions (MWDs), typically $M_w/M_n \le 1.1$ or even $\le 1.05$, are attained; and (3) chain-end anions are highly reactive, but stable under the appropriate conditions.[9–11]

The synthetic strategy of the iterative methodology is divided into the following two categories, as shown in Schemes 2.1 and 2.2. Scheme 2.1 shows the iterative methodology using a difunctional compound, **1**, having two functionalities, **X** and **F**. In the first step, a living anionic polymer reacts with **X** in **1** to introduce **F** at the chain-end. **F** then reacts with another living polymer in a 1:1 addition manner to link two polymer chains, and the **F**$^-$ anion newly generated by the linking reaction reacts *in situ* with **1** to reintroduce **F**. By repeating this reaction sequence involving 'the arm introduction by the linking reaction of a living polymer with **F**′ and 'the reintroduction of the same reaction site of **F**′ *via* the generated **F**$^-$, more polymer chains can be successively linked. Since the entire reaction system is designed to repeatedly reintroduce the same **F** reaction site after introduction of an arm during the synthesis, the arm number can, in principle, be limitlessly increased.

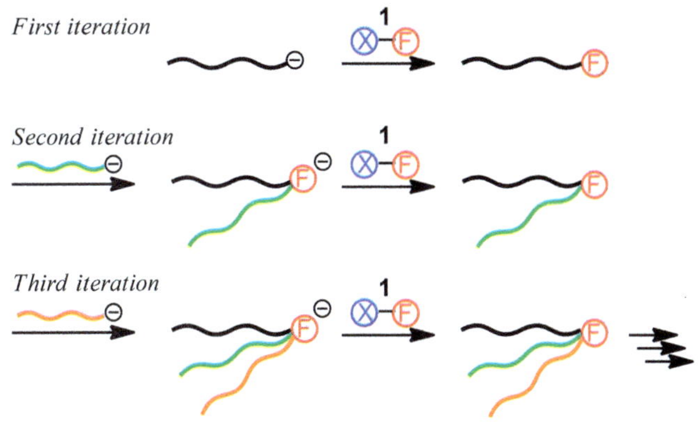


**Scheme 2.1** Iterative methodology using difunctional compound **1**.

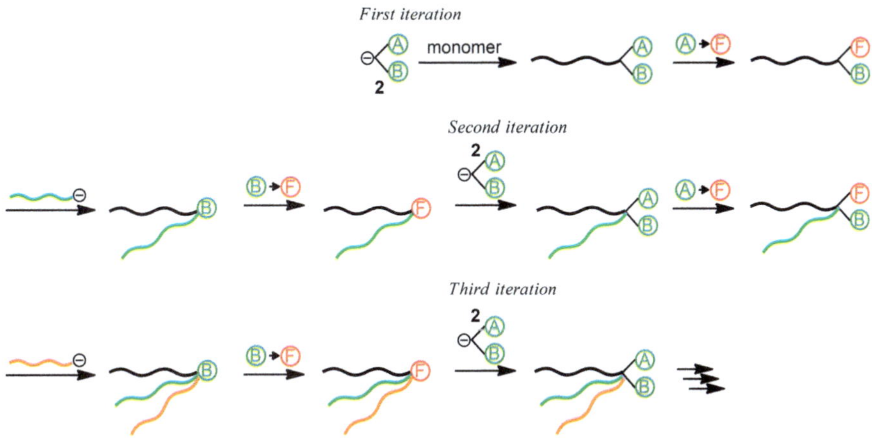


**Scheme 2.2** Iterative methodology using difunctional anionic species **2**.

The advantage of this system is that each arm is introduced *via* the same reaction site in each reaction step, thus avoiding the use of selective multistep reactions.

Scheme 2.2 shows the other iterative methodology with a difunctional anionic species, **2**, possessing two functionalities, **A** and **B**, which are stable toward anions. A monomer is first polymerized with **2** to introduce both **A** and **B** at the same polymer chain-end. **A** is then converted to a reaction site, **F**, which is used for the introduction of an arm by a linking reaction. The remaining **B** between the two polymer segments is next converted to **F**, which is used for the reintroduction of the same **A** and **B** upon reaction with **2**. Since the two functionalities **A** and **B** are convertible to **F** reaction sites and reintroduced, the reaction sequence can be repeated. Again, it should be mentioned that each arm is introduced *via* **F** in each reaction and, therefore, selective multistep reactions are not needed in this methodology.

## 2.2 Synthesis of Multi-Arm and Multi-Component Miktoarm Star Polymers by the Iterative Methodology Using a Difunctional Compound X-F

### 2.2.1 Iterative Methodology Using 1-(4-(3-Bromopropyl)phenyl)-1-phenylethylene

The iterative methodology using 1-(4-(3-bromopropyl)phenyl)-1-phenylethylene, **3**, as the **X-F** compound is shown in Scheme 2.3.[12] The DPE (1,1-diphenylethylene) and propyl bromide moieties in **3** correspond to the **F** and **X** functionalities in Scheme 2.1, respectively.

Since both the DPE and propyl bromide moieties are basically capable of reacting with a living anionic polymer, a specially designed reaction is needed to selectively promote the reaction of only **X** and not **F**. For this purpose, a living polymer 'A' is first end-capped with DPE, followed by reaction with **3**. Since DPE is a non-homopolymerizable monomer, the DPE-end-capped living polymer cannot be further reacted with the DPE moiety in **3**, but selectively and quantitatively reacts with the propyl bromide of **3** in THF at −78 °C, resulting in the introduction of the DPE moiety at the chain-end.

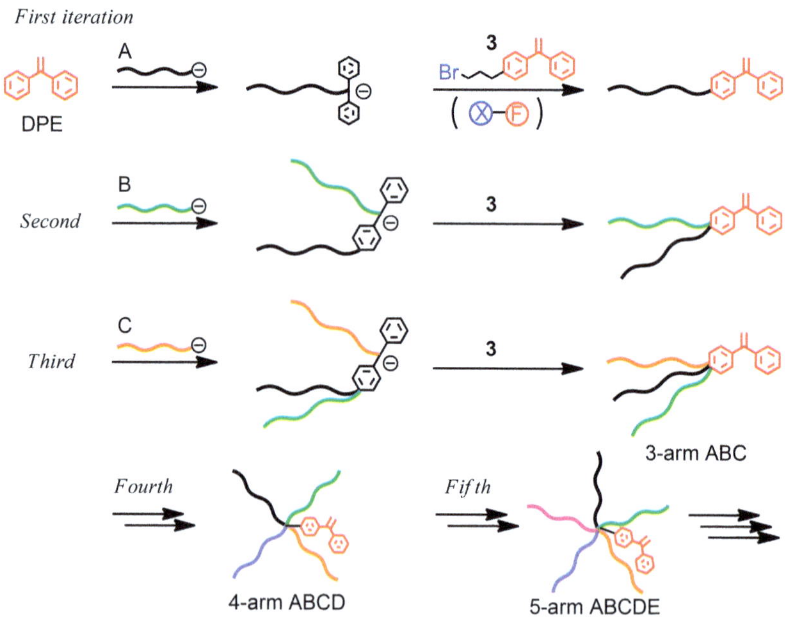


**Scheme 2.3** Iterative methodology using **3**. Adapted from *Progress in Polymer Science*, **36**(3), T. Higashihara, M. Hayashi, A. Hirao. Synthesis of well-defined star-branched polymers by stepwise iterative methodology using living anionic polymerization, 323–375, copyright (2011) with permission from Elsevier.

A living polymer 'B' is then reacted with the DPE moiety in 'A' to afford an in-chain-(DPE anion)-functionalized AB diblock copolymer, followed by reaction of the generated in-chain DPE anion with **3**. Thus, the DPE reaction site is reintroduced. By repeating the same reaction sequence, a different arm can be introduced in the star polymer in each reaction sequence. In practice, the reaction sequence can be iterated three more times to successively synthesize 3-arm ABC, 4-arm ABCD, and 5-arm ABCDE type μSPs. With this methodology, highly reactive living anionic polymers derived from styrene, α-methylstyrene, 1,3-butadiene, isoprene, and their derivatives have been used to introduce arm segments with quantitative efficiency, based on the complete linking reaction of each living polymer with the DPE reaction site.

The well-defined structures of the obtained μSPs were confirmed by $^1$H nuclear magnetic resonance (NMR), static light scattering (SLS), and size exclusion chromatography (SEC) analyses, as listed in Table 2.1, demonstrating their narrow MWDs ($M_w/M_n \leq 1.03$), predictable MWs, and composition. Since the 5-arm ABCDE μSP still possesses a DPE reaction site at the core, the reaction sequence could potentially be repeated to synthesize μSPs with more arms.

### 2.2.2 Iterative Methodology Using 3 and Either 1,3-Bis(1-phenylethenyl)benzene or 1,1-Bis(3-(1-phenylethenyl)phenyl)ethylene

During the iterative methodology mentioned in Section 2.2.1, the number of arm polymers increases one by one by repeating a reaction sequence. In this section, we show an extended iterative methodology, in which 2- or 3-arm polymers are simultaneously introduced in a star polymer at each reaction sequence. Scheme 2.4 illustrates the first example of such a methodology using 1,3-bis(1-phenylethenyl)benzene (**4**) and **3**.[13]

In the first step, two equivalents of a living polymer react with the two DPE moieties in **4** to link the two polymer chains, followed by reaction with **3** to reintroduce two DPE reaction sites. By repeating the reaction sequence

**Table 2.1** Successive synthesis of μSPs by the iterative methodology using **3**.

| | $M_w$ (kg mol$^{-1}$) | | $M_w/M_n$ | Composition (wt%) | |
|---|---|---|---|---|---|
| Type[a] | Calcd | SLS | SEC | Calcd | $^1$H NMR analysis |
| A | 10.5 | 10.7 | 1.03 | 100 | 100 |
| AB diblock | 21.3 | 22.0 | 1.02 | 48/52 | 48/52 |
| 3-Arm ABC | 32.0 | 34.4 | 1.02 | 32/35/33 | 31/35/34 |
| 4-Arm ABCD | 42.8 | 43.0 | 1.02 | 24/29/23/24 | 23/28/23/26 |
| 5-Arm ABCDE | 59.3 | 57.0 | 1.02 | 17/21/17/19/26 | 17/19/18/19/27 |

[a]A, B, C, D, and E are poly(4-trimethylsilylstyrene), poly(4-methoxystyrene), poly(4-methylstyrene), PS, and poly(4-*tert*-butyldimethylsilyloxystyrene), respectively.

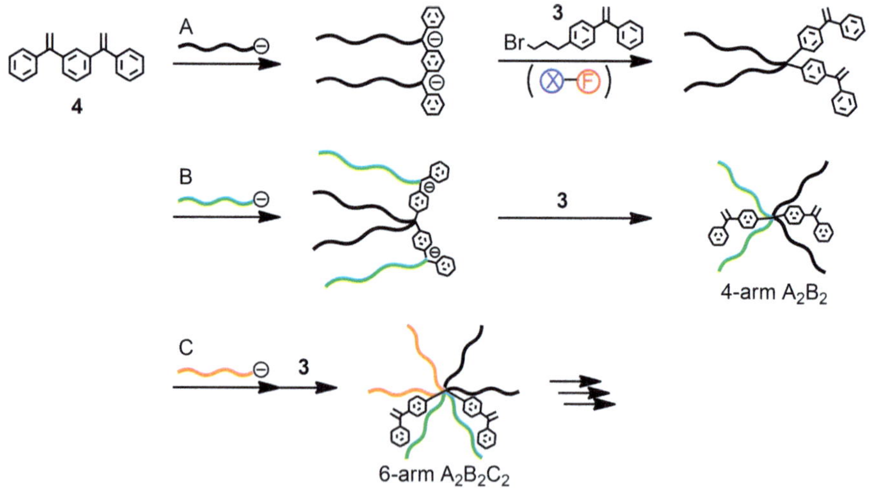


**Scheme 2.4** Iterative methodology using **3** and **4**. Adapted from ref. 13, with permission from John Wiley and Sons.

**Table 2.2** Successive synthesis of μSPs by the iterative methodology using **3** and **4**, and **3** and **5**.

| | $M_w$ (kg mol$^{-1}$) | | $M_w/M_n$ | Composition (wt%) | |
|---|---|---|---|---|---|
| Type | Calcd | SLS | SEC | Calcd | $^1$H NMR analysis |
| 4-Arm $A_2B_2$[a] | 41.1 | 41.8 | 1.02 | 55/45 | 54/46 |
| 6-Arm $A_2B_2C_2$[a] | 61.7 | 62.4 | 1.02 | 37/30/33 | 38/30/32 |
| 6-Arm $A_3B_3$[b] | 67.8 | 66.7 | 1.03 | 50/50 | 52/48 |
| 9-Arm $A_3B_3C_3$[b] | 97.0 | 97.1 | 1.02 | 34/35/31 | 35/34/31 |

[a]A, B, and C are PS, poly(α-methylstyrene), and poly(4-methoxystyrene), respectively.
[b]A, B, and C are PS, poly(4-methoxystyrene), and poly(4-trimethylsilylstyrene), respectively.

shown in Scheme 2.4, two arm polymers are successively introduced at each reaction sequence. In fact, 4-arm $A_2B_2$ and 6-arm $A_2B_2C_2$ type μSPs are readily synthesized by repeating the reaction sequence two more times. The resulting μSPs were found to possess well-controlled MWs and narrow MWDs, as listed in Table 2.2. Thus, the proposed methodology was found to work satisfactorily.

As can be seen in Scheme 2.5, three arms are simultaneously introduced by the iterative methodology using **3** and 1,1-bis(3-(1-phenylethenyl)phenyl) ethylene (**5**) instead of **4**.[14] First, a 3-arm $A_3$ type star polymer having three DPE reaction sites at the core is prepared by linking three equivalents of a living anionic polymer with **5**, followed by reaction with **3**. Using this material as the starting polymer, a similar reaction sequence involving 'introduction of the three arms by a linking reaction' and 'reintroduction of three DPE

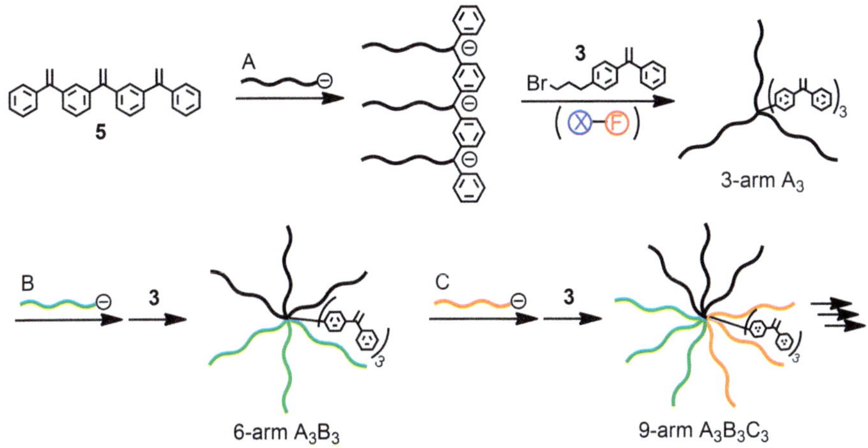


**Scheme 2.5** Iterative methodology using **3** and **5**.

reaction sites at the core by reacting the generated three DPE anions with **3'** is repeated to afford 6-arm $A_3B_3$ and 9-arm $A_3B_3C_3$ μSPs. Their well-defined structures were also confirmed, as listed in Table 2.2.

### 2.2.3 Iterative Methodology Using 3,5-Bis(3-(4-(1-phenylethenyl)phenyl)propoxy)benzyl Bromide

In Sections 2.2.1 and 2.2.2, **3** was used at the reintroduction step of the DPE reaction site by reaction of the DPE anion generated by the preceding linking reaction. Here, 3,5-bis(3-(4-(1-phenylethenyl)phenyl)propoxy)benzyl bromide (**6**), a new compound bearing one benzyl bromide and two DPE moieties, was prepared and used in the iterative methodology.[15]

As observed in Scheme 2.6, the starting chain-end-$(DPE)_2$-difunctionalized polymer 'A' was prepared by reacting a DPE-end-capped living polymer with **6**. Another living polymer 'B' was then reacted with the two DPE termini to link the two polymer chains, followed by reaction with **6** to reintroduce the DPE reaction sites, resulting in a 3-arm $AB_2$ μSP with a core functionalized with four DPE reaction sites. By repeating the same reaction sequence three more times, new and very complex 7-arm $AB_2C_4$, 15-arm $AB_2C_4D_8$, and 31-arm $AB_2C_4D_8E_{16}$ type μSPs were successively synthesized without any difficulty. Unexpectedly, no steric hindrance arose even in the synthesis of the 31-arm star polymer. Thus, by repeating the reaction sequence and using **6**, the number of DPE reaction sites and the corresponding arms increase two-fold at each reaction sequence. Their well-defined structures were confirmed, as listed in Table 2.3.

Both **3** or **6** are reacted with DPE anion(s) for reintroduction of the DPE reaction site(s) in Schemes 2.3 and 2.6. The use of **3** results in the reintroduction

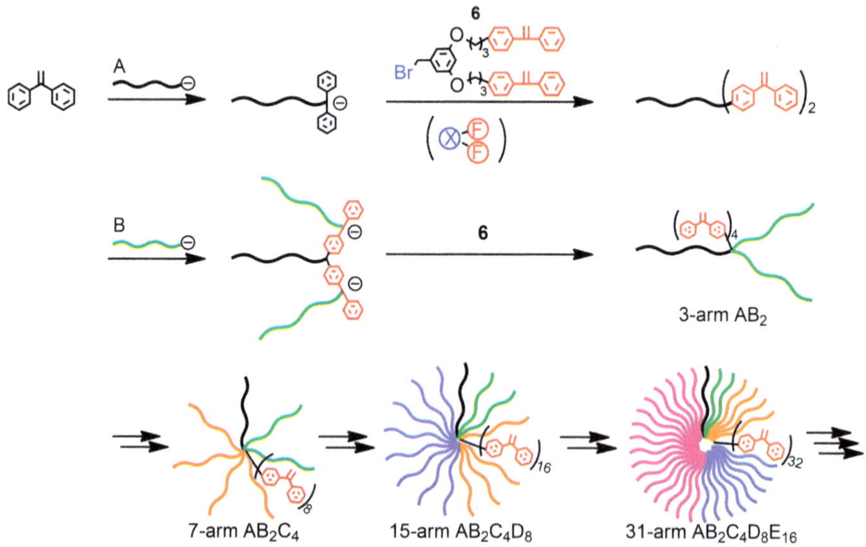


**Scheme 2.6** Iterative methodology using **6**. Adapted with permission from ref. 15, copyright (2006) American Chemical Society.

**Table 2.3** Successive synthesis of μSPs by the iterative methodology using **6**.

| | $M_w$ (kg mol$^{-1}$) | | $M_w/M_n$ | Composition (wt%) | |
|---|---|---|---|---|---|
| Type[a] | Calcd | SLS | SEC | Calcd | $^1$H NMR analysis |
| 3-Arm $AB_2$ | 32.1 | 35.7 | 1.02 | 36/64 | 36/64 |
| 7-Arm $AB_2C_4$ | 77.0 | 79.2 | 1.02 | 16/28/56 | 17/27/56 |
| 15-Arm $AB_2C_4D_8$ | 162 | 172 | 1.02 | 8/12/27/53 | 8/10/30/52 |
| 31-Arm $AB_2C_4D_8E_{16}$ | 339 | 363 | 1.02 | 4/6/13/27/50 | 4/7/15/28/46 |

[a]A, B, C, D, and E are PS, poly(α-methylstyrene), poly(4-methylstyrene), poly(4-methoxystyrene), and poly(4-trimethylsilylstyrene), respectively.

of the same number of DPE reaction sites as that of the DPE anion generated in the linking reaction, while the reintroduced DPE is doubled in number when using **6**. As shown in Scheme 2.7, systematic repetition of the reaction sequence using a combination of **3** and **6** provides a wide variety of μSPs.[16] In the second reaction sequence, for example, 3-arm $AB_2$ type μSPs possessing two and four DPE reaction sites at the core are synthesized using **3** or **6**, respectively. In the next reaction sequence, 5-arm $AB_2C_2$ and 7-arm $AB_2C_4$ with either two or four DPE reaction sites, respectively, are obtained. Further repetition of the reaction sequence affords eight types of μSPs, from 9-arm $AB_2C_2D_2E_2$ to 31-arm $AB_2C_4D_8E_{16}$ stars. As listed in Table 2.4, all the obtained polymers possessed the expected well-defined structures. This clearly indicated that all the reaction sequences proceeded smoothly and quantitatively. Thus, the iterative methodology using **3** and **6** provides access to a series of

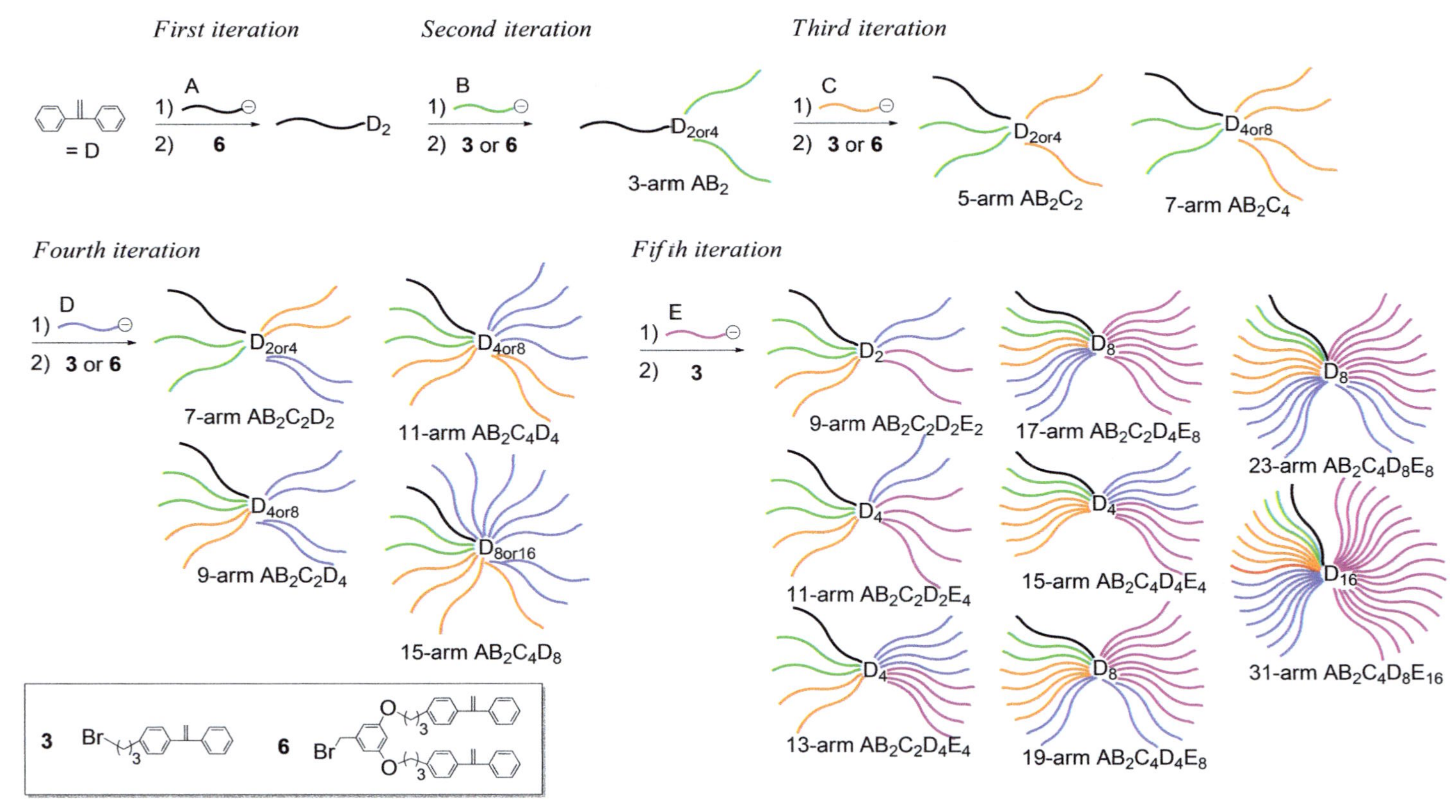


**Scheme 2.7** Iterative methodology using **3** and **6**. Adapted with permission from ref. 16, copyright (2009) American Chemical Society.

**Table 2.4** Successive synthesis of μSPs by the iterative methodology using **3** and **6**.

| Type[a] (Number of DPE moieties at the core) | $M_w$ (kg mol$^{-1}$) | | $M_w/M_n$ | Composition (wt%) | |
|---|---|---|---|---|---|
| | Calcd | SEC-LS | SEC | Calcd | $^1$H NMR analysis |
| 3-Arm $AB_2$ (2) | 33.7 | 32.6 | 1.04 | 31/69 | 30/70 |
| 3-Arm $AB_2$ (4) | 31.5 | 35.0 | 1.02 | 36/64 | 38/62 |
| 5-Arm $AB_2C_2$ (2) | 54.6 | 55.9 | 1.04 | 19/42/39 | 21/39/40 |
| 5-Arm $AB_2C_2$ (4) | 56.5 | 58.0 | 1.02 | 19/41/40 | 17/42/41 |
| 7-Arm $AB_2C_4$ (4) | 74.4 | 74.4 | 1.02 | 15/27/58 | 16/26/58 |
| 7-Arm $AB_2C_4$ (8) | 75.5 | 77.6 | 1.04 | 16/28/56 | 17/27/56 |
| 7-Arm $AB_2C_2D_2$ (2) | 72.3 | 78.4 | 1.05 | 14/32/30/24 | 20/29/29/22 |
| 7-Arm $AB_2C_2D_2$ (4) | 73.2 | 83.1 | 1.05 | 14/32/29/25 | 17/28/29/26 |
| 9-Arm $AB_2C_2D_4$ (4) | 98.4 | 100 | 1.02 | 10/24/23/43 | 10/25/23/42 |
| 9-Arm $AB_2C_2D_4$ (8) | 98.2 | 106 | 1.05 | 11/24/23/42 | 10/25/22/43 |
| 9-Arm $AB_2C_2D_2E_2$ (2) | 87.5 | 92.9 | 1.05 | 12/26/24/20/18 | 14/23/25/18/20 |
| 11-Arm $AB_2C_4D_4$ (4) | 116 | 119 | 1.02 | 10/17/37/36 | 10/17/37/36 |
| 11-Arm $AB_2C_4D_4$ (8) | 118 | 121 | 1.04 | 9/17/37/37 | 10/16/33/41 |
| 11-Arm $AB_2C_2D_2E_4$ (4) | 116 | 109 | 1.04 | 9/20/18/16/37 | 7/18/18/16/41 |
| 13-Arm $AB_2C_2D_4E_4$ (4) | 136 | 150 | 1.05 | 8/17/17/29/29 | 6/19/19/28/28 |
| 15-Arm $AB_2C_4D_8$ (8) | 159 | 163 | 1.04 | 7/9/27/57 | 6/9/28/57 |
| 15-Arm $AB_2C_4D_8$ (16) | 159 | 168 | 1.02 | 8/12/27/53 | 7/10/31/52 |
| 15-Arm $AB_2C_4D_4E_4$ (4) | 158 | 154 | 1.04 | 7/13/27/26/27 | 7/13/27/26/27 |
| 17-Arm $AB_2C_2D_4E_8$ (8) | 171 | 178 | 1.04 | 6/14/13/24/43 | 9/15/15/24/37 |
| 19-Arm $AB_2C_4D_4E_8$ (8) | 188 | 181 | 1.04 | 6/11/23/23/37 | 7/14/25/20/34 |
| 23-Arm $AB_2C_4D_8E_8$ (8) | 243 | 260 | 1.04 | 4/6/18/37/35 | 3/11/21/33/32 |
| 31-Arm $AB_2C_4D_8E_{16}$ (32) | 332 | 356 | 1.02 | 4/6/13/27/50 | 4/7/15/28/46 |

[a] A, B, C, D, and E are PS, poly(α-methylstyrene), poly(4-methylstyrene), poly(4-methoxystyrene), and poly(4-trimethylsilylstyrene), respectively. SEC-LS: size exclusion chromatography with light scattering.

multi-arm and multi-component μSPs, most of which are difficult to synthesize by other methods.

### 2.2.4 Synthesis of Miktoarm Star Polymers Using Intermediate Polymer Anions Prepared by the Iterative Methodology

In Sections 2.2.1 and 2.2.2, a DPE anion at the core of the μSP, which was prepared by a linking reaction, that is, the arm introduction step, was always used for the reintroduction of the DPE reaction site(s) by reaction with **3** or **6**. In this section, the successful utility of such μSPs possessing DPE anion(s) at the core, hereafter called the 'intermediate polymer anion', will be demonstrated for the synthesis of μSPs.

The first example is to use the intermediate polymer anion as a macroinitiator. As shown in Scheme 2.8(a), the intermediate polymer anion, although an A-*b*-B diblock copolymer in-chain anion in this case, was used to initiate the living anionic polymerization of 4-methylphenyl vinyl sulfoxide (MVS) to afford a 3-arm ABC μSP with a poly(MVS) segment.[17] The PMVS segment was then converted to poly(acetylene) by thermal treatment at 150 °C. Thus, the above star polymer was dramatically converted to a new and interesting ABC μSP composed of coil and rigid rod-like conductive arms. With the use of sequentially prepared intermediate polymer anions, 4-arm ABCD and 5-arm ABCDE μSPs bearing a poly(acetylene) segment were also synthesized. These are the first successfully synthesized μSPs composed of coil and rod arm scgmcnts. Nccdlcss to say, their structures are well-controlled, as listed in Table 2.5.

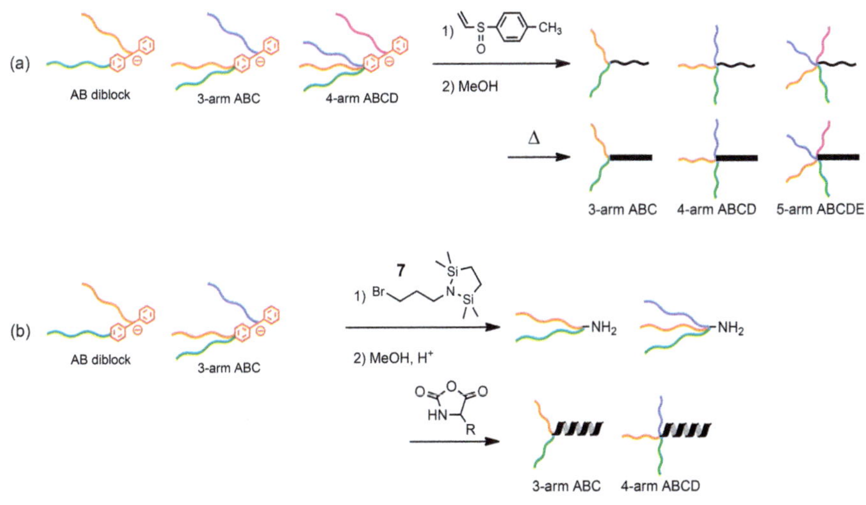


**Scheme 2.8** Synthesis of μSPs containing (a) poly(acetylene) or (b) poly(NCA) segments using intermediate polymer anions.

In a similar manner, intermediate polymers possessing two DPE anions at the core also worked as macroinitiators during the living polymerization of MVS to afford well-defined 3-arm $A_2B$, 4-arm $A_2BC$, 5-arm $A_2BCD$ and $A_2BE_2$, 6-arm $A_2BCE_2$, and 7-arm $A_2BCDE_2$ type rod-coil μSPs (see also Table 2.5).[18]

The intermediate polymer anion is also readily convertible to another function, for example, the amino group, which is an anionic initiator for the living polymerization of α-amino acid *N*-carboxy anhydrides (NCAs), as shown in Scheme 2.8(b).[19] The intermediate polymer anion is first reacted with a terminating agent bearing a protected amino functionality, 7, followed by deprotection. The regenerated amino group then initiates the living anionic polymerization of the γ-benzyl-L-glutamate NCA, resulting in the synthesis of polypeptide-containing μSPs, such as the 3-arm $AB_2$ and ABC and 4-arm $A_2B_2$ and $AB_2C$ μSPs, in which the A segments are helical poly(γ-benzyl-L-glutamate)s. The results are listed in Table 2.6. These miktoarm stars may be interesting from the viewpoints of helical conformation and compatibility with biomaterials.

**Table 2.5** Synthesis of μSPs containing poly(acetylene) segment(s) using intermediate polymer anions.

| | $M_w$ (kg $mol^{-1}$) | | $M_w/M_n$ | Composition (wt%) | |
|---|---|---|---|---|---|
| Type | Calcd | SLS | SEC | Calcd | $^1$H NMR analysis |
| 3-Arm ABC[a] | 31.8 | 32.7 | 1.02 | 38/31/31 | 38/31/31 |
| 4-Arm ABCD[a] | 43.3 | 43.6 | 1.03 | 26/24/24/26 | 29/23/23/25 |
| 5-Arm ABCDE[a] | 51.5 | 53.6 | 1.03 | 17/19/20/23/21 | 18/19/20/23/20 |
| 3-Arm $A_2B$[b] | 33.6 | 29.7 | 1.04 | 73/27 | 70/30 |
| 4-Arm $A_2BC$[b] | 43.1 | 44.5 | 1.02 | 53/23/24 | 53/23/24 |
| 5-Arm $A_2BCD$[b] | 52.5 | 54.3 | 1.03 | 41/21/19/19 | 43/21/18/18 |
| 5-Arm $A_2BE_2$[b] | 51.5 | 53.8 | 1.03 | 42/41/17 | 44/19/37 |
| 6-Arm $A_2BCE_2$[b] | 62.6 | 63.4 | 1.03 | 35/17/16/32 | 22/18/17/43 |
| 7-Arm $A_2BCDE_2$[b] | 75.6 | 75.9 | 1.03 | 31/16/13/13/27 | 32/15/14/14/25 |

[a]A, B, C, D, and E are poly(MVS), PS, poly(α-methylstyrene), poly(4-trimethylsilylstyrene), and poly(4-methoxystyrene), respectively.
[b]A, B, C, D, and E are poly(MVS), PS, poly(α-methylstyrene), poly(4-methoxystyrene), and poly(4-trimethylsilylstyrene), respectively.

**Table 2.6** Synthesis of μSPs containing a poly(NCA) segment using intermediate polymer anions.

| | $M_w$ (kg $mol^{-1}$) | | $M_w/M_n$ |
|---|---|---|---|
| Type[a] | Calcd | SEC-LS | SEC |
| 3-Arm $AB_2$ | 32.0 | 32.6 | 1.05 |
| 3-Arm ABC | 35.0 | 33.3 | 1.09 |
| 4-Arm $A_2B_2$ | 40.2 | 40.4 | 1.20 |
| 4-Arm $AB_2C'$ | 40.0 | 38.9 | 1.08 |

[a]A, B, C, and C′ are poly(γ-benzyl-L-glutamate), PS, PI, and poly(α-methylstyrene), respectively.

## 2.2.5 Iterative Methodology Using 6-Bromo-3-methylene-1-hexene

Throughout the previous sections, the dual-functionality of the DPE reaction site has been highlighted as the key to achieve the iterative methodology. The DPE reaction site first works in the arm introduction step to link the polymer chains, and the DPE anion generated by the linking reaction is used to reintroduce the next DPE reaction site. Herein, a 1,3-butadiene function (Bd) is presented as another reaction site that can be used in the iterative methodology. We found that a Bd moiety introduced into the polymer chain-end shows a dual-functionality similar to that of DPE, as confirmed by the linking reaction of living polystyrene (PS) with chain-end-Bd-functionalized PS in THF at −78 °C, as shown in Scheme 2.9.[20] Under these conditions, the Bd moiety was found to undergo 1 : 1 addition in the linking reaction to afford PS with an in-chain 1,3-butadienyl anion, which *in situ* reacted with 6-bromo-3-methylene-1-hexene, **8**, to reintroduce the same Bd moiety. Interestingly and unexpectedly, neither polymerization nor oligomerization of the Bd moiety occurred at all during the linking reaction in THF at −78 °C.

Based on these results, an iterative methodology using **8** was developed, as shown in Scheme 2.10.[21] The starting chain-end-Bd-functionalized polymer was prepared by the reaction of a DPE-end-capped living polymer with **8**. It was then reacted with another living polymer to link the two polymer chains, followed by reaction of the generated Bd anion with **8** to reintroduce the Bd moiety. This reaction sequence was repeated to successively synthesize 3-arm ABC, 4-arm ABCD, 5-arm ABCDE, and 6-arm ABCDEF type μSPs. Although further repetition of the reaction sequence might have been possible, the utility of the intermediate ABCDEF type μSP anion as a macroinitiator was demonstrated by initiating the living anionic polymerization of 2-vinylpyridine (2VP) to afford a 7-arm ABCDEFG type μSP, as shown in Scheme 2.10. The effectiveness of **8** in the iterative methodology was thus demonstrated by the quantitative synthesis of μSPs with well-defined structures (see Table 2.7).

An additional advantage of the Bd moiety incorporated at the core is its high reactivity through its transformation into epoxide and acid anhydride

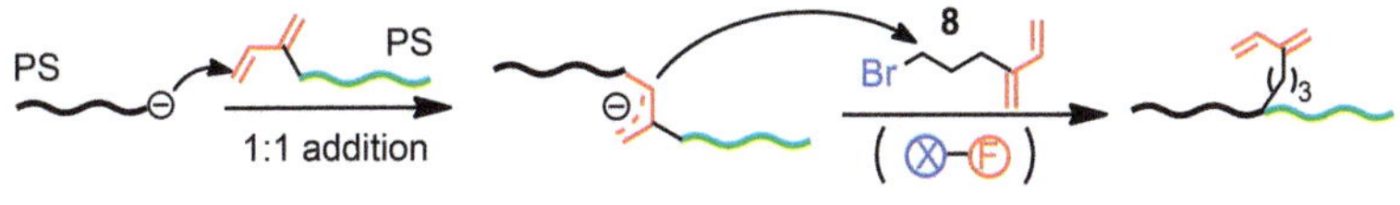


**Scheme 2.9** The 1 : 1 addition reaction of a living PS with a chain-end-Bd-functionalized PS, followed by reintroduction of a Bd moiety using **8**. Adapted from *Progress in Polymer Science*, **36**, 3, T. Higashihara, M. Hayashi, A. Hirao, Synthesis of well-defined star-branched polymers by stepwise iterative methodology using living anionic polymerization, 323–375, copyright (2011) with permission from Elsevier.

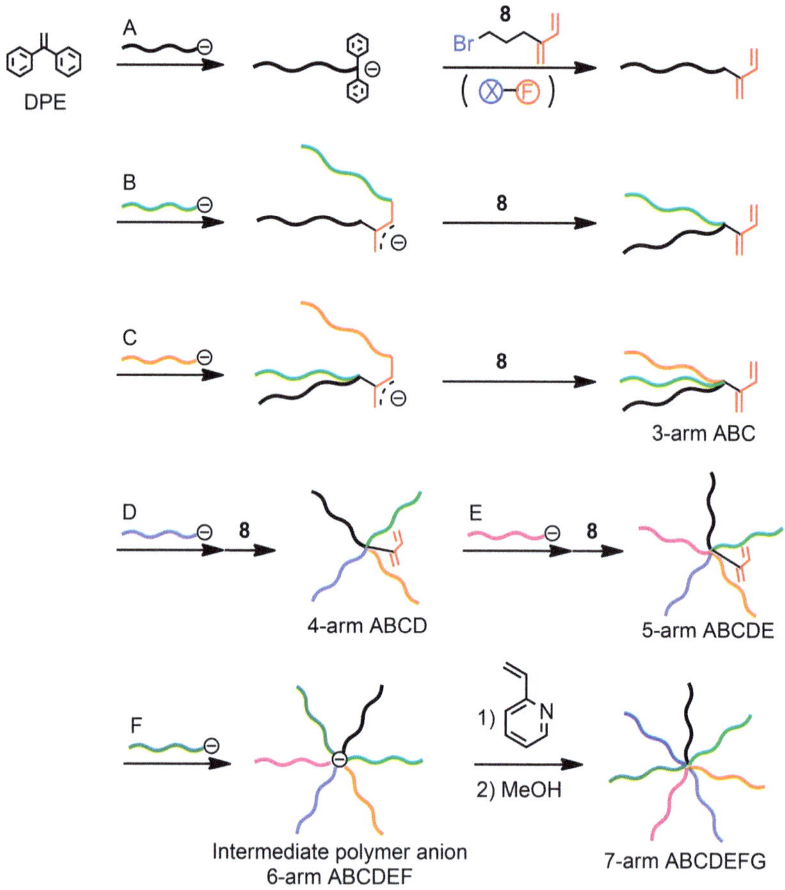


**Scheme 2.10** Iterative methodology using **8**, and utility of the 6-arm ABCDEF type intermediate polymer anion for the polymerization of 2VP. Adapted with permission from ref. 21, copyright (2008) American Chemical Society.

functions by oxidation and Diels–Alder reactions.[20] As shown in Scheme 2.11(a), a chain-end-(epoxide)$_4$-functionalized polymer was prepared by oxidation of a chain-end-(Bd)$_2$-fuctionalized polymer, which then reacted with another living polymer to afford a 5-arm star polymer. The Bd moiety could also be transformed into an anhydride function by the Diels–Alder reaction with maleic anhydride, as shown in Scheme 2.11(b). The anhydride functions are reacted with a chain-end-amino-functionalized polymer to quantitatively afford a 3-arm μSP with two carboxylic acid functions at the core. Since the resulting star polymers have hydroxyl or carboxylic acid functions at their cores, they are potential sources for further functionalization of the star polymers.

**Table 2.7** Successive synthesis of μSPs by the iterative methodology using **8**.

| | $M_w$ (kg mol$^{-1}$) | | $M_w/M_n$ | Composition (wt%) | |
|---|---|---|---|---|---|
| Type[a] | Calcd | SLS | SEC | Calcd | $^1$H NMR analysis |
| A | 11.2 | 11.5 | 1.03 | 100 | 100 |
| AB diblock | 24.2 | 25.3 | 1.04 | 47/53 | 49/51 |
| 3-Arm ABC | 34.8 | 35.9 | 1.01 | 32/36/32 | 33/34/33 |
| 4-Arm ABCD | 47.1 | 47.2 | 1.01 | 24/26/24/26 | 25/27/25/23 |
| 5-Arm ABCDE | 58.8 | 58.5 | 1.02 | 19/22/19/21/19 | 20/21/20/21/18 |
| 6-Arm ABCDEF | 71.8 | 71.5 | 1.03 | 16/18/16/17/16/17 | 17/18/17/17/15/16 |
| 7-Arm ABCDEFG | 77.6 | 76.5 | 1.03 | 15/15/14/15/15/13/13 | 16/16/14/15/15/14/10 |

[a]A, B, C, D, E, F, and G are PS, poly(α-methylstyrene), poly(4-methylstyrene), poly(4-methoxystyrene), poly(4-trimethylsilylstyrene), poly(4-*tert*-butyldimethylsilyloxystyrene), and P2VP, respectively.

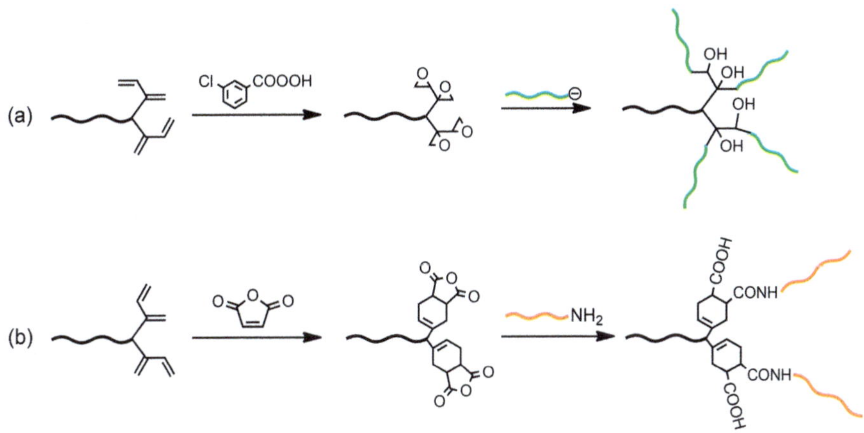


**Scheme 2.11** Transformation of chain-end-Bd-functionalities into (a) epoxide or (b) anhydride moieties, followed by reaction with (a) a living anionic polymer or (b) a chain-end-amino-functionalized polymer to afford μSPs.

## 2.3 Second-Generation Iterative Methodology

In Section 2.2, we described an iterative methodology by which a variety of new multi-arm and multi-component μSPs can be successfully synthesized. However, applicable monomers for the iterative methodology are rather limited since the DPE and Bd functions used as the reaction sites can only react with highly reactive living anionic polymers derived from styrene, 1,3-butadiene, isoprene, and their derivatives. As exceptions, other monomers such as phenyl vinyl sulfoxide, 2VP, alkyl methacrylates (RMAs), and NCAs can be directly or indirectly polymerized with intermediate polymer anions, as already mentioned in Section 2.2.4. After polymerization, however, the iterative methodology can no longer be continued since no reaction site is reintroduced.

Among the monomers feasible for living anionic polymerization, 2VP and RMAs are especially interesting as their polymers have attractive functionalities.[22,23] For example, P2VP can be converted into water-soluble polyelectrolytes by quarternization of the pyridine ring. PRMAs have reactive ester functions as side groups and are capable of carrying various functionalities as their ester groups, as shown in Figure 2.1.

Very unfortunately, both living anionic P2VP and PRMAs are not very reactive with the DPE and Bd reaction sites used in the above-mentioned iterative methodology. In order to apply such less reactive living P2VP and PRMAs to this methodology, a more reactive reaction site is needed. After examining many agents for this purpose, we finally found the α-phenylacrylate (PA) functionality to be the most suitable reaction site.[24] In the model reaction shown in Scheme 2.12, the PA reaction site introduced at the chain-end was observed to completely react with the living P2VP and PRMAs.[25] Furthermore, as expected, this reaction site also quantitatively reacted with living PS, poly(1,3-butadiene) (PBd), and polyisoprene (PI) end-functionalized with DPE.[25] Very importantly, the PA enolate anion generated by the linking reaction did not react with another PA terminus. As with the DPE and Bd reaction

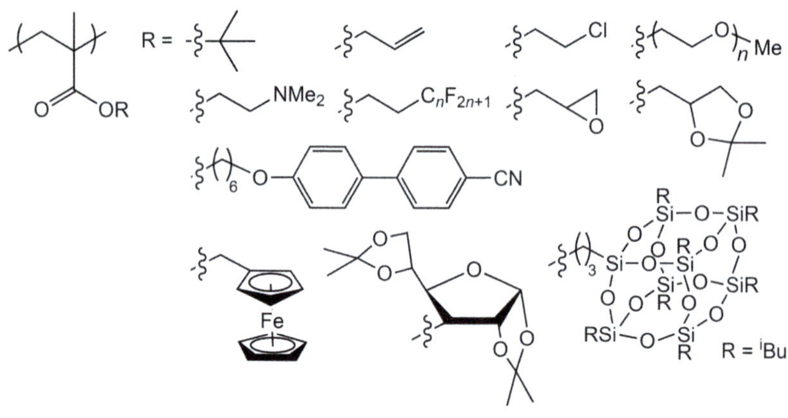


**Figure 2.1** Functional poly(methacrylate)s prepared by living anionic polymerization.

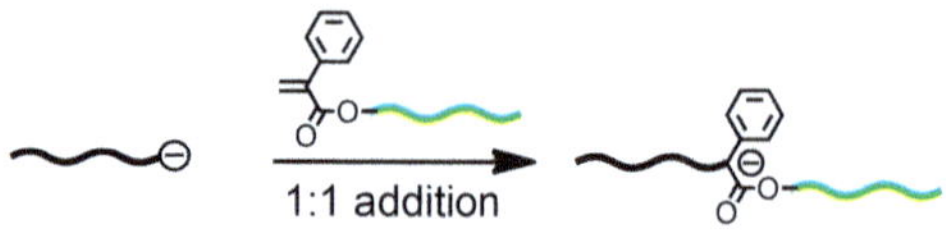


**Scheme 2.12** The 1:1 addition reaction of a living anionic polymer with a chain-end-PA-functionalized polymer. Adapted with permission from ref. 25, copyright (2012) American Chemical Society.

sites, neither the polymerization nor the oligomerization of the PA function occurred at all. Such reactivity of the PA function is expected to be useful as a reaction site for the iterative methodology, making it possible to synthesize μSPs. The methodology in which the PA reaction site is used is called the 'second-generation (2G) iterative methodology'. The synthetic outline has already been illustrated in Scheme 2.2 in the Introduction section and will be developed in the next section.

### 2.3.1 Second-Generation Iterative Methodology Using a Difunctional DPE Anion Bearing Trimethylsilyl and *Tert*-Butyldimethylsilyl Ethers

The first example of 'the 2G iterative methodology' is shown in Scheme 2.13.[25] In this methodology, a new difunctional DPE anion, **9**, possessing trimethylsilyl (TMS) and *tert*-butyldimethylsilyl (TBS) ethers is used as the core agent. These two silyl ethers are protected hydroxyl functionalities. Compound **9** is prepared by the addition reaction of a difunctional DPE derivative bearing TMS and TBS ethers with *sec*-BuLi and used as initiator for the living anionic polymerization of methyl methacrylate (MMA) to introduce both the TMS and TBS ethers at the PMMA chain-end. The TMS ether is then selectively cleaved by treatment with $K_2CO_3$ in MeOH, and the regenerated hydroxyl group is

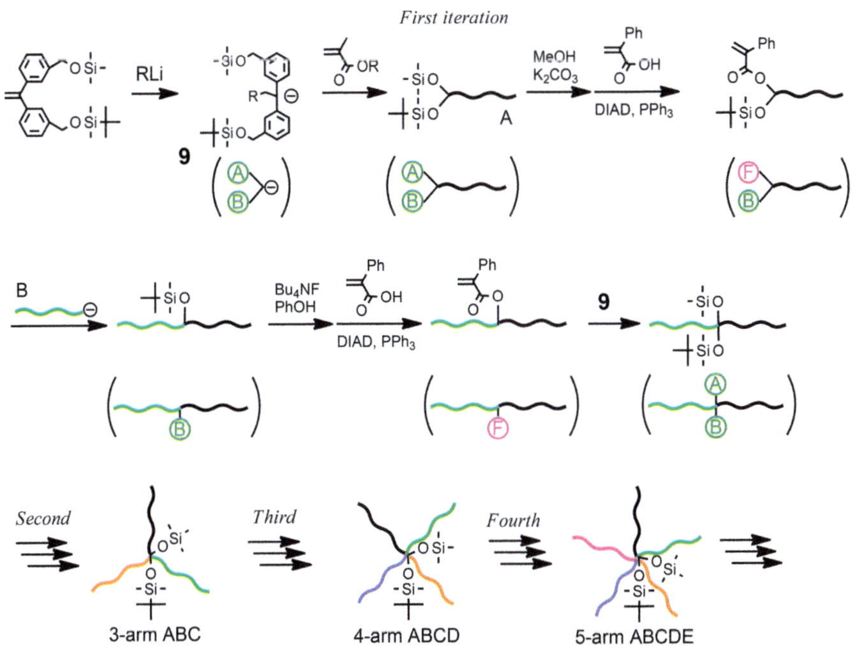


**Scheme 2.13** Iterative methodology using **9**. Adapted with permission from ref. 25, copyright (2012) American Chemical Society.

subsequently transformed into a PA reaction site by a Mitsunobu esterification reaction with α-phenylacrylic acid. Under these conditions, the TBS ether was observed to remain intact. Next, a living poly(ethyl methacrylate) (PEMA) is reacted with the resulting chain-end-(PA and TBS ether)-functionalized PMMA to quantitatively link the two polymer chains, resulting in an in-chain-(TBS ether)-functionalized PMMA-*b*-PEMA diblock copolymer. The remaining TBS ether is transformed into a PA reaction site by deprotection with $Bu_4NF$ and phenol, followed by a Mitsunobu esterification reaction. The functional anion of **9** is then reacted with the PA reaction site to introduce both silyl ethers.

Since the two TMS and TBS ethers transformable into PA reaction sites are reintroduced, the reaction sequence can be repeated three more times to successively synthesize 3-arm ABC, 4-arm ABCD, and 5-arm ABCDE type μSPs. The synthesized μSPs are exclusively composed of all the methacrylate-based polymers, in which the A, B, C, D, and E arms are PMMA, PEMA, poly(*tert*-butyl methacrylate) (P*t*BMA), poly(benzyl methacrylate) (PBnMA), and poly(2-methoxyethyl methacrylate), respectively. As expected, the living anionic polymers derived from the RMAs can be quantitatively linked with the PA reaction sites for introduction in the star polymers. Thus, the proposed 2G iterative methodology using PA reaction sites shown in Scheme 2.13 was found to work satisfactorily. All the μSPs were well-defined in structure, as listed in Table 2.8. Further repetition of the reaction sequence may afford

**Table 2.8** Successive synthesis of μSPs by the iterative methodology using **9** or **10**.

| | $M_w$ (kg mol$^{-1}$) | | $M_w/M_n$ | Composition (wt%) | |
|---|---|---|---|---|---|
| Type | Calcd | SEC-LS | SEC | Calcd | $^1$H NMR analysis |
| A[a] | 11.8 | 13.2 | 1.03 | 100 | 100 |
| AB diblock[a] | 23.6 | 23.1 | 1.03 | 55/45 | 54/46 |
| 3-Arm ABC[a] | 32.2 | 33.8 | 1.03 | 38/31/31 | 37/32/31 |
| 4-Arm ABCD[a] | 43.4 | 43.0 | 1.03 | 30/25/22/23 | 30/26/20/24 |
| 5-Arm ABCDE[a] | 52.0 | 51.5 | 1.03 | 24/21/18/19/18 | 26/20/16/20/18 |
| A[b] | 8.95 | 11.2 | 1.05 | 100 | 100 |
| AB diblock[b] | 23.1 | 22.6 | 1.03 | 47/53 | 48/52 |
| 3-Arm ABC[b] | 35.0 | 34.9 | 1.03 | 30/35/35 | 31/35/34 |
| 4-Arm ABCD[b] | 45.9 | 47.1 | 1.03 | 23/26/27/25 | 24/25/27/24 |
| 5-Arm ABCDE[b] | 56.6 | 53.5 | 1.04 | 20/21/22/20/17 | 20/23/21/20/16 |
| 6-Arm ABCDEF[b] | 62.8 | 64.9 | 1.05 | 17/18/19/17/15/14 | 18/20/18/17/13/14 |
| 7-Arm ABCDEFG[b] | 73.8 | 74.2 | 1.04 | 14/16/17/15/13/13/12 | 16/17/16/14/12/12/13 |

[a]A, B, C, D, and E are PMMA, PEMA, P*t*BMA, PBnMA, and poly(2-methoxyethyl methacrylate), respectively.
[b]A, B, C, D, E, F, and G are poly(cyclohexyl methacrylate), PS, poly(4-methoxystyrene), poly(4-methylstyrene), PMMA, PEMA, and poly(2-methoxyethyl methacrylate), respectively.

μSPs with more arm segments, since the final 5-arm polymer still possesses a PA reaction site at the core.

## 2.3.2 Second-Generation Iterative Methodology Using a Trifunctional DPE Anion Bearing Trimethylsilyl, *Tert*-Butyldimethylsilyl, and 2-Tetrahydropyranyl Ethers

The 2G iterative methodology using a trifunctional anion (**10**) is shown in Scheme 2.14 as an extension of the above-mentioned methodology.[26] The trifunctional anion **10**, prepared from 3-*tert*-butyldimethylsilyloxy-1-propyllithium and substituted DPE with TMS and 2-tetrahydropyranyl (THP) ethers, possesses three ethers, that is, TMS, TBS, and THP ethers, which are selectively transformed into three PA reaction sites one by one at different reaction stages. The first and second PA reaction sites arising from the TMS and TBS ethers are used for arm introduction upon reaction with two different living anionic polymers in different reaction steps, while the third PA obtained from the THP ether is used for the reintroduction of the three different ethers converted into PA reaction sites. Accordingly, two different polymer segments are introduced into a μSP in each reaction sequence *via* this methodology. Furthermore, 3-arm ABC, 4-arm ABCD, 5-arm ABCDE, 6-arm ABCDEF, and even 7-arm ABCDEFG type μSPs can be successively synthesized by repeating the reaction sequence three times. Thus, by using **10** instead of **9**, multi-arm and multi-component μSPs are readily synthesized in fewer reaction steps. The resulting star polymers are all well-defined in structure, as listed in Table 2.8.

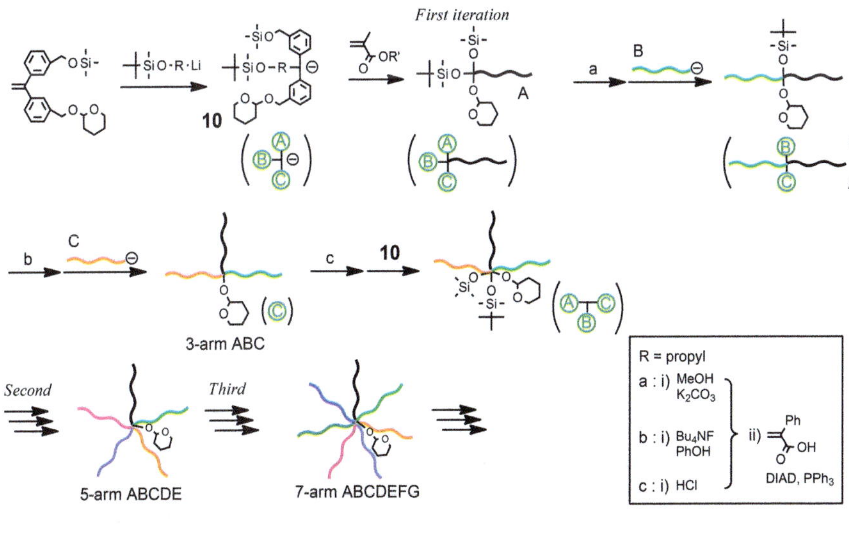


**Scheme 2.14** Iterative methodology using **10**. Adapted with permission from ref. 26, copyright (2013) American Chemical Society.

### 2.3.3 Second-Generation Iterative Methodology Using a DPE Anion Bearing a 1,3-Dioxolane Group

In the 2G iterative methodology, the successful use of a functional DPE anion possessing either two or three different ethers (protected hydroxyl functionalities) sequentially transformable into PA reaction sites has been demonstrated. With these transformation reactions, each of the different arms is step by step introduced in the μSP. In this section, we will introduce another functional DPE anion that works in a similar manner in the 2G iterative methodology.[27]

As shown in Scheme 2.15, a new functional DPE anion, **11**, is first prepared by reaction of DPE substituted with a 1,3-dioxolane group (protected formyl functionality) with *sec*-BuLi and used as the initiator for the living anionic polymerization of MMA to introduce a 1,3-dioxolane at the PMMA chain-end. After deprotection of the 1,3-dioxolane, the same anion, **11**, reacts with the generated formyl function to reintroduce the 1,3-dioxolane along with the production of a hydroxyl group. The hydroxyl group is then transformed into a PA reaction site in a manner similar to that mentioned above, followed by reacting with living PBnMA to link the two polymer chains. Thus, an in-chain-1,3-dioxolane-functionalized PMMA-*b*-PBnMA is quantitatively obtained.

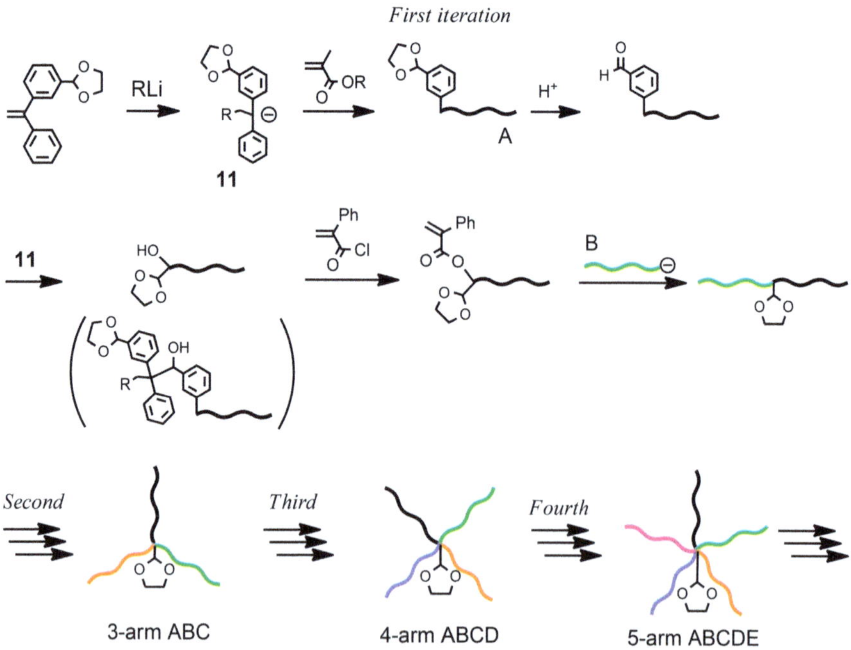


**Scheme 2.15** Iterative methodology using **11**. Adapted from *Polymer*, **54**, 8, R. Goseki, Y. Ozama, E. Akemine, S. Ito, S. Ehara, A. Hirao, Precise synthesis of poly(methacrylate)-based miktoarm star polymers by a new stepwise iterative methodology using a formyl-functionalized 1,1-diphenylethylene, **11**, copyright (2013) with permission from Elsevier.

Since the 1,3-dioxolane can be utilized for 'the arm introduction by a linking reaction with the PA reaction site' and 'the reintroduction of a 1,3-dioxolane group transformable into a PA reaction site', the same reaction sequence can be repeated.

By iterating the reaction sequence three more times, well-defined 3-arm ABC, 4-arm ABCD, and 5-arm ABCDE type μSPs composed of methacrylate-based arms are successively synthesized. The A, B, C, D, and E arms are PMMA, PBnMA, poly(allyl methacrylate), poly(6-(4-(4-cyanophenyl)phenoxy)hexyl methacrylate), and poly(2-*tert*-butyldimethylsilyloxy)ethyl methacrylate, respectively, convertible to poly(2-hydroxyethyl methacrylate). The results are summarized in Table 2.9. The success of the 2G iterative methodology shown in Scheme 2.15 using **11** is thus established.

## 2.3.4 Second-Generation Iterative Methodology Using 9 and In-Chain Block Copolymer Anions

We previously reported that an in-chain block copolymer anion enables the simultaneous introduction of two different polymer chains in a μSP by reaction with a chain-end-(benzyl bromide)$_2$-functionalized polymer, as shown in Scheme 2.16.[28,29] If an in-chain block copolymer anion is used instead of the

**Table 2.9** Successive synthesis of μSPs by the iterative methodology using **11**.

| | $M_w$ (kg mol$^{-1}$) | | $M_w/M_n$ | Composition (wt%) | |
|---|---|---|---|---|---|
| Type | Calcd | SEC-LS | SEC | Calcd | $^1$H NMR analysis |
| A[a] | 13.3 | 13.3 | 1.03 | 100 | 100 |
| AB diblock[a] | 23.4 | 22.8 | 1.07 | 58/42 | 61/39 |
| 3-Arm ABC[a] | 32.4 | 32.1 | 1.03 | 40/29/31 | 41/30/29 |
| 4-Arm ABCD[a] | 40.3 | 39.4 | 1.04 | 33/24/23/20 | 33/24/24/19 |
| 5-Arm ABCDE[a] | 52.2 | 50.6 | 1.05 | 26/18/18/14/24 | 26/19/18/15/22 |

[a]A, B, C, D, and E are PMMA, PBnMA, poly(allyl methacrylate), poly(6-(4-(4-cyanophenyl) phenoxy)hexyl methacrylate) and poly(2-*tert*-butyldimethylsilyloxyethyl methacrylate), respectively.

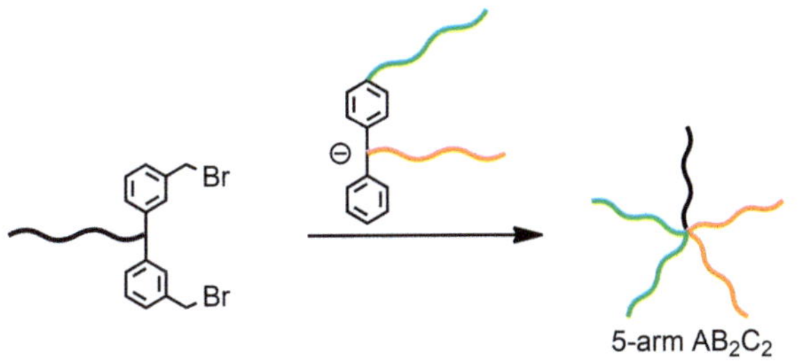


**Scheme 2.16** Linking reaction of an in-chain block copolymer anion with chain-end-(benzyl bromide)$_2$-functionalized polymer to afford 5-arm $AB_2C_2$ type μSPs. Adapted with permission from ref. 29, copyright (2002) American Chemical Society.

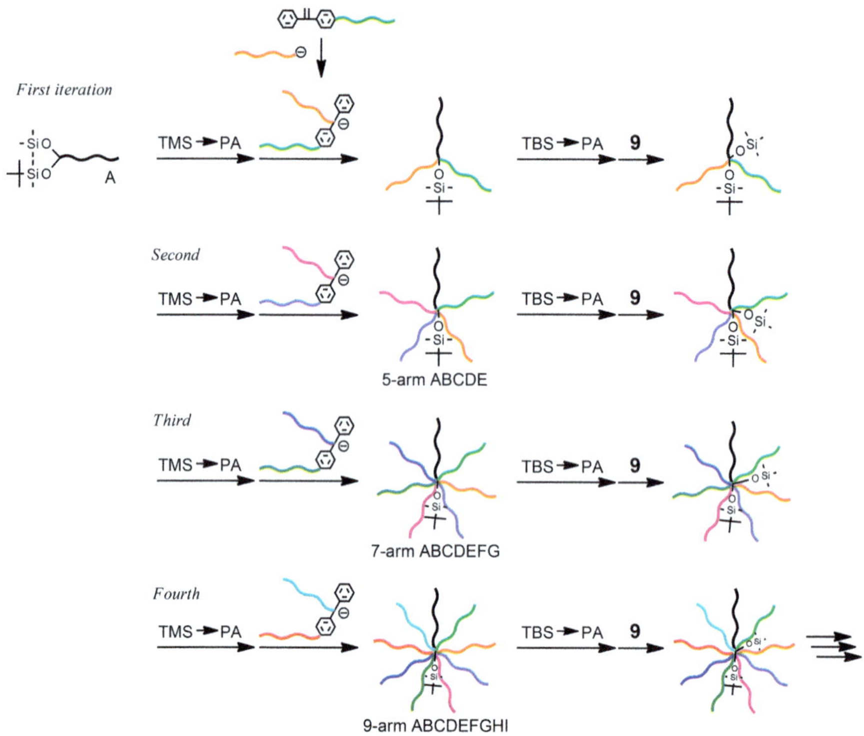


**Scheme 2.17** Iterative methodology using **9** and in-chain block copolymer anions. Adapted from ref. 30 with permission from John Wiley and Sons.

**Table 2.10** Successive synthesis of μSPs by the iterative methodology using **9** and in-chain block copolymer anions.

| | $M_w$ (kg mol$^{-1}$) | | $M_w/M_n$ | Composition (wt%) | |
|---|---|---|---|---|---|
| Type[a] | Calcd | SEC-LS | SEC | Calcd | $^1$H NMR analysis |
| 3-Arm ABC | 27.9 | 27.9 | 1.02 | 35/32/33 | 33/33/34 |
| 5-Arm ABCDE | 49.8 | 51.6 | 1.02 | 20/18/18/22/22 | 20/18/18/21/23 |
| 7-Arm ABCDEFG | 71.5 | 70.3 | 1.03 | 13/13/13/15/15/16/14 | 14/12/12/15/16/18/13 |
| 9-Arm ABCDEFGHI | 88.9 | 89.7 | 1.03 | 11/10/10/12/12/13/12/9/11 | 11/10/9/12/13/14/10/9/12 |

[a]A, B, C, D, E, F, G, H, and I are PBnMA, PS, poly(4-methylstyrene), poly(α-methylstyrene), P*t*BMA, poly(4-octylstyrene), poly(4-methoxystyrene), PI, and poly(3-*tert*-butyldimethylsilyloxymethylstyrene), respectively.

living anionic polymer in the iterative methodology shown in Scheme 2.13 (see Section 2.3.1), two different arm segments can be simultaneously introduced at each reaction sequence.

As shown in Scheme 2.17, the in-chain block copolymer anion is prepared by the reaction of a living polymer with a chain-end-DPE-functionalized polymer and *in situ* reacted with the chain-end-(PA and TBS ether)-functionalized polymer.[30] Indeed, this linking reaction quantitatively proceeds to synthesize a core-(TBS ether)-functionalized 3-arm ABC type μSP. After transforming the TBS ether into a PA reaction site, the iterative process using in-chain block copolymer anions is continued to synthesize well-defined 5-arm ABCDE, 7-arm ABCDEFG, and even 9-arm ABCDEFGHI type μSPs. These results are summarized in Table 2.10. The final 9-arm star polymer is obtained by repeating the reaction sequence only four times. Thus, the reaction steps are significantly reduced by using in-chain block copolymer anions for the synthesis of multi-arm and multi-component μSPs.

## 2.4 Conclusions

Throughout this chapter, we have described the effective utility of a methodology based on the novel concept of an 'iterative approach' for the precise synthesis of multi-arm and multi-component μSPs. The iterative methodology always consists of two steps, *i.e.*, 'arm introduction' and 'reintroduction of the same reaction site' at each reaction sequence. By repeating the reaction sequence, arm segments can be successively and, in principle, limitlessly introduced in a μSP, resulting in the formation of various types of complex μSPs. In this methodology, each arm segment can be introduced at each reaction step, thus avoiding the use of selective multistep reactions along with difficult and limited synthetic designs. Successful construction of such complex star-branched structures is undoubtedly attributed to the use of a living anionic polymerization system, which realizes the reaction sequence consisting of 'arm introduction' and 'reintroduction of the reaction site' steps in quantitative efficiencies and the introduction of precisely controlled arm segments.

DPE and Bd reaction sites are employed in the first iterative methodology mentioned in Section 2.2. Although they effectively work, the use of highly reactive living anionic polymers derived from styrene and 1,3-diene monomers is essential in order to react with the DPE or Bd reaction sites. As exceptions, other monomers, such as phenyl vinyl sulfoxide, 2VP, RMA, and NCA, can be directly or indirectly polymerized with the intermediate polymer anion, as shown in Section 2.2.4. After polymerization, however, the iterative methodology can be no longer continued, since no reaction site is reintroduced.

As mentioned in Section 2.3, we have developed a 2G iterative methodology using the PA functionality as a new reaction site, which allows the use of almost all types of living anionic polymers, such as PS, PBd, PI, P2VP, PRMAs, and their derivatives. Thus, not only structural variations of μSPs can be synthesized, but also the chemical composition of arm polymers is significantly broadened. For example, multi-component μSPs possessing five to nine arm polymers of PS, PBd, PI, P2VP, and/or PRMAs were successfully synthesized for the first time. Moreover, the synthesis of novel multi-component μSPs containing helical poly(NCA) or rod-like poly(acetylene) segments has been achieved using intermediate polymer anions generated during the iterative methodology, as shown in Scheme 2.8. Thus, the general synthesis of well-defined multi-arm and multi-component μSPs by the iterative methodology has been established.

## References


1. H. Hasegawa, *Complex Macromolecular Architectures: Synthesis, Characterization, and Self-Assembly*, ed. N. Hadjichristidis, A. Hirao, Y. Tezuka and F. Du Prez, John Wiley & Sons (Asia) Pte Ltd, 2011, pp. 569–591.
2. N. Hadjichristidis, H. Iatrou, M. Pitsikalis, S. Pispas and A. Avgeropoulos, *Prog. Polym. Sci.*, 2005, **30**, 725–782.
3. A. Hirao, M. Hayashi, T. Higashihara and N. Hadjichristidis, in *Complex Macromolecular Architectures: Synthesis, Characterization, and Self-Assembly*, ed. N. Hadjichristidis, A. Hirao, Y. Tezuka and F. Du Prez, John Wiley & Sons (Asia), Singapore, 2011, pp. 97–132.
4. N. Hadjichristidis, M. Pitsikalis, S. Pispas and H. Iatrou, *Chem. Rev.*, 2001, **101**, 3747–3792.
5. N. Hadjichristidis, H. Iatrou, M. Pitsikalis and J. Mays, *Prog. Polym. Sci.*, 2006, **31**, 1068–1132.
6. T. Higashihara, M. Hayashi and A. Hirao, *Prog. Polym. Sci.*, 2011, **36**, 323–375.
7. C. Ye, G. Zhao, M. Zhang, J. Du and Y. Zhao, *Macromolecules*, 2012, **45**, 7429–7439.
8. H. Liu, W. Pan, M. Tong and Y. Zhao, *Polym. Chem.*, 2016, 1603–1611.
9. *Anionic Polymerization: Principles and Practical Applications*, ed. H. Hsieh and R. P. Quirk, Marcel Dekker, Inc., New York, Basel, Hong Kong, 1996.
10. A. Hirao, R. Goseki and T. Ishizone, *Macromolecules*, 2014, **47**, 1883–1905.
11. *Anionic Polymerization: Principles, Practice, Strength, Consequences and Applications*, ed. N. Hadjichristidis and A. Hirao, Springer, Japan, Tokyo, 2015.
12. T. Higashihara, K. Inoue, M. Nagura and A. Hirao, *Macromol. Res.*, 2006, **14**, 287–299.
13. A. Hirao, M. Hayashi and T. Higashihara, *Macromol. Chem. Phys.*, 2001, **202**, 3165–3173.
14. A. Hirao and T. Higashihara, *Macromolecules*, 2002, **35**, 7238–7245.
15. A. Hirao, T. Higashihara, M. Nagura and T. Sakurai, *Macromolecules*, 2006, **39**, 6081–6091.
16. T. Higashihara, T. Sakurai and A. Hirao, *Macromolecules*, 2009, **42**, 6006–6014.
17. Y. Zhao, T. Higashihara, K. Sugiyama and A. Hirao, *J. Am. Chem. Soc.*, 2005, **127**, 14158–14159.
18. Y. Zhao, T. Higashihara, K. Sugiyama and A. Hirao, *Macromolecules*, 2007, **40**, 228–238.
19. A. Karatzas, H. Iatrou, N. Hadjichristidis, K. Inoue, K. Sugiyama and A. Hirao, *Biomacromolecules*, 2008, **9**, 2072–2080.

20. K. Sugiyama, K. Inoue, T. Higashihara, H. Hayashi and A. Hirao, *React. Funct. Polym.*, 2009, **69**, 480–492.
21. A. Hirao, T. Higashihara and K. Inoue, *Macromolecules*, 2008, **41**, 3579–3587.
22. A. Hirao, T. Ishizone and K. Sugiyama, in *Polymer Science: A Comprehensive Reference*, ed. K. Matyjaszewski and M. Möller, Elsevier BV, Amsterdam, 2012, pp. 591–621.
23. A. Hirao and T. Ishizone, in *Synthesis of Polymers: New Structures and Methods*, ed. D. Schlüter, C. J. Hawker and J. Sakamoto, Wiley-VCH Verlag GmbH&Co. KGaA, 2012, pp. 81–133.
24. K. Sugiyama, T. Oie, A. A. El-Magd and A. Hirao, *Macromolecules*, 2010, **43**, 1403–1410.
25. S. Ito, R. Goseki, S. Senda and A. Hirao, *Macromolecules*, 2012, **45**, 4997–5011.
26. S. Ito, R. Goseki, T. Ishizone, S. Senda and A. Hirao, *Macromolecules*, 2013, **46**, 819–827.
27. R. Goseki, Y. Ozama, E. Akemine, S. Ito, S. Ehara and A. Hirao, *Polymer*, 2013, **54**, 2049–2057.
28. A. Hirao, Y. Tokuda, K. Morifuji and M. Hayashi, *Macromol. Chem. Phys.*, 2001, **202**, 1606–1613.
29. A. Hirao and Y. Tokuda, *Macromolecules*, 2003, **36**, 6081–6086.
30. S. Ito, R. Goseki, I. Manners, T. Ishizone and A. Hirao, *Macromol. Chem. Phys.*, 2015, **216**, 1523–1533.

CHAPTER 3

# *Facile Synthesis of Multicomponent Star Copolymers* via *Controlled Polymerization and Click Chemistry*

YOULIANG ZHAO[a]

[a]Suzhou Key Laboratory of Macromolecular Design and Precision Synthesis, Jiangsu Key Laboratory of Advanced Functional Polymer Design and Application, State and Local Joint Engineering Laboratory for Novel Functional Polymeric Materials, College of Chemistry, Chemical Engineering and Materials Science, Soochow University, Suzhou 215123, China
*E-mail: ylzhao@suda.edu.cn

## 3.1 Introduction

The precise synthesis of complex topological polymers with versatile properties and multipurpose applications has been a recurring topic in current polymer science.[1–10] Polymers with low polydispersity (PDI) and well-defined architecture are essential to establish structure–property correlations, which allow us to rationally design macromolecules with predetermined functions and properties. The tremendous progress in living/controlled polymerizations

Polymer Chemistry Series No. 25
Miktoarm Star Polymers: From Basics of Branched Architecture to Synthesis, Self-assembly and Applications
Edited by Ashok Kakkar

Published by the Royal Society of Chemistry, www.rsc.org

and highly efficient linking reactions involving click chemistry has allowed for the facile construction of a wide range of macromolecules. Thus far, a number of living/controlled polymerization techniques, such as living anionic polymerization (LAP),[1–6] nitroxy-mediated stable free-radical polymerization (NMP or SFRP),[11,12] atom transfer radical polymerization (ATRP),[13–16] reversible addition–fragmentation chain transfer (RAFT) polymerization,[17–20] and ring-opening polymerization (ROP), have been developed, and some linking reactions have also been adopted for polymer synthesis. Among various coupling reactions, 'click reactions', a term coined by Sharpless and coworkers,[21] have attracted much attention due to their high specificity, quantitative yield, and tolerance toward most functional groups. Besides the most popular copper(I)-catalyzed azide–alkyne cycloaddition (CuAAC, Scheme 3.1a)

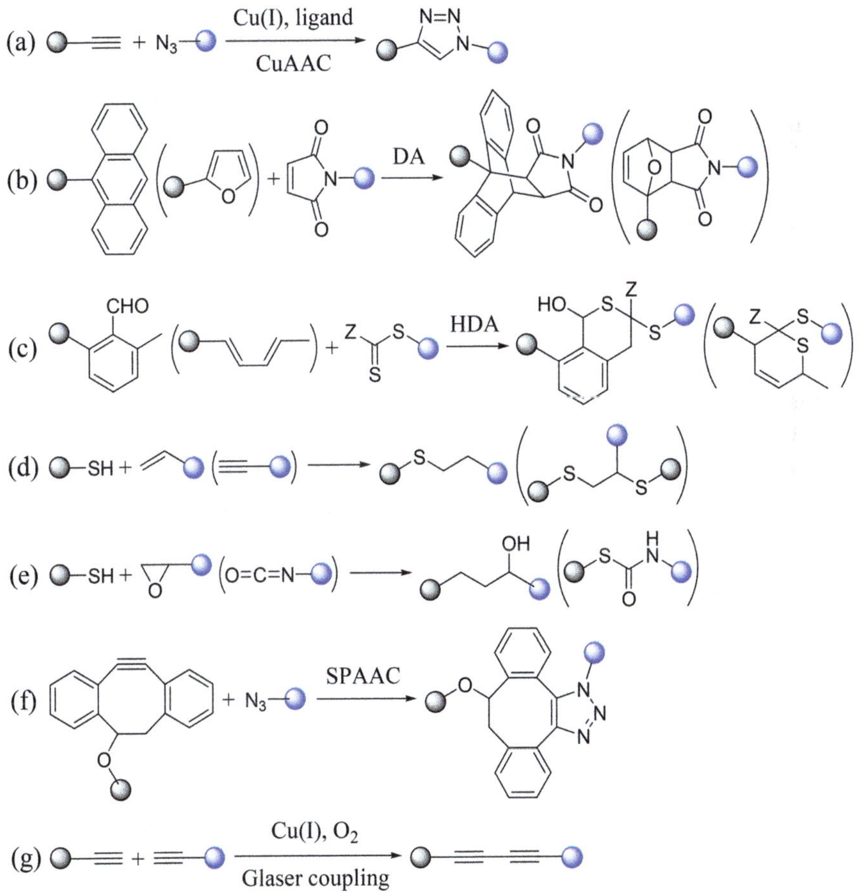


**Scheme 3.1** Schematic representation of the click reactions such as CuAAC (a), DA (b), HDA (c), thiol-ene/yne (d), thiol-epoxy/isocyanate (e), SPAAC (f) and Glaser coupling (g) exploited for the preparation of macromolecular architectures, in which the spherical symbol denotes a small molecular compound, linear (co)polymer, V-shaped (co)polymer, or star (co)polymer.

reactions,[21] other click or click-like reactions such as normal (DA, b) or hetero (HDA, c) Diels–Alder cycloaddition reactions, thiol-based reactions (d and e), strain promoted azide–alkyne cycloadditions (SPAAC, f),[22] and Glaser couplings (g) have been efficiently applied in polymer science. At present, the family of non-linear polymers has been drastically expanded owing to the rapid progress in synthetic methodologies. In addition to linear polymers, other non-linear polymers such as star, graft, branched, and dendrimer-like polymers and their derivatives have been successfully prepared. Because of the branching effect, non-linear polymers usually exhibit distinct properties and applications as compared to their linear analogues.

Star polymers, consisting of three or more linear arm segments and a central core, have attracted the interest of researchers because they constitute the simplest branched architecture.[1–7] The earliest attempt to synthesize star polymers was conducted by Schaefgen and Flory in 1948, and tetra- and octachain star-shaped polyamides were achieved by polymerizing ε-caprolactam in the presence of either cyclohexanonetetrapropionic or dicyclohexanoneoctacarboxylic acid.[23] In 1962, Morton and coworkers synthesized 4-arm star polystyrenes (PSt) *via* neutralization of living polystyryl lithium with tetrachlorosilane.[24] Although the products were mixtures of 3- and 4-arm PSt stars, this work eventually led to the synthesis of star polymers with arm numbers up to 128.[25] In 1989, Roovers and coworkers prepared star polybutadienes (PB) with 200 and 270 arms using hydrosilylated, low molecular-weight linear or star 1,2-polybutadiene as the linking agent.[26]

With increasing arm numbers and chemical compositions, star polymers can be roughly classified into different types. First, from the chemical compositions of arm segments in the outlayer, they can be subdivided into homoarm, miktoarm, and multicomponent stars. Second, the inner core may be comprised of a small molecule, linear oligomer, dendrimer, branched, and even a cross-linked polymer. Third, the linkages between the core and arms can be normal/dynamic covalent bonds, ionic bonds, and supramolecular interactions. Fourth, the core and arms can be inert, stimulus-responsive, or stimulus-cleavable linkages/segments. Last, according to the uniformity of the molecular weight and composition, they may be defined as well/ill-defined stars, symmetric/asymmetric stars, homo/heteroarm stars involving co-, ter-, quarter-, quintopolymers, *etc.*

Thus far, a wide range of star polymers with unequal arms (molecular weight asymmetry) or chemically different arms (chemical asymmetry) has been synthesized.[1–6] The term 'miktoarm stars' originates from the greek word μικτός (abbreviated as μ stars) and was adopted for stars with chemical asymmetry.[1] Compared to linear block copolymers, diverse architectures of star copolymers have been explored with the expectation of achieving different morphologies or physical properties induced by distinct molecular architectures.[27–36] Moreover, the introduction of star topology with enhanced compositions into multiphase materials further extends the types of morphological suprastructures and molecular assemblies formed in bulk and selected solvents, and the resultant nanoscale ordered materials may act as important devices with promising applications.

The synthesis of well-defined μ stars is extremely important since it can underline the advance of research on the elucidation of morphological evolution and special applications. To this end, living/controlled polymerization techniques are necessary for the growth of arm segments, and the linking reactions are also combined to bridge the core and arms in some cases. Considering the advantages, such as precise synthesis with molecular weights up to $10^6$ Da, extremely low polydispersity (PDI = 1.01–1.05), quantitative functionality of linking moieties, and high reactivity, LAP is the most preferred system in comparison with other living/controlled polymerization systems.[1–6] Compared to regular stars with the same types of arm segments, the preparation of μ stars is more challenging. The primary reasons lie in the selective multistep reactions required to introduce a series of different arms and the isolation of low-component polymers to obtain the target stars. To address this synthetic challenge, various methodologies have been adopted to generate miktoarm stars. In this chapter, the latest efforts leading to various types of μ stars with covalent linkages are summarized.

## 3.2 Miktoarm Stars Synthesized by Living/Controlled Polymerization

### 3.2.1 'Core First' Approach

In this method, a multifunctional agent (MFA) is initially synthesized and employed to initiate/mediate the polymerization of vinyl monomers, and the arms of the star polymer are formed during the polymerization process. The multifunctional initiators or chain transfer agents (CTAs) can be synthesized by organic reactions or living/controlled polymerization in the presence of a cross-linker/inimer, in which an inimer is widely used for the controlled synthesis of hyperbranched agents *via* self-condensing vinyl polymerization (SCVP).[37,38] As the MFA has defined functional reactive sites, it allows for the precise synthesis of well-defined multiarm and miktoarm stars with predetermined arm number ($f$), in which the $f$ value per star polymer is equal to the number of initiating functionalities if quantitative initiation is available. As small molecules are used as the initiator, $f$ is usually less than 10. In addition to homoarm stars, various polymerization techniques enable the precise synthesis of star block copolymers. For example, Aksakal *et al.* reported that 3-arm star-shaped copolymers comprising ABABC-type pentablocks could be efficiently prepared in less than 90 min *via* aqueous single electron transfer living radical polymerization (SET-LRP), in which the arm segments consisted of hydrophilic poly(*N*-isopropyl acrylamide) (PNIPAM, A), poly(*N*,*N*-dimethyl acrylamide) (PDMA, B), and poly(2-hydroxyethyl acrylate) (PHEA, C) segments.[39]

By increasing the functionality in initiators such as β-cyclodextrins (CDs)[40–42] and dendrimers,[43–46] the resultant star-like polymers exhibit a predetermined arm number and well-controlled arm length. For instance, we have previously synthesized a series of 6-, 12-, and 24-arm dendrimer-like star polymers using generation 1–3 Fréchet-type polyarylether dendrimer initiators *via* ATRP and ROP, in which the homoarms and diblock arms consisted of poly(L-lactide)s (PLLA), PSt, poly(methyl methacrylate) (PMMA), and poly(*tert*-butyl acrylate) (P*t*BA) segments.[45,46] Although ideal stars with well-controlled architecture and composition can be achieved, the polymerization conditions should be carefully optimized in some cases. Otherwise, undesired star polymers with reduced *f* and non-uniform chain length will remain in the last products, leading to more or less defects in the topology and composition. Meanwhile, the arm number of star polymers can be further enhanced as hyperbranched[47–53] and cross-linked polymers act as the inner core. With the aid of RAFT SCVP using functional inimers, we have synthesized hyperbranched polymers with multiple reactive sites, followed by chain extension polymerization and a linking reaction to generate multiarm and miktoarm stars with branched cores[51–53] (Scheme 3.2) and dendritic comb-like and toothbrush-like copolymers[54] with alternating poly(ethylene glycol) (PEG) and poly(ε-caprolactone) (PCL) grafts. More recently, Gao and coworkers have described synthesis of hyperstar polymers *via* sequential ATRP of inimers and functional monomers in aqueous dispersed media, in which core–shell structured hyperstar polymers containing a hyperbranched core and hundreds of radiating arms were synthesized in a one-pot process.[55] In addition, controlled radical polymerization (CRP) of a cross-linker under dilute conditions afforded a multifunctional nanogel bearing a statistical distribution of initiating sites, followed by polymerization of vinyl monomers to form core-cross-linked stars (CCS).[8,9,56–60]

Moreover, μ stars can be generated by tandem or stepwise polymerization processes using a heterofunctional initiator with several different initiating species based on CRP and ROP techniques. Under optimized conditions, the number and nature of arms are determined by those of the original initiator. Hedrick *et al.* reported the synthesis of 6-arm $A_3B_3$-type stars comprising PMMA and PCL arms by combining ATRP and ROP, in which a heterofunctional core compound carrying three bromoisobutylate and three hydroxyl (or silyl-protected hydroxyl) groups was used.[61] Similarly, Guo *et al.*

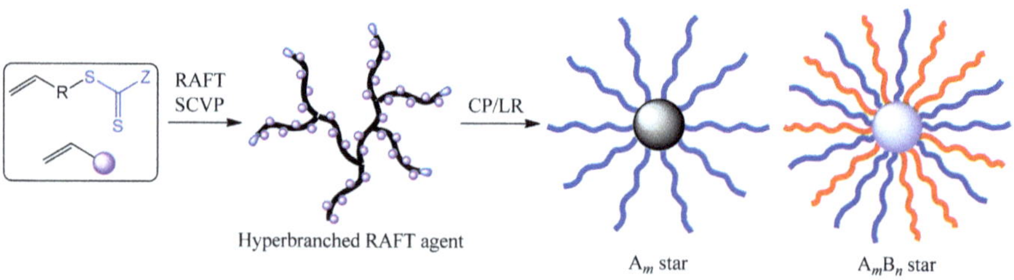


**Scheme 3.2** Synthesis of $A_m$-type multiarm and $A_mB_n$-type miktoarm stars by controlled polymerization (CP) and linking reaction (LR) based on a hyperbranched chain transfer agent generated by RAFT self-condensing vinyl copolymerization, in which the sphere connecting the arm segments denotes the hyperbranched core.

synthesized 8-arm $A_4B_4$-type stars containing PSt and polytetrahydrofuran (PTHF) arms through the combination of ATRP and cationic polymerization.[62] Furthermore, other $AB_x$ ($x = 2$–20),[63–72] $A_xB_x$ ($x = 2$–8),[73–76] and ABC[77–83] miktoarm stars were synthesized by various polymerization systems based on the core-first approach. In addition to directly synthesizing heterofunctional initiators lacking polymer segments, chain-end modification of a preformed linear polymer is an alternative strategy to prepare macroinitiators/macro-CTAs (MCs).[84–91] Multiple initiating sites are introduced into the chain end of a linear polymer, followed by chain extension polymerization of a second monomer to introduce other arm segments.

Although a series of ABC stars have been synthesized by the 'core first' method, they usually lack stimulus-cleavable linkages. To further extend their function and application, we designed and synthesized reduction-labile stimulus-responsive ABC-type μ stars with PCL (A), thermo-sensitive PNIPAM (B), and P*t*BA ($C_1$) or pH-responsive poly(acrylic acid) (PAA, $C_2$) segments (Scheme 3.3).[83] Starting from 2-((2-((2-hydroxymethyl-2-((2-bromo-2-methyl)propionyloxy)methyl)propionyloxy)ethyl)disulfanyl)ethyl 4-cyano-4-(phenylcarbonothioylthio) pentanoate (HBCP) with disulfide, bromide, dithiobenzoate, and hydroxyl functionalities, a core-first method involving ROP of CL, RAFT polymerization of NIPAM, and ATRP of *t*BA was used to generate $ABC_1$ stars, followed by selective hydrolysis to achieve $ABC_2$ stars. The disulfide

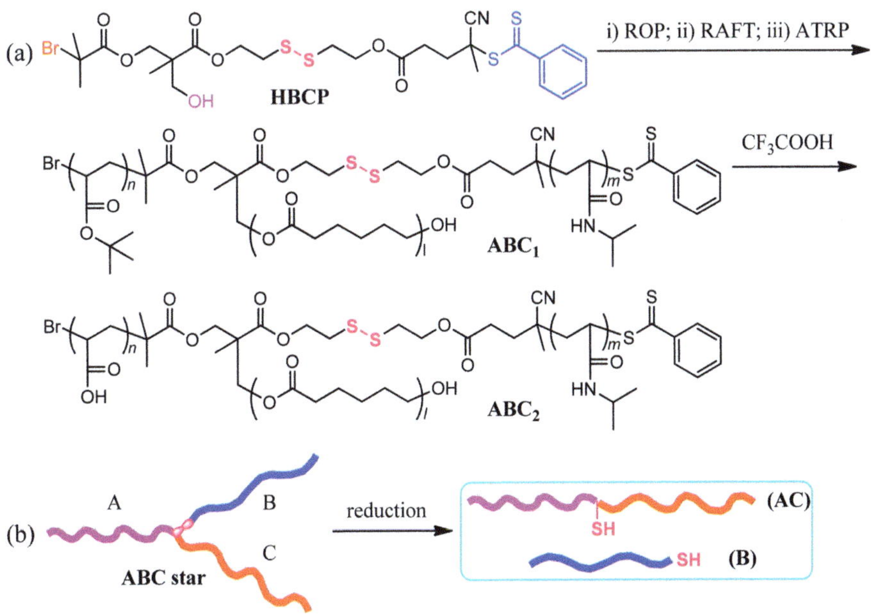


**Scheme 3.3** Straightforward synthesis of monocleavable ABC-type star terpolymers (a) and topological transformation upon reductive stimulus (b). Adapted from *Science China Chemistry*, Synthesis of multifunctional ABC stars with a reduction-labile arm by consecutive ROP, RAFT and ATRP processes, **58**, 2015, H. Liu, with permission from Springer.

linkage in the ABC stars could be efficiently cleaved upon reductive stimuli, during which the topology was converted from star terpolymers to mixtures of homopolymers (B) and diblock copolymers (AC). This research affords a straightforward route to construct stimulus-labile star terpolymers with great potential in versatile post-modification and topological transformation.

### 3.2.2 'Arm First' Approach

This approach involves the reaction of a living MC or macromonomer (MM) with a di- or multifunctional vinyl cross-linker to form a densely cross-linked core, in which the initial polymer segments act as the outlayer arms (Scheme 3.4). The crosslinking of MMs is similar to a coupling through process, whilst the crosslinking of MCs with cross-linkers is more similar to the combination of coupling onto and coupling through processes.

CRP in the presence of a cross-linker with an MC has been efficiently used to generate multiarm and miktoarm stars.[92–95] Moreover, the utilization of click reactions can be used for post-polymerization modification of star polymers. For instance, Cakir *et al.* synthesized α-fluoro and α-allyl functionalized PSt *via* the arm-first method, followed by successive thiol–*para*-fluro and thiol–ene click reactions to introduce heterofunctionalities in the periphery of multiarm stars.[96] Miktoarm stars can be readily obtained *via* one-pot

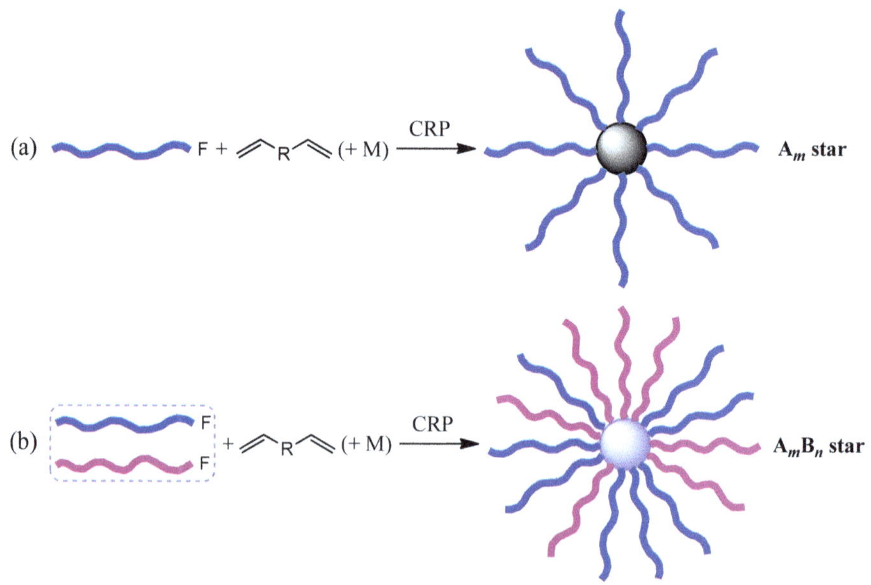


**Scheme 3.4** Synthesis of $A_m$-type multiarm stars (a) and $A_mB_n$-type miktoarm stars (b) with a cross-linked core *via* the 'arm first' method using a cross-linker, in which F denotes the initiating functionality for CRP (MC method) or a reactive C=C double bond (MM method). Vinyl monomers (M) are added in some cases.

crosslinking of a mixture of linear MCs. The composition of the arms in the μ star can be roughly tuned by simply changing the molar ratio of the initial MCs. For example, Gao *et al.* confirmed that ATRP experiments *via* a cross-linking reaction of MCs A and B using divinylbenzene (DVB) as the cross-linker could afford the desired μ star with high molecular weight and high star yield, in which A and B arms were concurrently incorporated into the star copolymer.[92] As confirmed from liquid chromatography under critical conditions, the resultant star copolymers prepared by one-pot copolymerization lacked any homoarm stars.

Meanwhile, crosslinking a mixture of linear MMs using a normal ATRP initiator can also give μ stars with high yield and low polydispersity. The utilization of a lower molar ratio of initiator to total MM is liable toward a decrease in the number of initiating sites in the inner core, and the probability of star–star coupling reactions is correspondingly reduced.[97] Since an extra cross-linker and initiator are adopted, both molecular weight and yield of star copolymers are enhanced, while the polydispersity remains relatively low. When two or more types of MMs are chosen for the tandem crosslinking reaction, multiarm μ stars with different arm segments can be potentially obtained.

More recently, Burts *et al.* reported the synthesis of miktoarm brush-arm star polymers from branched and linear norbornene-terminated MMs *via* the brush-first ring-opening metathesis polymerization (ROMP) method.[98] First, PSt-*branch*-PLA was synthesized *via* a combination of ATRP, ROP, and CuAAC. Graft-through ROMP of this macromonomer, followed by *in situ* crosslinking with a photo-cleavable bis-norbornene derivative, provided nanoscopic copolymers with photodegradable cores and precise arm composition. The intramolecular phase segregation of these copolymers was characterized, and a UV-induced structural rearrangement from three-faced Janus particles to micelles was observed.

### 3.2.3 'In–Out' Method

The 'in–out' approach represents an important route to synthesize μ stars with multiple A and B arms, and the number of B arms obtained by chain extension polymerization is always lower than that of the initially generated A arms due to the presence of side reactions, losing part of the reactive sites (Scheme 3.5).[99–101] Moreover, it is conceptually difficult to generate a multicomponent star unless different MCs are combined to prepare star-like copolymers with miktoarms and a cross-linked core in an initial CRP process using a cross-linker.

Star polymers formed *via* the 'arm first' method possess multiple initiating sites in the inner cross-linked core, and thus they can act as multifunctional MCs for CRP or ROP to generate μ stars. The word 'in' refers to the 'arm first' method to form a star-like polymer with an *in situ* formed cross-linked core, and the word 'out' denotes the growth of the second generation of arms from the inner core.[99] Various polymerization techniques can be used to

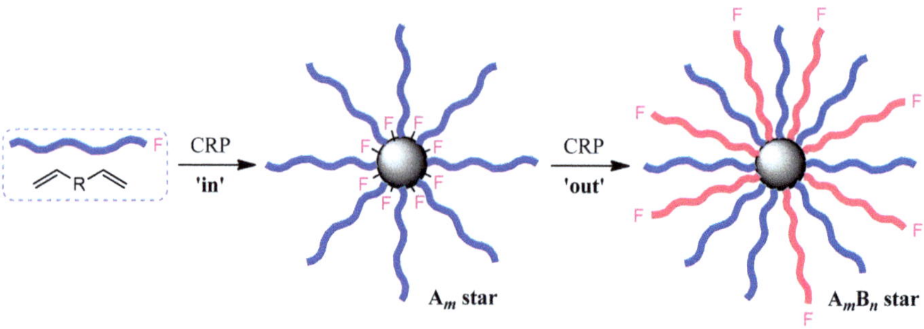


**Scheme 3.5** Illustration of the CRP-based 'in–out' method for the synthesis of multiarm and miktoarm stars, in which the initiating functionality (F) can be introduced *via* CRP or an end-group transformation, and some of the functionalities may be inert during the chain extension polymerization to construct the miktoarm stars.

prepare the initial MC, and post-modification is necessary in some cases. For instance, the terminal hydroxyl moiety of the polyether and polyester can be readily converted into alkyl bromide or dithiocarbonate groups *via* end-group transformation.[102] On this basis, living/controlled processes such as NMP, ATRP, RAFT, and ROP techniques are applied to achieve the desired μ stars *via* the 'in–out' route.

The 'in–out' strategy was first developed for the construction of $A_mB_m$-type miktoarm stars *via* LAP, in which the number of A and B miktoarms was assumed to be equal due to the lack of coupling termination originating from the electrostatic repulsions between carbanions.[1] As this method is extended to CRP systems, the radical–radical coupling becomes an inevitable side reaction, leading to a lower number of effective initiating sites in the star MC. Moreover, the remaining initiating functionalities are distributed throughout the inner cross-linked core, and they can participate in the CRP processes only if the catalyst and monomer are accessible to them due to the congested environment.[103] Therefore, the phenomenon of incomplete initiation of star MCs is usually observed in the synthesis of miktoarm stars *via* the 'in–out' method.[104]

### 3.2.4 LAP-Based General and Iterative Methodologies

With the aid of LAP, a series of methodologies involving the utilization of special linking functions, such as silyl chloride, benzyl halide, 1,1-diphenylethylene (DPE), 1,3-butadiene, and α-phenylacrylate, have been developed to synthesize many well-defined star copolymers,[1–6] and some typical methods are introduced below.

The first method lies in the utilization of silyl chloride for the coupling reaction. Pennisi and Fetters reported the synthesis of a 3-arm asymmetric $AA'_2$-type star PSt with different arm lengths in 1988,[105] and a similar method was adopted by Mays to synthesize a 3-arm $PSt(PI)_2$ star bearing two polyisoprene

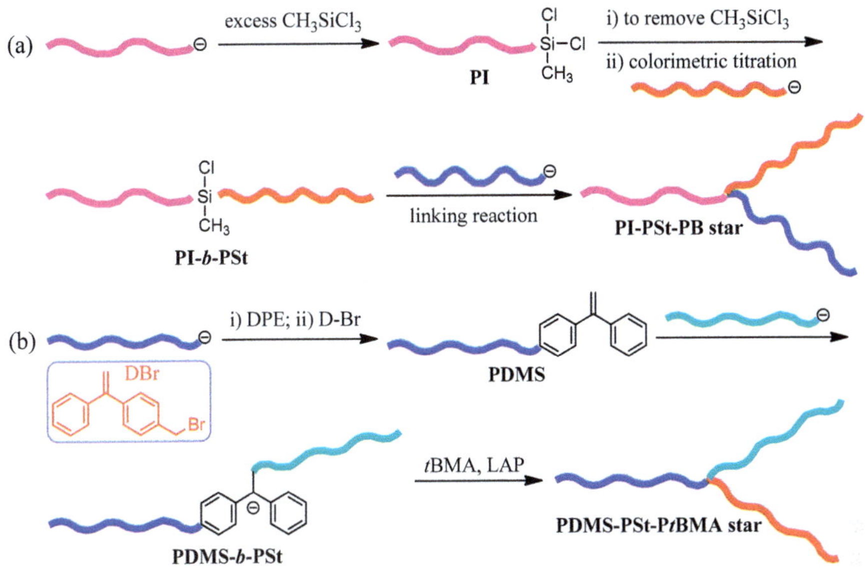


**Scheme 3.6** Synthesis of ABC star terpolymers *via* selective linking reaction (a) and DPE chemistry (b) using the LAP technique.

(PI) arms.[106] The method was subsequently extended to the synthesis of a 3-arm ABC star terpolymer comprising PB, PI, and PSt segments, in which Hadjichristidis *et al.* utilized the different anionic reactivity toward the Si–Cl reaction site caused by the chain-end steric bulkiness of the polymer anions (Scheme 3.6a).[107] The linking reaction was initially conducted between PILi and a large excess of $MeSiCl_3$ to prepare PI-$SiCl_2$, and the least reactive PStLi was then stoichiometrically reacted with PI-$SiCl_2$ *via* colorimetric titration to obtain PI-SiCl-PSt. Finally, it was reacted with the least sterically hindered PBLi to generate the desired star with precisely controlled molecular weight and composition and low polydispersity (PDI = 1.03). On this basis, they synthesized some other $A_xB_y$, ABC,[108,109] and ABCD[110] miktoarm stars using a similar methodology.

The second method is based on DPE chemistry. As well-documented, highly reactive living polymers of styrene, 1,3-butadiene, and isoprene can quantitatively react with DPE to form a polymer carbanion bearing one alkyl and two phenyl substituents. Due to the high steric hindrance, the resultant carbanion cannot react with the DPE moiety, although it is able to initiate the LAP of various monomers to form a diblock copolymer. Since a chain-end-DPE-functionalized polymer is used for successive addition and polymerization processes, a 3-arm ABC star can be readily formed. The synthetic strategy was first realized by Fujimoto *et al.* in 1992 (Scheme 3.6b),[111] in which DPE-functionalized poly(dimethylsiloxane) (PDMS) was reacted with PStLi to form a polymer anion with PDMS and PSt chains, followed by anionic polymerization of *tert*-butyl methacrylate (*t*BMA) to generate PDMS-PSt-P*t*BMA stars.

In this method, the 'core first' method is ingeniously combined with the 'arm first' route to grow the final arm segment. The versatile method involving DPE chemistry has been applied to synthesize many star copolymers such as $A_2B_2$, ABC,[112–117] and ABCD[118,119] stars.

The third method aims at linking reactions using benzyl bromide. Hirao and coworkers prepared a variety of in-chain- and chain-end-functionalized polymers with a plural number of benzyl bromide moieties by reaction of living anionic polymers with *tert*-butyldimethylsilyl-protected benzyl alcohol derivatives, followed by direct conversion to benzyl bromides by treatment with $Me_3SiCl$/LiBr. On this basis, some well-defined star copolymers including $A_xB_y$, $ABC_2$, and $ABC_4$ (A = PSt, B = poly(α-methylstyrene) (PαMS), C = PI) stars[120–124] were prepared, and this method was also generalized to construct high generation dendrimer-like (co)polymers with precise architectures and tunable compositions.

Finally, a more general and versatile methodology to synthesize multiarm μ stars was proposed by Hirao and coworkers.[6] An iterative methodology was designed in such a way that the same reaction site is regenerated after the arm introduction in each reaction sequence and this process is repeatable; the related functions are primarily composed of DPE, 1,3-butadiene, and α-phenylacrylate moieties (Scheme 3.7).[125–134] The development of this

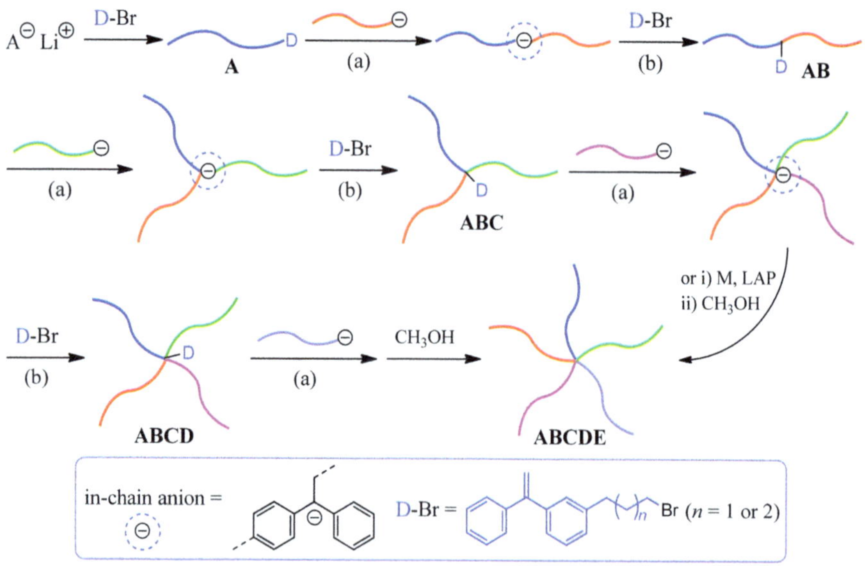


**Scheme 3.7** Synthesis of ABC, ABCD, and ABCDE miktoarm stars *via* an iterative methodology comprising LAP and DPE-based linking reactions, in which route (a) is used to introduce an arm segment, route (b) is adopted to introduce the DPE functionality, and the in-chain anion can be utilized for both the linking reaction and *in situ* anionic polymerization of a second monomer. Adapted with permission from ref. 126, copyright (2005) American Chemical Society and ref. 127, copyright (2005) American Chemical Society.

methodology allows successive and theoretically limitless introduction of arm segments in the target μ stars. Thus far, a variety of novel multiarm and multicomponent star copolymers with precise architectures and chemical compositions up to nine different arm segments have been successfully achieved. This perfect methodology stands at the highest level to construct multicomponent stars with quite low polydispersity in polymer science, and it is extremely difficult to surpass by any other methods involving CRP and click chemistry at present.

## 3.3 Synthesis of Miktoarm Stars *via* Combinatorial Approaches Involving Click Chemistry

In a 'coupling onto' approach, controlled polymerization/end-group transformation and coupling reactions are usually combined to synthesize various stars, in which preformed arms are tethered onto an MFA, and the latter forms the core of the star polymers obtained.[135–150] The most popular click reaction is CuAAC, leading to formation of a 1,2,3-triazole linkage.[151–153] However, some other linking reactions have been explored in polymer chemistry, most of them are conducted metal-free, and have succeeded under suitable conditions. The active terminal functionality can be available from either the original initiator/RAFT agent or post-modification of the MCs.

### 3.3.1 Combinational Methods Involving CuAAC/SPAAC

The first synthesis of star polymers using a combination of ATRP and CuAAC was accomplished by coupling azide-terminated PSt with 2–4 compounds bearing alkyne functionalities.[154] All coupling reactions were completed within 3 h, and the presence of small amounts of Cu(0) enhanced the coupling efficiency. Although the efficiency of the click reaction decreases with the increasing molecular weight of the arm precursors due to the high steric congestion and less precisely balanced stoichiometry between the azide and alkyne groups, this 'coupling onto' strategy has proven a versatile method to prepare star polymers in high yield. Similar methodologies have been utilized for the synthesis of various homoarm stars,[155–157] star cyclic polymers,[158,159] star block copolymers,[160–162] and $A_xB_y$-type μ stars.[163–174]

Meanwhile, SPAAC reactions have been developed for polymer synthesis, which can be performed at low temperatures and in the absence of a copper catalyst since the driving force lies in the ring strain of cyclooctyne. Besides labeling living cells and even animals,[175–178] the robustness and orthogonality of SPAAC are also demonstrated by its application in the controlled synthesis of 3- and 4-arm PSt and PEG stars,[179] 3-arm $(PSt)_2PEO$ (PEO = poly(ethylene oxide)) miktoarm stars,[180] and $(PSt)_2PEO(PSt)_2$ H-shaped copolymers.[180]

The synthesis of ABC-type μ stars is more complex than that of normal star copolymers, in which an MFA bearing three different functionalities is usually adopted, and living/controlled polymerization and linking reaction such as CuAAC are combined to achieve the desired star terpolymers (Scheme 3.8).[181–197] In some cases, functional group transformations (FGTs) are involved in order to generate the initiating functionality or clickable moiety. In addition to generate two different arm segments *via* controlled polymerization, the third functionality can act as a linking moiety to form the core of star copolymers during the click process. Monteiro reported the synthesis of 3-arm miktoarm star polymers and the first-generation of polymeric miktodendrimers comprising PSt, P*t*BA, and poly(methyl acrylate) (PMA) arms by combining ATRP and CuAAC.[181] Tunca prepared ABC miktoarm stars through the combination of ATRP, NMP, and CuAAC,[182] in

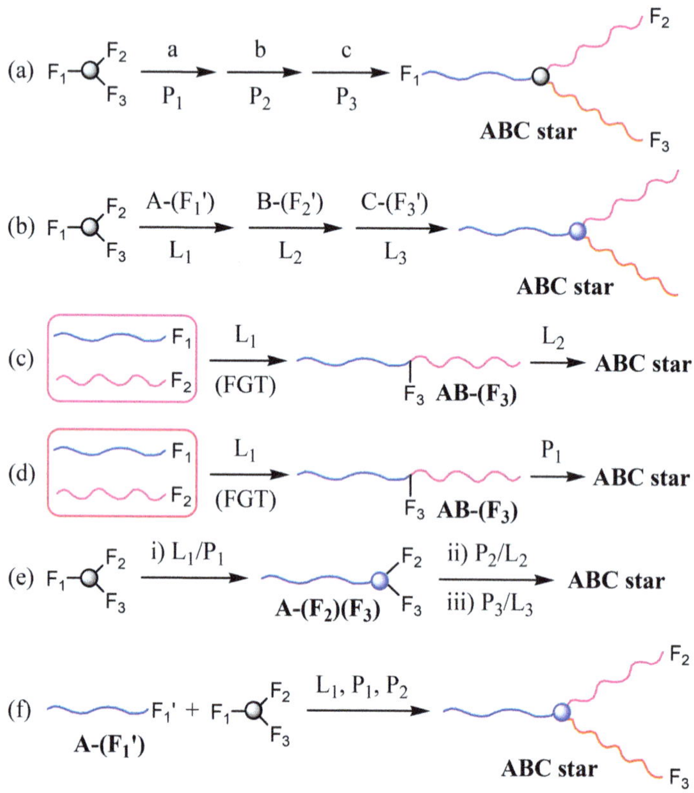


**Scheme 3.8** Typical synthetic routes to ABC star terpolymers by 'core first' (a), 'coupling onto' (b and c), and combinatorial (d–f) methods, in which the letters a, b, and c above the arrows are monomers, F stands for coupling or initiating functionality, L means linking reaction, P is living/controlled polymerization, and the sphere and wavy lines denote the core reagent and the arms of the star polymers, respectively. Functional group transformation (FGT) may be involved in routes (c) and (d) to regenerate a new reactive moiety. In route (e), both polymerization and coupling reaction are involved, and precursors to produce the MFA can also be applied in some cases.

which an MFA containing an ATRP initiating moiety, an alkoxyamine, and an alkyne group was sequentially used for ATRP of MMA and NMP of St, followed by click coupling with azide-terminated P*t*BA or PEO to generate the desired star. By combination of ATRP, ROP, and CuAAC, Deng *et al.* synthesized PEO-PSt-PCL miktoarm stars, in which an azide-functional PEO was initially clicked onto propargyl 2-hydroxylmethyl-2-(α-bromoisobutyraloxymethyl)propionate, followed by ATRP of St and ROP of CL.[183] By combination of controlled polymerization and CuAAC, Liu and coworkers prepared a series of stimulus-sensitive 3-arm μ stars involving PEG-PDEA-PM (PM = P*t*BMA, poly(methacrylic acid) (PMAA), PNIPAM),[184,185] PSt-PNIPAM-PCL,[186] PEG-P*t*BMA-PNIPAM,[187] and PCL-PM-PDMAEMA (PM = PSt, PEO; PDMAEMA = poly(2-(dimethylamino)ethyl methacrylate)).[188] Apart from successive processes, the versatile synthesis of star terpolymers may be conducted in one-pot.[188,189] Based on one-pot or two-step reactions involving NMP, ROP, and CuAAC, Altintas *et al.* synthesized three types of PCL-PSt-PM (PM = PEG, P*t*BA, PMMA) star terpolymers.[189]

It should be mentioned that the multifunctional core of star copolymers can be readily formed *via* click-like multicomponent reactions. For instance, Li and coworkers demonstrated that the Passerini three-component reaction could be a simple and effective method for the synthesis of polymers with double functional end-groups, which were then key precursors for the preparation of ABC-type μ stars (Scheme 3.9).[196] Using the one-step Passerini reaction of monomethoxy poly(ethylene glycol)-propionaldehyde (PEG-CHO) with 2-bromo-2-methylpropionic acid and propargyl isocyanoacetamide, the PEG chain end was simultaneously functionalized with one ATRP initiating site and one alkynyl group. The resulting PEG(-alkynyl)-Br was then used for the synthesis of three types of ABC

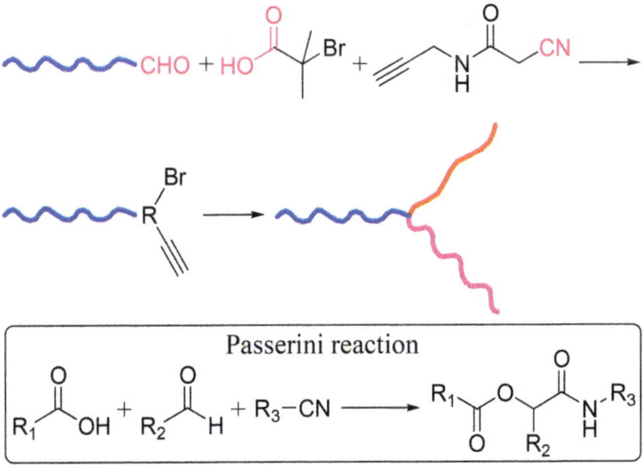


**Scheme 3.9** Synthesis of dual end-functionalized polymers *via* the Passerini reaction and the subsequent synthesis of ABC miktoarm terpolymers *via* successive ATRP/CuAAC or one-pot SET-LRP/CuAAC. Adapted from ref. 196, with permission from John Wiley and Sons.

miktoarm terpolymers *via* two approaches: ATRP of NIPAM and a subsequent click reaction with azido-terminated PSt or P*t*BA afforded PEG-PNIPAM-PSt and PEG-PNIPAM-P*t*BA stars, and one-pot SEP-LRP of *t*BA and CuAAC with PSt-$N_3$ gave PEG-P*t*BA-PSt stars.

The synthesis of 4-arm ABCD star quaterpolymers is more challenging due to the numerous reaction steps required to introduce an increasing variety of compositions. Thus far, the examples of ABCD stars generated by click chemistry are relatively scarce.[198–201] A number of synthetic examples of ABCD miktoarm stars by a similar approach appeared in 2008.[198–200] Tunca *et al.* synthesized an ABCD star by a [3 + 2] click reaction between an in-chain-azide-functionalized AB diblock copolymer (A = PSt, B = PCL) and an in-chain-alkyne-functionalized CD diblock copolymer (C = PEO, D = PMMA or P*t*BMA).[198] Wang *et al.* synthesized ABCD miktoarm stars *via* a combination of anionic polymerization, ATRP, ROP, and CuAAC.[199] PSt and PI were anionically polymerized using butyl lithium. PSt was functionalized with propargyl and 2-bromoisobutyryl groups, and the subsequent ATRP of butyl acrylate (BA) afforded PSt-*b*-PBA with a propargyl group at the junction. PI was functionalized with a hydroxyl group and an ethoxyethyl-protected hydroxyl group, followed by ROP of EO to yield PI-*b*-PEO, and the protected hydroxyl group was further converted into an azide functionality. The two block copolymers were clicked together to form an ABCD star. Other than the normal '2 + 2' strategy, Pan and coworkers reported the controlled synthesis of PSt-PCL-PMA-PEO stars *via* a '3 + 1' route (Scheme 3.10).[200] Using a mechanism transformation strategy, PSt with hydroxyl, alkyne, and trithiocarbonate functional groups was synthesized by reaction of the RAFT-prepared PSt with 2-hydroxyethyl-3-(4-(prop-2-ynyloxy) phenyl) acrylate (HEPPA), followed by ROP, RAFT, and CuAAC processes to generate the desired star. More recently, we have adopted an alternative '3 + 1' strategy to achieve a dual-sensitive monocleavable ABCD star

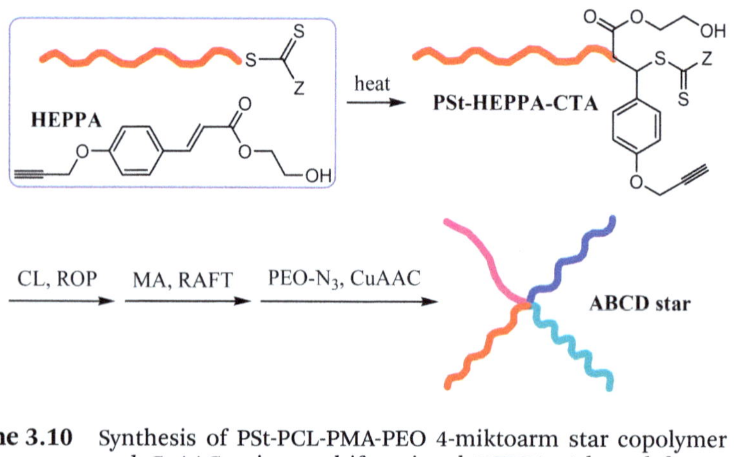


**Scheme 3.10** Synthesis of PSt-PCL-PMA-PEO 4-miktoarm star copolymer *via* CP and CuAAC using multifunctional HEPPA. Adapted from ref. 200 with permission from John Wiley and Sons.

(Scheme 3.11).[201] To this end, controlled polymerization and click chemistry were combined to generate the desired star comprising PNIPAM (A), poly(2-diisopropylaminoethyl methacrylate) (PDPA, B), PCL (C), and acetal-linked poly(ethylene glycol) (*a*PEG, *D*) segments. Starting from a special MFA bearing alkyne, bromide, dithiobenzoate, and hydroxyl functionalities, an alkyne-core-functionalized ABC star terpolymer was obtained *via* successive RAFT, ATRP, and ROP, followed by CuAAC between the ABC star and azide-functionalized *a*PEG to obtain the target star. Upon acid stimulus, the ABCD star was liable to 'degradation' into a mixture of PEG and an AB'C star with a protonated PDPA (B') segment, and the differences in topology and composition further induced time-dependent morphological transformations of the copolymer aggregates. With extended time, various morphologies including large compound micelles, flower-like micelles, small micelles and their aggregates, and compound micelles with reduced size were observed. Drug-loaded aggregates exhibited accelerated drug release kinetics upon thermo and pH stimuli, and the micellar system thus holds great promise for biomedical applications.

Thus far, the synthesis of ABCDE star quintopolymers remains extremely challenging. Although Hirao and coworkers have systematically developed iterative methodologies by introducing individually various arms to achieve μ stars with 5–9 chemical compositions,[126–134] so far, ABCDE stars obtained *via* combinatorial controlled polymerization and click chemistry are very scarce.[202–204] Considering some monomers are difficult to polymerize directly *via* living anionic polymerization, we developed a novel '1 + 2 + 2' synthetic strategy to generate 5-arm ABCDE stars *via* selective and

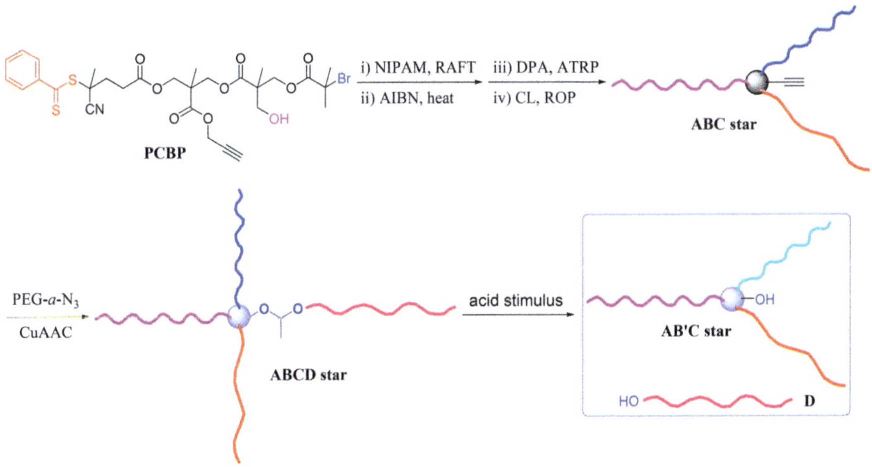


**Scheme 3.11** Synthesis of an acetal-functionalized ABCD (A = PNIPAM, B = PDPA, C= PCL, and D = *a*PEG) star quaterpolymer *via* successive RAFT, ATRP, ROP, and CuAAC processes and acid-triggered topological transformation. Adapted from ref. 201 with permission from the Royal Society of Chemistry.

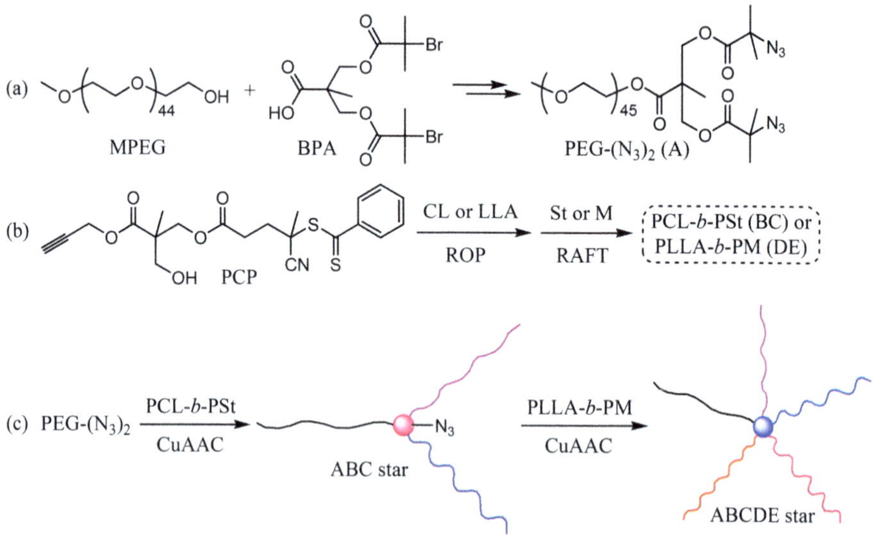


**Scheme 3.12** Synthesis of PEG-PCL-PSt-PLLA-PM (PM = PDMAEMA, PMMA, PMA, or P*t*BA) star quintopolymers by combination of end group transformation (a), ROP and RAFT process (b), and CuAAC reaction (c). Adapted with permission from ref. 202 and 203, copyright (2012) American Chemical Society and the Royal Society of Chemistry.

successive CuAAC reactions, in which '1' and '2' stand for the segment number of building blocks (Scheme 3.12).[202,203] Based on successive click reactions, a few PEG-PCL-PSt-PLLA-PM (PM = PDMAEMA, PMMA, PMA, P*t*BA, and PAA) star copolymers were successfully prepared. This approach is efficient as the selective reaction between A and BC occurs smoothly; however, there are two remaining limitations for it to be a robust synthetic method. First, excess diazide-functionalized A should be efficiently removed from the mixture of A and the ABC star *via* fractional precipitation or high performance liquid chromatography (HPLC) separation. Second, two azide moieties of A may simultaneously participate in the initial click reaction to give low-component $AB_2C_2$ stars[205] due to steric hindrance, leading to the eventual production of $AB_2C_2$ and ABCDE mixed stars. Therefore, it is extremely urgent to further explore more versatile routes to create well-defined star quintopolymers. More recently, we have chosen orthogonal chemistry to address the limitations remaining in the traditional modular '1 + 2 + 2' strategy using the same type of ligation method, in which CuAAC and DA click reactions are combined to prepare ABCDE stars (Scheme 3.13).[204] The target stars comprise PCL (A), P*t*BA (B), PLLA (C), PNIPAM (D), and poly(5-methyl-5-allyloxycarbonyl-1,3-dioxane-2-one) (PMAC, E) segments. The integration of multiple segments endows the star copolymer with some unique properties including biodegradability, stimulus-responsiveness, and versatile post-modification. Subsequent hydrolysis afforded the PAA (B′)-bearing AB′CDE star with clickable and dual-sensitive segments. The star copolymers could be self-assembled into intriguing morphologies

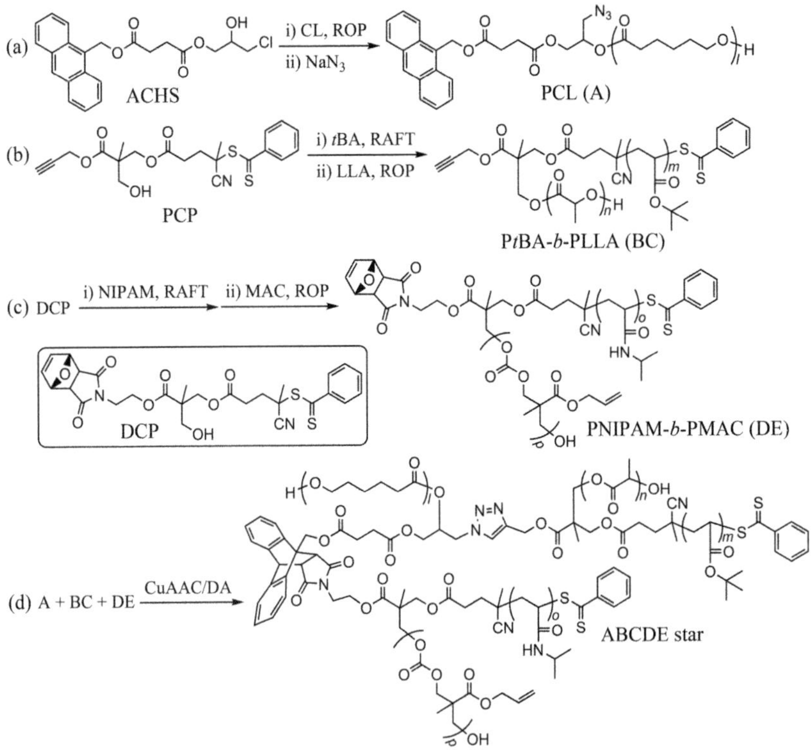


**Scheme 3.13** Synthesis of clickable PCL-P*t*BA-PLLA-PNIPAM-PMAC star copolymer by one-pot CuAAC and DA click reactions, in which ROP (a), RAFT process and ROP (b, c), and click chemistry (d) are adopted. Adapted from ref. 204 with permission from the Royal Society of Chemistry.

including vesicles, spherical micelles, and large compound micelles, which could be tuned by thermo and pH stimuli. This general modular synthesis using orthogonal chemistry allowed us to prepare various star quintopolymers and their derivatives.

### 3.3.2 Combinatorial Approaches Involving DA/HDA Reactions

Since its first report in 1928, Diels–Alder cycloaddition reactions have been listed as some of the most fascinating organic reactions involving the formation of new carbon–carbon bonds. In combination with the hetero DA approach, these reactions cover a wide range of possible starting materials and reaction conditions, including widely tunable reaction temperatures.[206–212] Therefore, DA and HDA reactions have been readily adopted to synthesize distinct polymers in polymer science.

DA reactions can be an appropriate route for the conjugation of different polymer segments. Dag *et al.* efficiently utilized the reaction of maleimide and anthracene to construct diverse star block copolymers.[213,214] In an 'arm first' approach, polymeric precursors were prepared using α-functionalized

ATRP initiators and esterification reactions to prepare maleimide or anthracene end-functionalized polymers. The DA reaction was performed in toluene under reflux, and tri-anthracene functional linking agents and cross-linked PSt-cores with anthracene periphery were used as building blocks. More recently, Bahadori *et al.* reported on the synthesis of Y-shaped PEG$(PCL)_2$ star copolymers *via* DA reactions.[215] A furan protected maleimide end-functionalized PEG and an anthracene in-chain-functionalized PCL were subjected to a DA click reaction to obtain star copolymers. The star copolymer micelles were used as drug delivery vehicles, which showed satisfactory results and enhanced the cytotoxic effect of the anticancer drug vinorelbine.

Meanwhile, the HDA approach was adopted by Barner-Kowollik and coworkers to synthesize star polymers.[216] PSt homopolymers with electron-deficient dithioester end-groups were synthesized by a RAFT process using the functionalized RAFT agent diethoxyphosphoryldithioformate or pyridine-2-yldithioformate, followed by HDA reactions with diene functionalized compounds to generate the desired stars. For the synthesis of 2–4 arm star polymers, di-, tri- and tetrafunctional diene cores were reacted with both types of PSt-RAFT precursors. Depending on the choice of the RAFT agent and the number of arms, yields between 65% and 91% were achieved, in which the pyridine-2-yldithioformate PSt showed the highest yields in each case. The HDA coupling reaction was monitored *via* UV-vis spectroscopy of the RAFT moiety and by $^1$H NMR spectroscopy of the diene functionality. The results indicated that the phosphoryldiethoxydithioformate-terminated polymer reached 92% conversion within 24 h and the pyridin-2-yldithioformate-terminated polymer reached 96% conversion in 10 h. As the 4-arm stars were subjected to high-temperature treatment, gel permeation chromatography (GPC) measurements indicated that complete cleavage of all four arms from the core was achieved in 24 h at 160 °C.

Moreover, DA reactions were extended to construct a series of ABC,[217] ABCD,[218] and ABCDE[204] multicomponent stars. Durmaz *et al.* prepared PEG-PSt-P*t*BA miktoarm stars *via* 'core-in' and 'core-out' methods by combining a DA reaction, SFRP, and ATRP.[217] A PEG-maleimide precursor and an MFA bearing anthracenylmethyl, alkyl bromide, and 2,2,6,6-tetramethylpiperidinyl-1-oxy moieties were subjected to a DA click reaction, and the resulting DA adduct displayed appropriate functional groups for SFRP and ATRP. On this basis, successive SFRP of St and ATRP of *t*BA were performed to generate the desired stars with controlled molecular weight and low polydispersity (PDI < 1.27). A DA reaction was adopted by Altintas *et al.* to synthesize ABCD 4-miktoarm stars comprising PCL, P*t*BA, PSt, and PMMA arms.[218] Firstly, PCL with anthracene and P*t*BA with furan-protected maleimide end-functionalities were synthesized separately and then subjected to a DA click reaction to give a PCL-*b*-P*t*BA copolymer. Next, the diblock copolymer was utilized successively as macroinitiator in the NMP of styrene and in the free radical photopolymerization of MMA to achieve the target star quaterpolymer.

### 3.3.3 Combinatorial Approaches with Thiol-Based Click Reactions

Thiol–ene reactions are important due to the wide range of commercially available thiols and alkenes and the two efficient routes *via* either anti-Markovnikov radical addition or base-catalyzed Michael addition.[219–225] Other thiol-based 'click' reactions such as thiol–maleimide addition, thiol–isocyanate addition, pyridyl–disulfide exchange, and thiol-*para*-fluoro have also been efficiently developed.[226–231] Thiol-terminated polymers can be easily obtained by RAFT polymerization of a wide range of monomers and the subsequent cleavage of the RAFT moiety *via* aminolysis. Alternatively, disulfide containing bifunctional ATRP initiators can be used for the preparation of thiol functionalized polymers.[232,233]

Chan *et al.* synthesized a 3-arm star poly(*N*,*N*-diethylacrylamide) by a thiol–ene reaction.[234] In a one-pot reaction, a RAFT polymer with a thiocarbonylthio end-group was reacted with a trifunctionalized acrylate coupling agent under phoshine catalysis for 5 min. The reduction of the RAFT end-group was managed simultaneously by a primary amine, and an excess of polymer was used to guarantee the full conversion of the double bonds. Pounder *et al.* used a thiol–maleimide reaction for the synthesis of a 3-arm star polylactide.[235] A trifunctional thiol compound was conjugated to a maleimide-functionalized PLA in 2 h under triethylamine catalysis with a slight excess of the thiol.

Using the 'core first' strategy, a series of amphiphilic $A_4B_4$-type (A = PEO, B = PCL, PSt, P*t*BA) star-shaped copolymers were synthesized by mechanism transformation combined with a thiol–ene reaction.[236] First, using a designed MFA with four active hydroxyl groups and four allyl groups, 4-arm star PEOs with active hydroxyl moieties at the core position were obtained by sequential ROP, capping reaction of the living oxyanion with benzyl chloride, and transformation of the allyl groups into hydroxyl functionalities by a thiol–ene reaction. On this basis, star copolymers were obtained by ROP of CL or ATRP of vinyl monomers from a macroinitiator bearing alkyl bromide functionalities.

More recently, thiol–ene chemistry has been applied to synthesize ABC star terpolymers. Ding and coworkers reported on the synthesis of ABC miktoarm star terpolyphosphoesters (PPEs) *via* combination of ROMP, ROP, and acyclic diene metathesis (ADMET) polymerization.[237] Acrylate-terminated monotelechelic PPE was first prepared through ROMP of a 7-membered cyclic phosphate monomer in the presence of a terminating agent, and then the terminal acrylate group was readily converted to a heterodifunctional PPE with hydroxyl and acrylate groups *via* a thiol–Michael addition click reaction and esterification. ROP of the cyclic phosphoester monomer initiated by this PPE was successively carried out to generate the acrylate-functionalized block PPE, which was utilized as a selective macromolecular chain stopper in the subsequent ADMET polymerization of an α,ω-diene phosphate monomer, finally producing a miktoarm terpolymer. In addition, Patil *et al.* synthesized $A_2B$ and ABC-type miktoarm stars *via* thiol-based click reactions.[238] Two MFAs were used to generate well-defined α-allyl-α'-allyloxy and α-allyl-α'-propargyloxy bifunctionalized PCLs *via* ROP. On this basis, thiol–ene reaction with PEG thiol to obtain a $(PEG)_2PCL$ miktoarm star and orthogonal reactions *via* CuAAC with azido-functionalized PNIPAM, followed by thiol–ene reaction with PEG thiol, were used to synthesize a PCL-PNIPAM-PEG star terpolymer.

### 3.3.4 Combinatorial Approaches Involving Other Click Reactions

In theory, any click reaction can be used to construct multiarm and miktoarm stars. In this section, the application of Glaser coupling and aldehyde–aminoxy linking reactions in the preparation of star polymers is briefly summarized.

Glaser coupling was reported recently in the context of star copolymers. As a first example, Gungor *et al.* described the preparation of $A_2B_2$-type μ stars, $(PSt)_2(PMMA)_2$ and $(PSt)_2(PCL)_2$, exploiting the Glaser coupling reaction as an efficient reaction,[239] in which terminal alkyne homocoupling (oxidative alkyne coupling) was used to prepare the desired μ stars. The block copolymer with a terminal alkyne at the junction point prepared by NMP-ATRP and ROP-NMP sequential routes was coupled *via* diyne formation to afford two types of 4-miktoarm stars using a combination of $(PPh_3)_2PdCl_2/PPh_3/CuI$ in a solvent mixture of $Et_3N/CH_3CN$ at room temperature. This approach was subsequently further extended to the synthesis of other $A_2B_2$ miktoarm stars and barbwire-like graft copolymers comprising PEO, PSt, PI, P*t*BA, PCL, and PAA segments.[240–242]

Meanwhile, RAFT processes and the aldehyde–aminoxy click coupling reactions were combined by Wu *et al.* to generate μ stars comprising PEG and PNIPAM/PSt segments.[243,244] Take PNIPAM-PEG miktoarm stars with multiple arms, for example.[243] Star PNIPAM with an aldehyde functionalized core was initially prepared by the RAFT 'arm first' technique *via* crosslinking of the preformed linear macroRAFT agents using an aldehyde-containing divinyl compound, which then acted as a multifunctional coupling agent for the formation of second-generation PEG arms *via* a click coupling reaction between the aldehyde groups and aminooxy-terminated PEGs. On this basis, the thermo-responsive micellization behavior of the PNIPAM-PEG miktoarm star with different $f_{PEG}$ in water was investigated. Thermo-induced inter/intramolecular micellization of the miktoarm stars could be available *via* control of the weight composition and formation of multi/unimolecular micelles.

### 3.3.5 Combinatorial Methods Using Dual and Multiple Click Reactions

The combination of different types of click reactions allows for the preparation of distinct architectural polymers and their versatile post-modification.[147–150] The advantage of this strategy lies in the facile introduction of different functions and compositions that cannot be easily achieved using only one click reaction. The latest tendency aims at the rational synthesis of topological polymers *via* such combinatorial click reactions.

### 3.3.5.1 Combination of Thiol–Ene and CuAAC Reactions

Iskin *et al.* used this combination to prepare an ABC miktoarm star.[245] Thiol-terminated PSt (PSt-SH) was synthesized in a two-step protocol, in which the bromo-terminated PSt was first converted to a xanthate function, followed by reaction with 1,2-ethandithiol. PSt-SH was then grafted onto the 1-(allyloxy)-3-azidopropan-2-ol core by a UV-initiated radical thiol–ene reaction, followed by ROP to introduce a PCL segment. Finally, PEG-acetylene obtained by the esterification of methoxy-terminated PEG and 5-pentynoic acid was clicked onto the core *via* CuAAC to construct the star terpolymer.

### 3.3.5.2 Combination of CuAAC and DA/HDA Reactions

With the aid of one-pot and sequential double DA and CuAAC reactions, Hizal, Tunca, and coworkers synthesized a series of multiarm star triblock copolymers, as well as multi-miktoarm star block copolymers, using α,ω-functionalized polymers and cross-linked PSt-cores possessing anthracene and/or alkyne exteriors.[246–248] Xiong *et al.* synthesized well-defined dendritic star polymers with PCL on the periphery *via* one-pot CuAAC and DA reactions.[249] Barner-Kowollik and coworkers combined HDA and CuAAC reactions for the preparation of PCL-*b*-PSt star copolymers.[250] The diene-terminated PCL was converted through an esterification reaction with 4-pentynoic acid, and the resultant telechelic PCL was capable of reacting with an electron-deficient dithioester and an azide compound. A trifunctional azide compound and RAFT-prepared PSt were used to synthesize the star block copolymer.

### 3.3.5.3 Combination of CuAAC and Nitroxide Radical Coupling (NRC) Reactions

Fu *et al.* first synthesized an azide end-functional P*t*BA, TEMPO end-functional PEO or PCL, and alkyne end-functional PSt, and then combined click chemistry with atom transfer nitroxide radical coupling (NRC) in a one-pot procedure to yield ABC miktoarm stars.[251] Interestingly, Jia *et al.* combined CRP, NRC, and CuAAC to generate ABC star terpolymers comprising three different cyclic arms comprising PSt, P*t*BA, and PMA, whose synthesis could be performed in one pot at 25 °C by modulating the copper catalyst activity through different combinations of solvents and ligands (Scheme 3.14).[252]

### 3.3.5.4 Combination of CuAAC, DA, and NRC Reactions

Following this strategy, the group of Tunca and Hizal reported the synthesis of ABC miktoarm stars.[253] The linear precursors, furan-protected maleimide-terminated PEG (PEG-MI), tetramethylpiperidine-1-oxyl-terminated PCL (PCL-TEMPO), and azide-terminated PSt (PSt-$N_3$) or poly(*N*-butyl oxanorbornene imide) (PONB-$N_3$) were ligated to the core with anthracene, alkyne, and bromide functionalities *via* a 'coupling onto' approach. In a first step, DA

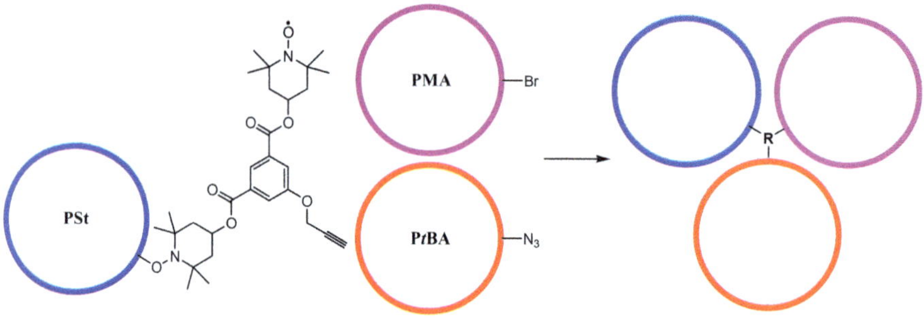


**Scheme 3.14** Synthesis of μ-ABC-tricyclic miktoarm star *via* NRC and CuAAC. Adapted with permission from ref. 252, copyright (2012) American Chemical Society.

reaction of PEG-MI with an MFA was conducted to yield an α-alkyne-α-bromide-terminated PEG precursor. On this basis, the precursor was further clicked with PCL-TEMPO and PSt/PONB-$N_3$ through CuAAC and NRC reactions to yield PEG-PCL-PSt and PEG-PCL-PONB stars in a one-pot fashion. Using similar triple click reactions, Dedeoglu *et al.* synthesized tadpole polymers with cyclic PSt and linear PEG/PCL blocks[254] and heterograft brush copolymers comprising a polyoxanorbornene backbone and PMMA/PEG/PCL grafts.[255]

### 3.3.5.5 *Combination of Thiol–Ene, CuAAC, and DA Reactions*

Yagci's group demonstrated that triple click reactions can be efficiently employed for the formation of ABC miktoarm stars containing PSt, PCL, and PEG segments.[256] An MFA with allyl, azide, and anthracene orthogonal functionalities was clicked with PSt-SH *via* a thiol–ene reaction to produce an azide- and anthracene-terminated PSt precursor. On this basis, linear PCL-alkyne and PEG-MI were reacted with the precursor in a sequential/two-step or a one-pot/one-step mode using CuAAC and DA reactions to yield a PSt-PCL-PEG star.

## 3.4 Conclusions and Outlook

In the past few decades, the tremendous progress in living/controlled polymerization and highly efficient linking reactions has enabled the precise synthesis of numerous multiarm and miktoarm stars with well-defined molecular weight, relatively low polydispersity, tunable parameters (*i.e.* chain length and arm number), rich composition, and on-demand functions. The high versatility of the different methodologies allows for the rational design and tailoring of the topology and composition of star polymers. The combination of controlled polymerization and click chemistry has revolutionized the facile construction of miktoarm stars, and thus many suitable routes can be chosen to construct the desired stars. Thus far, miktoarm stars with up to nine compositions have been achieved by LAP-based iterative methodologies, and multicomponent stars with up to five compositions have been obtained *via* successive or tandem click reactions using different building blocks. All these studies underlie the multipurpose applications of star polymers in various fields, and functional miktoarm stars hold great promise for next-generation smart materials. The latest advances in synthetic star polymers have further promoted the development of some related disciplines such as polymer physics, materials science, bioscience, and nanotechnology.

In my opinion, the following five aspects will attract increasing interest in the near future. First, special functions and parameters can be introduced into miktoarm stars, including sequence-controlled arm segments, nearly monodisperse molecular weight distributions, single chain folding, and stimulus-cleavable linkages/segments. Second, more advanced organic reactions and polymerization techniques are continuously developed and adopted for materials synthesis, and the multifunctional core of miktoarm stars can be formed *via* either linking reactions or post-polymerization modification. Among them, click-like multicomponent reactions may be the method of choice for straightforward syntheses. Third, exploration of the accelerated synthesis of miktoarm stars with fewer step reactions, high star yield, and satisfactory purity remains important due to the basic requirements for practical applications in the near future. Fourth, ingenious design and tailoring can endow multicomponent star copolymers with versatile functions and applications in heterophase materials, biomaterials, and self-healing materials. Last, with the extended scopes of linkages and interactions, more emphasis will be laid on the preparation and properties of other types of miktoarm stars, such as ionic and supramolecular stars, besides the traditional stars with covalent linkages.

## Acknowledgements

This work was financially supported by the National Natural Science Foundation of China (Grants 21074081 and 21274096) and the Project Funded by the Priority Academic Program Development (PAPD) of Jiangsu Higher Education Institutions.

## References

1. N. Hadjichristidis, *J. Polym. Sci., Part A: Polym. Chem.*, 1999, **37**, 857.
2. N. Hadjichristidis, M. Pitsikalis, S. Pispas and H. Iatrou, *Chem. Rev.*, 2001, **101**, 3747.
3. N. Hadjichristidis, H. Iatrou, M. Pitsikalis, S. Pispas and A. Avgeropoulos, *Prog. Polym. Sci.*, 2005, **30**, 725.
4. A. Hirao, M. Hayashi, S. Loykulnant, K. Sugiyama, S.-W. Ryu, N. Haraguchi, A. Matsuo and T. Higashihara, *Prog. Polym. Sci.*, 2005, **30**, 111.

5. T. Higashihara, M. Hayashi and A. Hirao, *Prog. Polym. Sci.*, 2011, **36**, 323.
6. S. Ito, R. Goseki, T. Ishizone and A. Hirao, *Eur. Polym. J.*, 2013, **49**, 2545.
7. K. Khanna, S. Varshney and A. Kakkar, *Polym. Chem.*, 2010, **1**, 1171.
8. H. F. Gao and K. Matyjaszewski, *Prog. Polym. Sci.*, 2009, **34**, 317.
9. A. Blencowe, J. F. Tan, T. K. Goh and G. G. Qiao, *Polymer*, 2009, **50**, 5.
10. C. Gao and D. Y. Yan, *Prog. Polym. Sci.*, 2004, **29**, 183.
11. M. K. Georges, R. P. N. Veregin, P. M. Kazmaier and G. K. Hamer, *Macromolecules*, 1993, **26**, 2987.
12. C. J. Hawker, A. W. Bosman and E. Harth, *Chem. Rev.*, 2001, **101**, 3661.
13. J. S. Wang and K. Matyjaszewski, *J. Am. Chem. Soc.*, 1995, **117**, 5614.
14. K. Matyjaszewski and J. Xia, *Chem. Rev.*, 2001, **101**, 2921.
15. M. Kato, M. Kamigaito, M. Sawamoto and T. Higashimura, *Macromolecules*, 1995, **28**, 1721.
16. M. Kamigaito, T. Ando and M. Sawamoto, *Chem. Rev.*, 2001, **101**, 3689.
17. J. Chiefari, Y. K. Chong, F. Ercole, J. Krstina, J. Jeffery, T. P. T. Le, R. T. A. Mayadunne, G. F. Meijs, C. L. Moad, G. Moad, E. Rizzardo and S. H. Thang, *Macromolecules*, 1998, **31**, 5559.
18. G. Moad, E. Rizzardo and S. H. Thang, *Aust. J. Chem.*, 2005, **58**, 379.
19. S. Perrier and P. Takolpuckdee, *J. Polym. Sci., Part A: Polym. Chem.*, 2005, **43**, 5347.
20. C. Barner-Kowollik and S. Perrier, *J. Polym. Sci., Part A: Polym. Chem.*, 2008, **46**, 5715.
21. H. C. Kolb, M. G. Finn and K. B. Sharpless, *Angew. Chem., Int. Ed.*, 2001, **40**, 2004.
22. J. F. Lutz, *Angew. Chem., Int. Ed.*, 2008, **47**, 2182.
23. J. R. Schaefgen and P. J. Flory, *J. Am. Chem. Soc.*, 1948, **70**, 2709.
24. M. Morton, T. E. Helminiak, S. D. Gadkary and F. Bueche, *J. Polym. Sci.*, 1962, **57**, 471.
25. J. Roovers, L. L. Zhou, P. M. Toporowski, M. Vanderzwan, H. Iatrou and N. Hadjichristidis, *Macromolecules*, 1993, **26**, 4324.
26. J. Roovers, P. M. Toporowski and J. Martin, *Macromolecules*, 1989, **22**, 1897.
27. S. Sioula, N. Hadjichristidis and E. L. Thomas, *Macromolecules*, 1998, **31**, 8429.
28. S. Pispas, Y. Poulos and N. Hadjichristidis, *Macromolecules*, 1998, **31**, 4177.
29. S. Pispas, N. Hadjichristidis, I. Potemkin and A. Khokhlov, *Macromolecules*, 2000, **33**, 1741.
30. Z. B. Li, E. Kesselman, Y. Talmon, M. A. Hillmyer and T. P. Lodge, *Science*, 2004, **306**, 98.
31. Z. B. Li, M. A. Hillmyer and T. P. Lodge, *Langmuir*, 2006, **22**, 9409.
32. A. Walther and A. H. E. Müller, *Chem. Commun.*, 2009, 1127.
33. J. Park, S. Jang and J. K. Kim, *J. Polym. Sci., Part B: Polym. Phys.*, 2015, **53**, 1.
34. D. Kuckling and A. Wycisk, *J. Polym. Sci., Part A: Polym. Chem.*, 2013, **51**, 2980.

35. Z. Iatridi and C. Tsitsilianis, *Polymers*, 2011, **3**, 1911.
36. L. Zhao and Z. Q. Lin, *Soft Matter*, 2011, **7**, 10520.
37. J. M. J. Fréchet, M. Henmi, I. Gitsov, S. Aoshima, M. R. Leduc and R. B. Grubbs, *Science*, 1995, **269**, 1080.
38. C. J. Hawker, J. M. J. Fréchet, R. B. Grubbs and J. Dao, *J. Am. Chem. Soc.*, 1995, **117**, 10763.
39. R. Aksakal, M. Resmini and C. R. Becer, *Polym. Chem.*, 2016, **7**, 171.
40. C.-G. Mu, X.-D. Fan, W. Tian, Y. Bai and X. Zhou, *Polym. Chem.*, 2012, **3**, 1137.
41. Y. Li, Y. F. Qian, T. Liu, G. Y. Zhang, J. M. Hu and S. Y. Liu, *Polym. Chem.*, 2014, **5**, 1743.
42. P.-F. Gou, W.-P. Zhu, N. Xu and Z.-Q. Shen, *J. Polym. Sci., Part A: Polym. Chem.*, 2010, **48**, 2961.
43. N. J. Hovestad, G. van Koten, S. A. F. Bon and D. M. Haddleton, *Macromolecules*, 2000, **33**, 4048.
44. A. Heise, S. Diamanti, J. L. Hedrick, C. W. Frank and R. D. Miller, *Macromolecules*, 2001, **34**, 3798.
45. Y. L. Zhao, X. T. Shuai, C. F. Chen and F. Xi, *Macromolecules*, 2004, **37**, 8854.
46. Y. L. Zhao, Y. M. Chen, C. F. Chen and F. Xi, *Polymer*, 2005, **46**, 5808.
47. X. Zhang, Y. M. Chen, A. J. Gong, C. F. Chen and F. Xi, *Polym. Int.*, 1999, **48**, 896.
48. S. Maier, A. Sunder, H. Frey and R. Mülhaupt, *Macromol. Rapid Commun.*, 2000, **21**, 226.
49. Z. Shen, Y. Chen, E. Barriau and H. Frey, *Macromol. Chem. Phys.*, 2006, **207**, 57.
50. H. Y. Hong, Y. Y. Mai, Y. F. Zhou, D. Y. Yan and J. Cui, *Macromol. Rapid Commun.*, 2007, **28**, 591.
51. C. B. Zhang, Y. Zhou, Q. Liu, S. X. Li, S. Perrier and Y. L. Zhao, *Macromolecules*, 2011, **44**, 2034.
52. M. J. Zhang, H. H. Liu, W. Shao, C. N. Ye and Y. L. Zhao, *Macromolecules*, 2012, **45**, 9312.
53. C. X. Li, H. H. Liu, D. D. Tang and Y. L. Zhao, *Polym. Chem.*, 2015, **6**, 1474.
54. M. J. Zhang, H. H. Liu, W. Shao, K. Miao and Y. L. Zhao, *Macromolecules*, 2013, **46**, 1325.
55. X. F. Wang, R. W. Graff, Y. Shi and H. F. Gao, *Polym. Chem.*, 2015, **6**, 6739.
56. H. Eschwey, M. Hallensleben and W. Burchard, *Makromol. Chem.*, 1973, **173**, 235.
57. Y. Gnanou, P. Lutz and P. Rempp, *Makromol. Chem.*, 1988, **189**, 2885.
58. H. F. Gao and K. Matyjaszewski, *Macromolecules*, 2008, **41**, 1118.
59. G. H. Deng and Y. M. Chen, *Macromolecules*, 2004, **37**, 18.
60. G. H. Deng, M. Cao, J. Huang, L. H. He and Y. M. Chen, *Polymer*, 2005, **46**, 5698.
61. A. Heise, M. Trollsås, T. Magbitang, J. L. Hedrick, C. W. Frank and R. D. Miller, *Macromolecules*, 2001, **34**, 2798.
62. Y. M. Guo, C. Y. Pan and J. Wang, *J. Polym. Sci., Part A: Polym. Chem.*, 2001, **39**, 2134.

63. T. Erdogan, Z. Ozyurek, G. Hizal and U. Tunca, *J. Polym. Sci., Part A: Polym. Chem.*, 2004, **42**, 2313.
64. A. E. Muftuoglu, I. Cianga, D. Colak and Y. Yagci, *Des. Monomers Polym.*, 2004, 7, 563.
65. S. Peleshanko, J. Jeong, V. V. Shevchenko, K. L. Genson, Y. Pikus, M. Ornatska, S. Petrash and V. V. Tsukruk, *Macromolecules*, 2004, **37**, 7497.
66. G. H. Deng, L. W. Zhang, C. D. Liu, L. H. He and Y. M. Chen, *Eur. Polym. J.*, 2005, **41**, 1177.
67. Y. Shi, Z. F. Fu, B. Y. Li, L. Q. Zhang, X. P. Cai and D. S. Zhang, *Eur. Polym. J.*, 2007, **43**, 2612.
68. X. J. Liao, H. L. Zhang, J. F. Chen and X. Y. Wang, *Polym. Bull.*, 2007, **58**, 819.
69. O. Glaied, C. Delaite and P. Dumas, *J. Polym. Sci., Part A: Polym. Chem.*, 2006, **44**, 1796.
70. O. Glaied, C. Delaite and P. Dumas, *J. Polym. Sci., Part A: Polym. Chem.*, 2007, **45**, 4179.
71. Z. S. Ge, Y. L. Cai, J. Yin, Z. Y. Zhu, J. Y. Rao and S. Y. Liu, *Langmuir*, 2007, **30**, 1114.
72. Y. Miura, A. Narumi, S. Matsuya, T. Satoh, Q. Duan, H. Kaga and T. Kakuchi, *J. Polym. Sci., Part A: Polym. Chem.*, 2005, **43**, 4271.
73. O. Glaied, C. Delaite and P. Dumas, *J. Polym. Sci., Part A: Polym. Chem.*, 2007, **45**, 968.
74. D. R. Lu, Y. Wang, T. Y. Wu, K. Tao, L. J. An and R. K. Bai, *J. Polym. Sci., Part A: Polym. Chem.*, 2008, **46**, 5805.
75. H. Durmaz, S. Aras, G. Hizal and U. Tunca, *Des. Monomers Polym.*, 2005, **8**, 203.
76. L. Zhong, Y. F. Zhou, D. Y. Yan and C. Y. Pan, *Macromol. Rapid Commun.*, 2008, **29**, 1385.
77. X. S. Feng and C. Y. Pan, *Macromolecules*, 2002, **35**, 2084.
78. T. He, D. J. Li, X. Sheng and B. Zhao, *Macromolecules*, 2004, **37**, 3128.
79. U. Tunca, Z. Ozyurek, T. Erdogan and G. Hizal, *J. Polym. Sci., Part A: Polym. Chem.*, 2004, **42**, 4228.
80. P. J. Shi, Y. G. Li and C. Y. Pan, *Eur. Polym. J.*, 2004, **40**, 1283.
81. C. Gordin, C. Delaite, H. Medlej, D. Josien-Lefebvre, K. Hariri and M. Rusu, *Polym. Bull.*, 2009, **63**, 789.
82. C. Liu, M. A. Hillmyer and T. P. Lodge, *Langmuir*, 2009, **25**, 13718.
83. H. H. Liu, D. D. Tang, R. P. Tang and Y. L. Zhao, *Sci. China: Chem.*, 2015, **58**, 1724.
84. S. Penczek, M. Cypryk, A. Duda, P. Kubisa and S. Slomkowski, *Prog. Polym. Sci.*, 2007, **32**, 247.
85. C. Celik, G. Hizal and U. Tunca, *J. Polym. Sci., Part A: Polym. Chem.*, 2003, **41**, 2542.
86. D. Priftis, M. Pitsikalis and N. Hadjichristidis, *J. Polym. Sci., Part A: Polym. Chem.*, 2007, **45**, 5164.
87. Y.-M. Guo, J. Xu and C.-Y. Pan, *J. Polym. Sci., Part A: Polym. Chem.*, 2001, **39**, 437.

88. R. Francis, B. Lepoittevin, D. Taton and Y. Gnanou, *Macromolecules*, 2002, **35**, 9001.
89. Y. L. Cai and S. P. Armes, *Macromolecules*, 2005, **38**, 271.
90. C. Liu, Y. Zhang and J. L. Huang, *Macromolecules*, 2008, **41**, 325.
91. J. Babin, D. Taton, M. Brinkmann and S. Lecommandoux, *Macromolecules*, 2008, **41**, 1384.
92. H. F. Gao and K. Matyjaszewski, *J. Am. Chem. Soc.*, 2007, **129**, 11828.
93. C. Boyer, M. Whittaker and T. P. Davis, *J. Polym. Sci., Part A: Polym. Chem.*, 2011, **49**, 5245.
94. X. F. Shi, W. Zhou, Q. Qiu and Z. S. An, *Chem. Commun.*, 2012, **48**, 7389.
95. X. H. Wei, G. Moad, B. W. Muir, E. Rizzardo, J. Rosselgong, W. T. Yang and S. H. Thang, *Macromol. Rapid Commun.*, 2014, **35**, 840.
96. N. Cakir, U. Tunca, G. Hizal and H. Durmaz, *Macromol. Chem. Phys.*, 2016, **217**, 636.
97. H. F. Gao and K. Matyjaszewski, *Macromolecules*, 2008, **41**, 4250.
98. A. O. Burts, A. X. Gao and J. A. Johnson, *Macromol. Rapid Commun.*, 2014, **35**, 168.
99. H. F. Gao, N. V. Tsarevsky and K. Matyjaszewski, *Macromolecules*, 2005, **38**, 5995.
100. J. Z. Du and Y. M. Chen, *Macromolecules*, 2004, **37**, 3588.
101. T. Tsoukatos, S. Pispas and N. Hadjichristidis, *J. Polym. Sci., Part A: Polym. Chem.*, 2001, **39**, 320.
102. J. Z. Du and Y. M. Chen, *J. Polym. Sci., Part A: Polym. Chem.*, 2004, **42**, 2263.
103. B. S. Sumerlin, D. Neugebauer and K. Matyjaszewski, *Macromolecules*, 2005, **38**, 702.
104. H. F. Gao and K. Matyjaszewski, *Macromolecules*, 2006, **39**, 7216.
105. R. W. Pennisi and L. J. Fetters, *Macromolecules*, 1988, **21**, 1094.
106. J. W. Mays, *Polym. Bull.*, 1990, **23**, 247.
107. H. Iatrou and N. Hadjichristidis, *Macromolecules*, 1992, **25**, 4649.
108. A. Zioga, S. Sioula and N. Hadjichristidis, *Macromol. Symp.*, 2000, **157**, 239.
109. V. Bellas, H. Iatrou and N. Hadjichristidis, *Macromolecules*, 2000, **33**, 6993.
110. H. Iatrou and N. Hadjichristidis, *Macromolecules*, 1993, **26**, 2479.
111. T. Fujimoto, H. Zhang, T. Kazama, Y. Isono, H. Hasegawa and T. Hashimoto, *Polymer*, 1992, **33**, 2208.
112. H. Hückstädt, V. Abetz and R. Stadler, *Macromol. Rapid Commun.*, 1996, **17**, 599.
113. H. Hückstädt, A. Göpfert and V. Abetz, *Macromol. Chem. Phys.*, 2000, **201**, 296.
114. A. Walther, J. Yuan, V. Abetz and A. H. E. Müller, *Nano Lett.*, 2009, **9**, 2026.
115. O. Lambert, P. Dumas, G. Hurtrez and G. Riess, *Macromol. Rapid Commun.*, 1997, **18**, 343.
116. M. Nasser-Eddine, S. Reutenauer, C. Delaite, G. Hurtrez and P. Dumas, *J. Polym. Sci., Part A: Polym. Chem.*, 2004, **42**, 1745.

117. S. Sioula, N. Hadjichristidis and E. L. Thomas, *Macromolecules*, 1998, **31**, 5272.
118. X. J. Wang, J. P. He and Y. L. Yang, *J. Polym. Sci., Part A: Polym. Chem.*, 2007, **45**, 4818.
119. A. Mavroudis and N. Hadjichristidis, *Macromolecules*, 2006, **39**, 535.
120. M. Hayashi, K. Kojima and A. Hirao, *Macromolecules*, 1999, **32**, 2425.
121. M. Hayashi, Y. Negishi and A. Hirao, *Proc. Jpn. Acad., Ser. B*, 1999, **75**, 93.
122. A. Hirao and N. Haraguchi, *Macromolecules*, 2002, **35**, 7238.
123. A. Hirao, M. Hayashi and A. Matsuo, *Polymer*, 2002, **43**, 7125.
124. A. Hirao and A. Matsuo, *Macromolecules*, 2003, **36**, 9742.
125. A. Hirao, M. Hayashi and T. Higashihara, *Macromol. Chem. Phys.*, 2001, **202**, 3165.
126. Y. L. Zhao, T. Higashihara, K. Sugiyama and A. Hirao, *J. Am. Chem. Soc.*, 2005, **127**, 14158.
127. T. Higashihara, M. Nagura, K. Inoue, N. Haraguchi and A. Hirao, *Macromolecules*, 2005, **38**, 4577.
128. A. Hirao, T. Higashihara, M. Nagura and T. Sakurai, *Macromolecules*, 2006, **39**, 6081.
129. A. Hirao, T. Higashihara and K. Inoue, *Macromolecules*, 2008, **41**, 3579.
130. T. Higashihara, T. Sakurai and A. Hirao, *Macromolecules*, 2009, **42**, 6006.
131. S. Ito, R. Goseki, S. Senda and A. Hirao, *Macromolecules*, 2012, **45**, 4997.
132. R. Goseki, Y. Ozama, E. Akemine, S. Ito, S. Ehara and A. Hirao, *Polymer*, 2013, **54**, 2049.
133. S. Ito, R. Goseki, T. Ishizone, S. Senda and A. Hirao, *Macromolecules*, 2013, **46**, 819.
134. S. Ito, R. Goseki, I. Manners, T. Ishizone and A. Hirao, *Macromol. Chem. Phys.*, 2015, **216**, 1523.
135. D. Fournier, R. Hoogenboom and U. S. Schubert, *Chem. Soc. Rev.*, 2007, **36**, 1369.
136. W. H. Binder and R. Sachsenhofer, *Macromol. Rapid Commun.*, 2007, **28**, 15.
137. J. F. Lutz, *Angew. Chem., Int. Ed.*, 2007, **46**, 1018.
138. C. R. Becer, R. Hoogenboom and U. S. Schubert, *Angew. Chem., Int. Ed.*, 2009, **48**, 4900.
139. A. B. Lowe, *Polym. Chem.*, 2010, **1**, 17.
140. U. Mansfeld, C. Pietsch, R. Hoogenboom, C. R. Becer and U. S. Schubert, *Polym. Chem.*, 2010, **1**, 1560.
141. B. S. Sumerlin and A. P. Vogt, *Macromolecules*, 2010, **43**, 1.
142. P. J. Roth, C. Boyer, A. B. Lowe and T. P. Davis, *Macromol. Rapid Commun.*, 2011, **32**, 1123.
143. N. Akeroyd and B. Klumperman, *Eur. Polym. J.*, 2011, **47**, 1207.
144. M. A. Harvison and A. B. Lowe, *Macromol. Rapid Commun.*, 2011, **32**, 779.
145. K. Kempe, A. Krieg, C. R. Becer and U. S. Schubert, *Chem. Soc. Rev.*, 2012, **41**, 176.

146. O. Altintas, A. P. Vogt, C. Barner-Kowollik and U. Tunca, *Polym. Chem.*, 2012, **3**, 34.
147. H. Durmaz, A. Sanyal, G. Hizal and U. Tunca, *Polym. Chem.*, 2012, **3**, 825.
148. U. Tunca, *Macromol. Rapid Commun.*, 2013, **34**, 38.
149. U. Tunca, *J. Polym. Sci., Part A: Polym. Chem.*, 2014, **52**, 3147.
150. C. Yi, Q. Han and X. Q. Xiong, *Acta Polym. Sin.*, 2014, **5**, 584.
151. R. Huisgen, *Angew. Chem., Int. Ed.*, 1963, **2**, 565.
152. V. V. Rostovtsev, L. G. Green, V. V. Fokin and K. B. Sharpless, *Angew. Chem., Int. Ed.*, 2002, **41**, 2596.
153. C. W. Tornoe, C. Christensen and M. Meldal, *J. Org. Chem.*, 2002, **67**, 3057.
154. H. F. Gao and K. Matyjaszewski, *Macromolecules*, 2006, **39**, 4960.
155. J. Xu and S. Y. Liu, *J. Polym. Sci., Part A: Polym. Chem.*, 2009, **47**, 404.
156. R. Hoogenboom, B. C. Moore and U. S. Schubert, *Chem. Commun.*, 2006, 4010.
157. O. Altintas, B. Yankul, G. Hizal and U. Tunca, *J. Polym. Sci., Part A: Polym. Chem.*, 2006, **44**, 6458.
158. Z. S. Ge, D. Wang, Y. M. Zhou, H. W. Liu and S. Y. Liu, *Macromolecules*, 2009, **42**, 2903.
159. J. Kulis, Z. F. Jia and M. J. Monteiro, *Macromolecules*, 2012, **45**, 5956.
160. H. F. Gao, K. Min and K. Matyjaszewski, *Macromol. Chem. Phys.*, 2007, **208**, 1370.
161. M. Lammens, D. Fournier, M. W. M. Fijten, R. Hoogenboom and F. Du Prez, *Macromol. Rapid Commun.*, 2009, **30**, 2049.
162. L. P. Yang, X. H. Dong and C. Y. Pan, *J. Polym. Sci., Part A: Polym. Chem.*, 2008, **46**, 7757.
163. L.-Y. Li, W.-D. He, J. Li, B.-Y. Zhang, T.-T. Pan, X.-L. Sun and Z.-L. Ding, *Biomacromolecules*, 2010, **11**, 1882.
164. J. Zhu, X. L. Zhu, E. T. Kang and K. G. Neoh, *Polymer*, 2007, **48**, 6992.
165. J. Y. Rao, Y. F. Zhang, J. Y. Zhang and S. Y. Liu, *Biomacromolecules*, 2008, **9**, 2586.
166. Z. S. Ge, J. Xu, J. M. Hu, Y. F. Zhang and S. Y. Liu, *Soft Matter*, 2009, **5**, 3932.
167. E. H. H. Wong, M. H. Stenzel, T. Junkers and C. Barner-Kowollik, *Macromolecules*, 2010, **43**, 3785.
168. C. Gao, Y. Wang, P. F. Gou, X. Cai, X. D. Li, W. P. Zhu and Z. Q. Shen, *J. Polym. Sci., Part A: Polym. Chem.*, 2013, **51**, 2824.
169. Y. Wang, B. Liu, X. Wang, W. Zhao, D. T. Liu, X. L. Liu and D. M. Cui, *Polym. Chem.*, 2014, **5**, 4580.
170. E. Doganci, M. Gorur, C. Uyanik and F. Yilmaz, *J. Polym. Sci., Part A: Polym. Chem.*, 2014, **52**, 3390.
171. E. Doganci, M. A. Tasdelen and F. Yilmaz, *Macromol. Chem. Phys.*, 2015, **216**, 1823.
172. B. Huang, M. Chen, S. X. Zhou and L. M. Wu, *Polym. Chem.*, 2015, **6**, 3913.

173. Y. Satoh, K. Miyachi, H. Matsuno, T. Isono, K. Tajima, T. Kakuchi and T. Satoh, *Macromolecules*, 2016, **49**, 499.
174. H. H. Liu, X. Jiang, R. J. Bian, M. Tong, D. D. Tang, X. D. Zhou and Y. L. Zhao, *Polymer*, 2015, **64**, 249.
175. P. V. Chang, J. A. Prescher, E. M. Sletten, J. M. Baskin, I. A. Miller, N. J. Agard, A. Lo and C. R. Bertozzi, *Proc. Natl. Acad. Sci. U. S. A.*, 2010, **107**, 1821.
176. N. J. Agard, J. A. Prescher and C. R. Bertozzi, *J. Am. Chem. Soc.*, 2004, **126**, 15046.
177. S. T. Laughlin and C. R. Bertozzi, *ACS Chem. Biol.*, 2009, **4**, 1068.
178. P. V. Chang, X. Chen, C. Smyrniotis, A. Xenakis, T. S. Hu, C. R. Bertozzi and P. Wu, *Angew. Chem., Int. Ed.*, 2009, **48**, 4030.
179. S. S. Wang, X. K. Yang, W. Zhu, L. Zou, K. Zhang, Y. M. Chen and F. Xi, *Polymer*, 2014, **55**, 4812.
180. W. Zhu, Z. Li, P. Sun, L. X. Ren and K. Zhang, *Polymer*, 2016, **86**, 1.
181. M. R. Whittaker, C. N. Urbani and M. J. Monteiro, *J. Am. Chem. Soc.*, 2006, **128**, 11360.
182. O. Altintas, G. Hizal and U. Tunca, *J. Polym. Sci., Part A: Polym. Chem.*, 2006, **44**, 5699.
183. G. H. Deng, D. Y. Ma and Z. Z. Xu, *Eur. Polym. J.*, 2007, **43**, 1179.
184. H. Liu, C. H. Li, H. W. Liu and S. Y. Liu, *Langmuir*, 2009, **25**, 4724.
185. Y. F. Zhang, H. Liu, J. M. Hu, C. H. Li and S. Y. Liu, *Macromol. Rapid Commun.*, 2009, **30**, 941.
186. Y. F. Zhang, H. Liu, H. F. Dong, C. H. Li and S. Y. Liu, *J. Polym. Sci., Part A: Polym. Chem.*, 2009, **47**, 1636.
187. C. H. Li, Z. S. Ge, H. W. Liu and S. Y. Liu, *J. Polym. Sci., Part A: Polym. Chem.*, 2009, **47**, 4001.
188. Y. F. Zhang, C. H. Li and S. Y. Liu, *J. Polym. Sci., Part A: Polym. Chem.*, 2009, **47**, 3066.
189. O. Altintas, B. Yankul, G. Hizal and U. Tunca, *J. Polym. Sci., Part A: Polym. Chem.*, 2007, **45**, 3588.
190. A. Vora, K. Singh and D. C. Webster, *Polymer*, 2009, **50**, 2768.
191. A. Gozgen, A. Dag, H. Durmaz, O. Sirkecioglu, G. Hizal and U. Tunca, *J. Polym. Sci., Part A: Polym. Chem.*, 2009, **47**, 497.
192. S. Ozlem, B. Iskin, G. Yilmaz, M. Kukut, J. Hacaloglu and Y. Yagci, *Eur. Polym. J.*, 2012, **48**, 1755.
193. R. Alizadeh and M. Ghaemy, *Polymer*, 2015, **66**, 179.
194. K. Khanna, S. Varshney and A. Kakkar, *Macromolecules*, 2010, **43**, 5688.
195. Y. F. Chu, H. Yu, Y. T. Zhang, G. Y. Zhang, Y. Y. Ma, R. X. Zhuo and X. L. Jiang, *J. Polym. Sci., Part A: Polym. Chem.*, 2014, **52**, 3346.
196. L. Li, X.-W. Kan, X.-X. Deng, C.-C. Song, F.-S. Du and Z.-C. Li, *J. Polym. Sci., Part A: Polym. Chem.*, 2013, **51**, 865.
197. A. J. Zhu, K. Miao, Y. B. Deng, H. T. Ke, H. He, T. Yang, M. Guo, Y. L. Li, Z. Q. Guo, Y. Y. Wang, X. L. Yang, Y. L. Zhao and H. B. Chen, *ACS Nano*, 2015, **9**, 7874.

198. O. Altintas, G. Hizal and U. Tunca, *J. Polym. Sci., Part A: Polym. Chem.*, 2008, **46**, 1218.
199. G. W. Wang, X. L. Luo, C. Liu and J. L. Huang, *J. Polym. Sci., Part A: Polym. Chem.*, 2008, **46**, 2154.
200. L. P. Yang, H. X. Zhou, G. Y. Shi, Y. Wang and C. Y. Pan, *J. Polym. Sci., Part A: Polym. Chem.*, 2008, **46**, 6641.
201. W. D. Pan, H. H. Liu, H. C. Zhang and Y. L. Zhao, *Polym. Chem.*, 2016, **7**, 2870.
202. C. N. Ye, G. D. Zhao, M. J. Zhang, J. Z. Du and Y. L. Zhao, *Macromolecules*, 2012, **45**, 7429.
203. H. H. Liu, K. Miao, G. D. Zhao, C. X. Li and Y. L. Zhao, *Polym. Chem.*, 2014, **5**, 3071.
204. H. H. Liu, W. D. Pan, M. Tong and Y. L. Zhao, *Polym. Chem.*, 2016, **7**, 1603.
205. K. Miao, H. H. Liu and Y. L. Zhao, *Polym. Chem.*, 2014, **5**, 3335.
206. T. Dispinar, R. Sanyal and A. Sanyal, *J. Polym. Sci., Part A: Polym. Chem.*, 2007, **45**, 4545.
207. M. L. Szalai, D. V. McGrath, D. R. Wheeler, T. Zifer and J. R. McElhanon, *Macromolecules*, 2007, **40**, 818.
208. B. Gacal, H. Akat, D. K. Balta, N. Arsu and Y. Yagci, *Macromolecules*, 2008, **41**, 2401.
209. A. J. Inglis, M. H. Stenzel and C. Barner-Kowollik, *Macromol. Rapid Commun.*, 2009, **30**, 1792.
210. G. Franc and A. K. Kakkar, *Chem.–Eur. J.*, 2009, **15**, 5630.
211. J. A. Syrett, G. Mantovani, W. R. S. Barton, D. Price and D. M. Haddleton, *Polym. Chem.*, 2010, **1**, 102.
212. M. A. Tasdelen, *Polym. Chem.*, 2011, **2**, 2133.
213. A. Dag, H. Durmaz, U. Tunca and G. Hizal, *J. Polym. Sci., Part A: Polym. Chem.*, 2009, **47**, 178.
214. A. Dag, H. Durmaz, G. Hizal and U. Tunca, *J. Polym. Sci., Part A: Polym. Chem.*, 2008, **46**, 302.
215. F. Bahadori, A. Dag, H. Durmaz, N. Cakir, H. Onyuksel, U. Tunca, G. Topcu and G. Hizal, *Polymers*, 2014, **6**, 214.
216. A. J. Inglis, S. Sinnwell, T. P. Davis, C. Barner-Kowollik and M. H. Stenzel, *Macromolecules*, 2008, **41**, 4120.
217. H. Durmaz, F. Karatas, U. Tunca and G. Hizal, *J. Polym. Sci., Part A: Polym. Chem.*, 2006, **44**, 499.
218. O. Altintas, G. Hizal and U. Tunca, *Des. Monomers Polym.*, 2009, **12**, 83.
219. J. W. Chan, B. Yu, C. E. Hoyle and A. B. Lowe, *Polymer*, 2009, **50**, 3158.
220. C. E. Hoyle and C. N. Bowman, *Angew. Chem., Int. Ed.*, 2010, **49**, 1540.
221. M. J. Kade, D. J. Burke and C. J. Hawker, *J. Polym. Sci., Part A: Polym. Chem.*, 2010, **48**, 743.
222. B. Yu, J. W. Chan, C. E. Hoyle and A. B. Lowe, *J. Polym. Sci., Part A: Polym. Chem.*, 2009, **47**, 3544.
223. A. B. Lowe, C. E. Hoyle and C. N. Bowman, *J. Mater. Chem.*, 2010, **20**, 4745.

224. G. Z. Li, R. K. Randev, A. H. Soeriyadi, G. Rees, C. Boyer, Z. Tong, T. P. Davis, C. R. Becer and D. M. Haddleton, *Polym. Chem.*, 2010, **1**, 1196.
225. A. H. Soeriyadi, G. Z. Li, S. Slavin, M. W. Jones, C. M. Amos, C. R. Becer, M. R. Whittaker, D. M. Haddleton, C. Boyer and T. P. Davis, *Polym. Chem.*, 2011, **2**, 815.
226. B. M. Rosen, G. Lligadas, C. Hahn and V. Percec, *J. Polym. Sci., Part A: Polym. Chem.*, 2009, **47**, 3931.
227. C. R. Becer, K. Kokado, C. Weber, A. Can, Y. Chujo and U. S. Schubert, *J. Polym. Sci., Part A: Polym. Chem.*, 2010, **48**, 1278.
228. C. R. Becer, K. Babiuch, D. Pilz, S. Hornig, T. Heinze, M. Gottschaldt and U. S. Schubert, *Macromolecules*, 2009, **42**, 2387.
229. M. Li, P. De, S. R. Gondi and B. S. Sumerlin, *J. Polym. Sci., Part A: Polym. Chem.*, 2008, **46**, 5093.
230. H. B. Li, B. Yu, H. Matsushima, C. E. Hoyle and A. B. Lowe, *Macromolecules*, 2009, **42**, 6537.
231. C. E. Hoyle, A. B. Lowe and C. N. Bowman, *Chem. Soc. Rev.*, 2010, **39**, 1355.
232. A. Klaikherd, S. Ghosh and S. Thayumanavan, *Macromolecules*, 2007, **40**, 8518.
233. J. A. Syrett, C. R. Becer and D. M. Haddleton, *Polym. Chem.*, 2010, **1**, 978.
234. J. W. Chan, B. Yu, C. E. Hoyle and A. B. Lowe, *Chem. Commun.*, 2008, 4959.
235. R. J. Pounder, M. J. Stanford, P. Brooks, S. P. Richards and A. P. Dove, *Chem. Commun.*, 2008, 5158.
236. Q. Q. Guo, C. Y. Liu, T. T. Tang, J. Huang, X. G. Zhang and G. W. Wang, *J. Polym. Sci., Part A: Polym. Chem.*, 2013, **51**, 4572.
237. J. Qiu, J. W. Zhang, F. L. Yu, J. Wei and L. Ding, *J. Polym. Sci., Part A: Polym. Chem.*, 2016, **54**, 692.
238. S. S. Patil, B. V. Tawade and P. P. Wadgaonkar, *J. Polym. Sci., Part A: Polym. Chem.*, 2016, **54**, 844.
239. E. Gungor, G. Hizal and U. Tunca, *J. Polym. Sci., Part A: Polym. Chem.*, 2008, **46**, 6703.
240. G. W. Wang, B. Hu and J. L. Huang, *Macromolecules*, 2010, **43**, 6939.
241. G. W. Wang, X. S. Fan and J. L. Huang, *J. Polym. Sci., Part A: Polym. Chem.*, 2010, **48**, 5313.
242. X. Y. Liang, Y. J. Liu, J. Huang, L. H. Wei and G. W. Wang, *Polym. Chem.*, 2015, **6**, 466.
243. Z. M. Wu, H. Liang and J. Lu, *Macromolecules*, 2010, **43**, 5699.
244. Z.-M. Wu, H. Liang, J. Lu and W.-L. Deng, *J. Polym. Sci., Part A: Polym. Chem.*, 2010, **48**, 3323.
245. B. Iskin, G. Yilmaz and Y. Yagci, *J. Polym. Sci., Part A: Polym. Chem.*, 2011, **49**, 2417.
246. H. Durmaz, A. Dag, A. Hizal, G. Hizal and U. Tunca, *J. Polym. Sci., Part A: Polym. Chem.*, 2008, **46**, 7091.
247. A. Dag, H. Durmaz, V. Kirmizi, G. Hizal and U. Tunca, *Polym. Chem.*, 2010, **1**, 621.

248. H. Durmaz, A. Dag, D. Gursoy, A. L. Demirel, G. Hizal and U. Tunca, *J. Polym. Sci., Part A: Polym. Chem.*, 2010, **48**, 1557.
249. X. Q. Xiong and Y. H. Xu, *Polym. Bull.*, 2010, **65**, 455.
250. S. Sinnwell, A. J. Inglis, M. H. Stenzel and C. Barner-Kowollik, *Macromol. Rapid Commun.*, 2008, **29**, 1090.
251. Q. Fu, G. W. Wang, W. C. Lin and J. L. Huang, *J. Polym. Sci., Part A: Polym. Chem.*, 2009, **47**, 986.
252. Z. F. Jia, D. E. Lonsdale, J. Kulis and M. J. Monteiro, *ACS Macro Lett.*, 2012, **1**, 780.
253. U. S. Gunay, H. Durmaz, E. Gungor, A. Dag, G. Hizal and U. Tunca, *J. Polym. Sci., Part A: Polym. Chem.*, 2012, **50**, 729.
254. T. Dedeoglu, H. Durmaz, H. Hizal and U. Tunca, *J. Polym. Sci., Part A: Polym. Chem.*, 2012, **50**, 1917.
255. N. Cakir, M. Yavuzarslan, H. Durmaz, G. Hizal and U. Tunca, *J. Polym. Sci., Part A: Polym. Chem.*, 2013, **51**, 899.
256. B. Iskin, G. Yilmaz and Y. Yagci, *Polym. Chem.*, 2011, **2**, 2865.

CHAPTER 4

# *Use of Click Chemistry as a Coupling Strategy for the Synthesis of Miktoarm Star Polymers*

ERDINC DOGANCI[a], MESUT GORUR[b] AND FARUK YILMAZ*[c]

[a]Kocaeli Vocational School, Kocaeli University, 41380, Kocaeli, Turkey; [b]Department of Chemistry, Istanbul Medeniyet University, 34700, Istanbul, Turkey; [c]Department of Chemistry, Gebze Technical University, 41400, Kocaeli, Turkey
*E-mail: fyilmaz@gyte.edu.tr

## 4.1 Introduction

The fast development of technology requires the design and synthesis of new polymers (or materials) with well-defined molecular properties, such as their topology, molecular weight and distribution, and functional groups. Profound investigation and use of elegant organic reactions to obtain tailor-made polymers for multipurpose applications are essential for the development of macromolecular chemistry.[1] In this perspective, the combination of controlled polymerization techniques and effective ligation reactions affords an effective platform for the synthesis of delicate macromolecular structures with different topologies.[2] Cu(I)-catalyzed 1,3-dipolar azide–alkyne

Polymer Chemistry Series No. 25
Miktoarm Star Polymers: From Basics of Branched Architecture to Synthesis, Self-assembly and Applications
Edited by Ashok Kakkar

Published by the Royal Society of Chemistry, www.rsc.org

cycloaddition (CuAAC) reactions, termed as Sharpless-type click reactions, are the most widely used ligation technique in polymer chemistry.[3]

According to Sharpless, there are some basic requirements for reactions to be regarded as click chemistry methods.[4,5] The reactions should

- be modular,
- give high yields at moderate temperatures with high stereo- and regio-selectivity,
- display high tolerance to various functional groups and solvents,
- not require laborious purification procedures.

Later, Barner-Kowollik *et al.* revised the definition of click chemistry by adding some other basic requirements. According to their definition, equimolar amounts of macromolecular building blocks must be coupled together at a reasonable time-scale with high yields and the main product must be isolated using a simple large-scale purification process.[6,7] However, in some cases, a stoichiometric excess of one building block can be used where simple large-scale purification procedures work, such as selective precipitation.

Other types of click reaction methods used in the synthesis of miktoarm star polymers are Diels–Alder (DA), thiol–ene, atom transfer nitroxide radical coupling (ATNRC), and aldehyde–aminooxy click reactions (Figure 4.1).[7]

The commonly used polymerization methods for the synthesis of miktoarm star polymers are atom transfer radical polymerization (ATRP), reversible addition–fragmentation chain transfer (RAFT), nitroxide-mediated polymerization (NMP), and ring opening polymerization (ROP) methods. By combining these polymerization methods with click chemistry techniques, a wide variety of miktoarm polymers have been prepared and reported in the literature. The topologies of the miktoarm stars, together with the methods used for their synthesis, are summarized in Table 4.1.

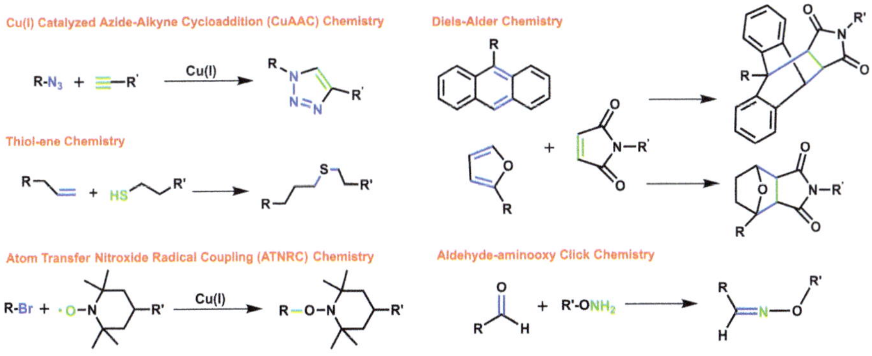


**Figure 4.1** Click chemistry methods used in the synthesis of miktoarm star polymers.

**Table 4.1** Miktoarm star topologies and their preparation methods.

| Reagent A for click | Reagent B for click | Click | Type of $A_nB_n$ |
|---|---|---|---|
| Alkyne-(PMMA-*b*-PS) | $N_3$-PEG, $N_3$-P*t*BA | CuAAC | ABC[8] |
| Tripropargyl-amine | X-$N_3$, Y-$N_3$ (X,Y = PS, P*t*BA, PMA) | CuAAC | $AB_2$[9] |
| PEG-maleimide | Anthracene-compound | DA | ABC[10] |
| PEO-$N_3$ | PHBP | CuAAC | ABC[11] |
| PMMA-$N_3$, PEG-$N_3$ | Alkyne-compound | CuAAC | ABC[12] |
| PS-$N_3$ | 3-(Prop-2-ynyloxy)-2,2-bisPPP-Br | CuAAC | $A_3B$[13] |
| (P*t*BA-$N_3$)$_3$, (PS-$N_3$)$_3$ | PEG-*b*-P*t*BA-alkyne, PEG-*b*-P*t*BA-alkyne | CuAAC | $A_3$-$(B\text{-}C)_3$[14] |
| X-$N_3$, Y-$N_3$ (X,Y = PS, P*t*BA, PMA) | Tripropagylamine | CuAAC | $AB_2$[15] |
| PS-*b*-PCL-$N_3$ | P*t*BA-*b*-PEG-alkyne, PMMA-*b*-PEG-alkyne | CuAAC | ABCD[3] |
| PBLG-$N_3$ | Dialkynyl-PZLL | CuAAC | $AB_2$[16] |
| Alkyne-(PS-*b*-P*t*BA) | $N_3$-(PI-*b*-PEO) | CuAAC | ABCD[17] |
| ($N_3$-PS)$_2$-core-(OH)$_2$ | Alkyne-PCL | CuAAC | $(AB)_2$-$B_2$ $(AB)_2$-$(BA)_2$[18] |
| PEO-$N_3$ | Alkyne-(PS-*b*-PCL-*b*-PMA) | CuAAC | ABCD[19] |
| MPEG-$N_3$ | Alkyne-PCL | CuAAC | ABC[20] |
| Maleimide-P*t*BA | Anthracene-PCL | Diels–Alder | ABC, ABCD[21] |
| P*t*BA-$N_3$ TEMPO-PEO | Alkyne-PS-Br | ATNRC | ABC[22] |
| PCL-$N_3$ | Alkyne-PEG | CuAAC | $A_2B_2$[23] |
| ($N_3$)$_7$-(CD-PNIPAM)$_{14}$ | Alkyne-PDEA | CuAAC | $A_7B_{14}$[24] |
| Alkyne-P*t*BMA | PEG(-$N_3$)-*b*-PNIPAM | CuAAC | ABC[25] |
| PS-*b*-PMMA-alkyne | PNBONI-$N_3$ | CuAAC | $AB_2$[26] |
| PS-PNIPAM-alkyne | PDMAEMA-$N_3$ | CuAAC | ABC[27] |
| Alkyne-(PLA)$_2$ | PBA-$N_3$, PPEG-$N_3$, PNIPAM-$N_3$ | CuAAC | $AB_2$[28] |
| $N_3$-(PEG-*b*-PNIPAM) | Alkyne-PLys | CuAAC | ABC[29] |
| PS-$N_3$ | Alkynyl(-OH)-Br | CuAAC | ABC[30] |
| PDEA-$N_3$ | PEG(-alkynyl)-*b*-PNIPAM | CuAAC | ABC[31] |
| Alkynyl-PDEA | PEG(-$N_3$)-*b*-P*t*BMA | CuAAC | ABC[32] |
| PS-$N_3$, PDMAEMA-$N_3$ | TPTTA | CuAAC | $AB_2$[33] |
| PAA-$N_3$, PS-$N_3$ | Tripropargylamine | CuAAC | $AB_2$[34] |
| PNIPAM-$N_3$ | PS(-alkynyl)-*b*-PCL | CuAAC | ABC[35] |
| Alkyne-PCL | PEG-$N_3$ | CuAAC | $A_2B_2$[36] |
| (Alkyne-PS)m–polyDVB–(PS-anthracene)m | P*t*BA-$N_3$ PMMA-maleimide | CuAAC, DA | Multi-$A_nB_n$[37] |
| Alkyne-PZLL | PNIPAM-$N_3$ | CuAAC | $AB_2$[38] |
| 17α-Ethynylestradiol, ethynylferrocene | PCL-$N_3$ | CuAAC | CCS[39] |
| CDS(PCL)$_{14}$-D-$N_3$ | Alkyne-PEG | CuAAC | $A_{14}B_7$[40] |
| PEG-$N_3$, PS-$N_3$ | 3-(Triisopropylsilyl)-5-ethynylbenzyl alcohol | CuAAC | ABC[41] |
| PEG-$N_3$ | 3,5-Bis(prop-2-ynyloxy)-phenyl methanol | CuAAC | $A_2B$[42] |
| Aminooxy-PEG | (PS)*n*-polyEVBA-aldehyde | Aldehyde–aminooxy | Multi-$A_nB_n$[43] |

| | | | |
|---|---|---|---|
| PS-$N_3$, PiBoA-$N_3$ | Alkyne-bearing nitrone | CuAAC | $AB_2$[44] |
| Aminooxy-PEG | (PS)*n*-polyEVBA-aldehyde | Aldehyde–aminooxy | Multi-$A_nB_n$[45] |
| PS-$N_3$-Ant<br>PS-PCL-Ant | Alkyne-PCL, Me-PEG-MIm | CuAAC, DA (one pot) | ABC[46] |
| PEG-$N_3$, 11-Azido-undecan-1-ol | Alkyne-compound, alkyne-PEG | CuAAC | ABC[47] |
| PS-$N_3$ | Alkyne-compound | CuAAC | $A_2B$[48] |
| PNIPAAm-$N_3$ | Alkyne-oligonucleotide | CuAAC | $A_3B$[49] |
| MPEG-$N_3$ | Alkyne-PCL | CuAAC | ABC[50] |
| 1-(Allyloxy)-3-azidopropan-2-ol, (PS-PCL)-$N_3$ | PS-SH, alkyne-PEG | Thiol–ene, CuAAC | ABC[51] |
| 2,2-Bis(azidomethyl)-1,3-propanediol | Alkyne-PE | CuAAC | $A_2B_2$[52] |
| c-PS-$N_3$, c-PS-alkyne | P*t*BA-alkyne, P*t*BA-$N_3$ | CuAAC | $AB_2$, $A_2B$[53] |
| Pyrene-PS-$N_3$ | Alkyne-PEG-PMMA | CuAAC | ABC[54] |
| PEG-MIm<br>PEG-alkyne/Br | Anthracene/alkyne/bromo-compound, PS-$N_3$,PONB-$N_3$, PCL-TEMPO | DA, CuAAC, ATNRC | ABC[55] |
| Alkyne mid-(PBu-P*t*BMA or PDMAEMA) | PS-$N_3$,PEO-$N_3$, P*t*BMA-$N_3$,PDMAEMA-$N_3$ | CuAAC | ABC[56] |
| Compound with alkyne and two TEMPO | X-Br (PS, cPS, P*t*BA, PEG)<br>Y-$N_3$(PS, cPS, P*t*BA, PEG) | CuAAC, ATNRC | $AB_2$[57] |
| Alkyne-PEG | β-CD-$N_3$-OTs | CuAAC | ABC[58] |
| Alkyne-PEG | CD-$N_3$ | CuAAC | $AB_x$[59] |
| Alkyne-P(DEAAm) | b-CD-$N_3$ | CuAAC | $AB_2$[60] |
| PNIPAAm-$N_3$ | Alkyne-oligonucleotide | CuAAC | $A_3B$[61] |
| PS-$N_3$, PLA-$N_3$ | Alkyne-carboxylic acid methyl ester | CuAAC | $A_2B$[62] |
| PEG-$(N_3)_2$<br>$N_3$-(PEG-PCL-PS) | Alkyne-(PCL-PS)<br>Alkyne-(PLLA-PM) | CuAAC | ABC, ABCDE[63] |
| mPEG-$N_3$ | Alkyne-dendron (D1) | CuAAC | $AB_2$[64] |
| PEO-$N_3$ | Alkyne-PS | CuAAC | $A_2B$ |
| PFS-$N_3$, PI-$N_3$ | Alkyne-PS | CuAAC | ABC[65–68] |
| PDLA-$N_3$<br>$(PDLA)_2$-$N_3$<br>$(PDLA)_3$-$N_3$ | Alkyne-PLA<br>Alkyne-$(PLA)_2$<br>Alkyne-$(PLA)_3$ | CuAAC | $A_xB_y$ ($x = 2, 3$; $y = 1–3$)[69] |
| $N_3$-$(PCL)_2$<br>$N_3$-$(PCL)_3$ | Alkyne-maltoheptaose | CuAAC | $AB_2$, $AB_3$[70] |
| $(N_3)_2$-PCL | Alkyne-maltoheptaose | CuAAC | $A_2B$[71] |
| R-8$N_3$-4OH | mPEG-alkyne | CuAAC | $A_8B_4$[72] |
| $(PEO)_2$-alkyne<br>cPEO-alkyne | $(PTHF)_2$-$N_3$<br>cPTHF-$N_3$ | CuAAC | $A_2B_2$, $A_2B$[73] |
| PEO-$N_3$<br>PEO(-alkynyl)-*b*-PCL | BPBA, PZLL-$N_3$ | CuAAC | ABC[74] |
| PS-$N_3$, P*t*BA-$N_3$ | PEG(-alkynyl)-*b*-PNIPAM | CuAAC | ABC[75] |
| PMMA-$(N_3)_2$ | Alkyne-P3HT | CuAAC | $A_2B$[76] |
| mPEG-$N_3$ | Alkyne-PS | CuAAC | ABC[77] |
| Alkyne-PS<br>PEO-PS-$N_3$<br>Alkyne-mid$(PEO\text{-}PS)_2$ | PEO-$N_3$<br>(3,5-Bis(prop-2-ynyloxy)phenyl)methanol<br>PLA-$N_3$ | CuAAC | $(AB)_2C$[78] |

(*continued*)

**Table 4.1** (*continued*)

| Reagent A for click | Reagent B for click | Click | Type of $A_nB_n$ |
|---|---|---|---|
| POSS-$(N_3)_8$ | Alkyne-PLA<br>Alkyne-PEO | CuAAC | $A_{8-x}B_x$[79] |
| Anth-$(PCL)_2$ | PEG-MIm | DA | $AB_2$[80] |
| Alkyne-PBLA | $N_3$-mid(mPEG-PCL) | CuAAC | ABC[81] |
| mPEG-$N_3$, PMMA-$N_3$, PS-$N_3$, PCL-$N_3$ | Alkyne-cholic acid | CuAAC | $AB_3$[82] |
| PHIC-$N_3$ | Alkyne-compound-$(OH)_2$, alkyne-compound-$(OH)_3$ | CuAAC | $AB_2$, $AB_3$[83–85] |
| Alkyne-(PCL-PM) | PEG-$(N_3)_2$ | CuAAC | $AB_2C_2$[86] |
| PEG-$N_3$, 11-azido-undecan-1-ol, $N_3$-(PEG-PCL) | Alkyne-compound, alkyne-PEG, alkyne-tetraiodo-fluorescein | CuAAC | ABC[87] |
| $N_3$-mPEG | Alkyne-$(PAMAM)_2$ | CuAAC | $AB_2$[88] |
| $N_3$-CD | Alkyne-mid(PDMA)<br>Alkyne-mid(PDEA) | CuAAC | $A_2B_2$[89] |
| POSS-$(N_3)_8$ | Alkyne-mPEG, alkyne-PS, alkyne-PMMA | CuAAC | $A_8B_8$[90] |
| $(Alken)_3$-core-$(OH)_1$<br>$(Alken)_2$-core-$(OH)_2$<br>$(Alken)_1$-core-$(OH)_3$ | PEO-SH | Thiol–ene | $A_3B$, $A_2B_2$, $AB_3$[91] |
| mPEG-$N_3$ | Alkyne-PCL | CuAAC | ABC[92] |
| β-CD-$N_3$ | (TMS-alkyne), PS-alkyne | CuAAC | $A(B\text{-}C)_{20}$[93] |
| PS-$N_3$, PB-$N_3$ | $(Alkyne)_4$-core | CuAAC | $A_2B_2$[94] |
| $(N_3)_2$-PS | Alkyne-MH | CuAAC | $A_2B$[95] |
| PEG-$N_3$ | $(Alkyne)_3$-benzene-OH | CuAAC | $AB_3$[96] |
| PMMA-$N_3$ | Tripropargylamine | CuAAC | $A_2B$[97] |
| PMMA-$(alkyne)_2$ | P3DDT-$N_3$ | | |
| $(Allyl)_2$-PCL | mPEG-SH | Thiol–ene | $A_2B$[98] |
| α-allyl, α′-alkyne -PCL | PNIPAAM-$N_3$ | CuAAC | |
| PNIPAM-$(alkyne)_2$ | mPEG-$N_3$ | CuAAC | $AB_2$[99] |
| $N_3$-compound-$(OH)_2$, $N_3$-compound-$(OH)_3$ | Alkyne-PF (poly[2,7-(9,9-dihexylfluorene)] | CuAAC | $AB_2$, $AB_3$[100] |
| Acrylate-PPE | 3-Mercapto-1,2-propane-diol | Thiol–ene | ABC[101] |
| Alkyne-(PMPC-PDPA) | $N_3$-PEG | CuAAC | ABC[102] |
| Aldehyde-PEG star | PBLG-$ONH_2$ | Aldehyde–aminooxy | $A_{16}B_{23}$[103] |
| $(N_3)_2$-P1, $N_3$-P4, $N_3$-P6 | Alkyne-P3, alkyne-P5, alkyne-P7 | CuAAC | $A_2B_2$, $AB_2$, $A_2B$[104] |
| Anthracene-PCL-$N_3$<br>Alkyne-(P*t*BA-PLLA) | Maleimide-mid-(PNIPAM-*b*-PMAC) | CuAAC, DA (one pot) | ABCDE[105] |
| Alkyne-(PMMA–PPEGMA) | $N_3$-$(PMMA\text{-}PNIPAm)_2$ | CuAAC | $(BA)(AC)_2$[106] |

## 4.2 Cu(I)-Catalyzed 1,3-Dipolar Azide–alkyne Cycloaddition Reaction (CuAAC)

1,3-Dipolar cycloaddition reactions between azide and alkyne functional groups were first reported by Huisgen.[107] These reactions are performed at considerably high temperatures and give racemic mixtures of 1,4- and

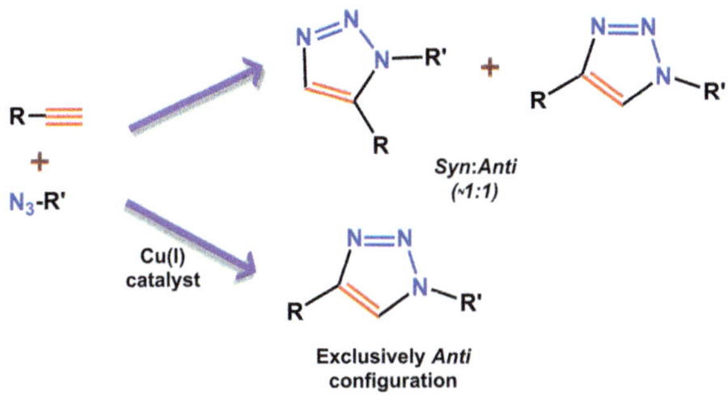


**Figure 4.2** 1,3-Dipolar cycloaddition reactions in the absence and presence of Cu(I) salts.

1,5-disubstituted 1,2,3-triazoles (Figure 4.2). Although alkyne and azide groups can be easily incorporated into polymers and they display kinetic stability and tolerance to various functional groups and diverse reaction conditions, the importance of these reactions could not be realized until the early 2000s. Medlar *et al.* and Sharpless *et al.* independently developed this coupling method and reported that 1,3-dipolar cycloaddition reactions between azides and alkynes in the presence of Cu(I) salts could be performed under very mild conditions affording exclusively 1,4-disubstituted 1,2,3-triazoles[108–110] (Figure 4.2). Since then, Cu(I)-catalyzed 1,3-dipolar azide–alkyne cycloaddition reactions (CuAAC) have been regarded as 'the cream of the crop' of click reactions[111] and widely used for the synthesis of numerous polymers with different topologies. A variety of miktoarm star polymers elegantly prepared using CuAAC click chemistry has been reported in the literature.

According to the mechanism of the Cu(I)-catalyzed reaction proposed by Sharpless, Cu(I) coordinates to alkynes, leading to the formation of a copper–acetylide intermediate through an exothermic reaction.[109] The azide group is attached to the copper–acetylide intermediate, yielding a six membered Cu(III)-metallacycle, and then a triazole ring is rapidly formed.[112] The obtained triazole ring has substantial chemical and redox stability, a large 5 D dipole moment and thus hydrogen bonding ability.[110,113,114]

Cu(I) species can be introduced in the reaction media in different ways. In one method, Cu(I) salts are added directly. CuBr is the most common one, although CuI is also used in the synthesis of miktoarm stars. This technique requires the use of one equivalent of nitrogen base, such as 2,6-lutidine, triethylamine, diisopropylethylamine, pyridine, 1,8-diazabicyclo[5.4.0]undec-7-ene (DBU), hexamethyltriethylenetetramine, (HMTETA), or *N,N,N′,N″,N″*-pentamethyldiethylenetriamine (PMDETA). On the other hand, the formation of trace amounts of undesired side products, such as diacetylenes, bistriazoles, and 5-hydroxytriazoles, has been encountered in some cases.[4,109,115] In another method, Cu(I) species are generated *in situ* by reduction of Cu(II) salts (*i.e.*, $CuSO_4$) in the presence of ascorbic acid sodium salt (Na-Asc). This method can be applied in aqueous media and does not require any cosolvents, amine bases, or additives.[109] In the third method,

Cu(I) species are produced by comproportionation of a $Cu^{2+}/Cu^{0}$ system and it is very useful when the substrates are incompatible with sodium ascorbate or its oxidation products.[4]

ATRP is one of the most widely used controlled polymerization methods for the preparation of miktoarm star polymers and the alkyl-halide end-groups of the obtained polymers are seamlessly transformed into azides. The functional groups of the polymers obtained *via* other type of controlled polymerization methods are converted into azide or alkyne groups through proper chemical transformation methods. Alternatively, various azide- and alkyne-functionalized initiators have been prepared and used in different controlled polymerization methods, such as reversible ATRP,[8,14,25,38,49,82,90,106] addition–fragmentation chain transfer polymerization (RAFT),[13,19,27,28,60,63,86,89,105] ROP,[18,20,23,36,71,81,105] and nitroxide-mediated polymerization (NMP),[8,14,26,116] due to their significantly high thermal and chemical stability.

One of the most prominent properties of click reactions is the use of equimolar amounts of the reacting functionalized polymers to obtain quantitative yields since, in most cases, highly tedious and time consuming purification procedures are required to isolate the targeted miktoarm star polymer from polymer mixtures by fractional precipitation. In contrast, if any of the reacting macromolecular species can be removed through simple purification procedures, these polymers can be used in excess. Moreover, the excess azide/alkyne functionalized polymers can be scavenged by reaction with alkyne/azide-functionalized insoluble structures, such as silica particles and Merrifield and Wang resins.[24,25,35,105]

ABC-type PMMA-PS-P*t*BA and PMMA-PS-PEG miktoarm star terpolymers were prepared *via* CuAAC, ATRP, and NMP methods (Figure 4.3).[8] PMMA-*b*-PS with an alkyne unit at the junction point was prepared *via* ATRP of MMA and NMP of styrene (S) using a trifunctional initiator. Then, azide-functionalized P*t*BA and PEG were reacted with PMMA-*b*-PS copolymer *via* a CuAAC reaction in the presence of CuBr/PMDETA in DMF at room temperature for 24 h. Azide-functionalized P*t*BA were used in equimolar amounts with respect to PMMA-*b*-PS, whereas the excess of PEG was removed by precipitating the final product in methanol. The gel permeation chromatography (GPC) traces were monomodal, indicating the absence of free linear polymers.

An alternative approach[13] describes the synthesis of well-defined $A_3B$ miktoarm star polymers in which azide end-functionalized PS ($PS\text{-}N_3$) is

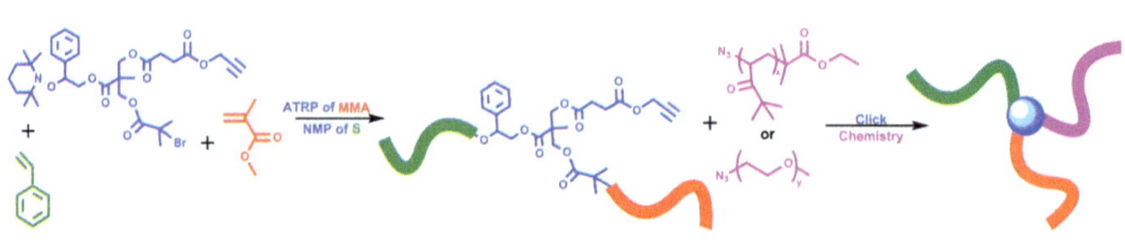


**Figure 4.3** Synthetic route for the preparation of hetero-arm star ABC-type terpolymers (PMMA-PS-P*t*BA and PMMA-PS-PEG) through a click chemistry strategy.

assembled with a bromide-containing trialkyne coupling agent *via* CuAAC in the presence of $CuSO_4$/Na-Asc to afford a three-armed PS star polymer with a bromide functionalized group at the core ($PS_3$-Br). Then, $PS_3$-Br is further employed as a macro-initiator in the ATRP of methyl methacrylate (MMA) to give a well-defined $A_3B$ miktoarm star polymer, containing three PS and one PMMA arms with low polydispersity index (Figure 4.4).

Dendrimer-like $A_3$-(B-C)$_3$ miktoarm star terpolymers, (P*t*BA)$_3$-(PS-PCL)$_3$ and (PS)$_3$-(P*t*BA-PEG)$_3$ were delicately prepared *via* ATRP and CuAAC methods (Figure 4.5).[14] First, azide end-capped three-armed P*t*BA and PS star polymers were synthesized *via* ATRP of *t*BA and St, respectively, and subsequent azidification of the bromide terminal groups of the obtained polymers.

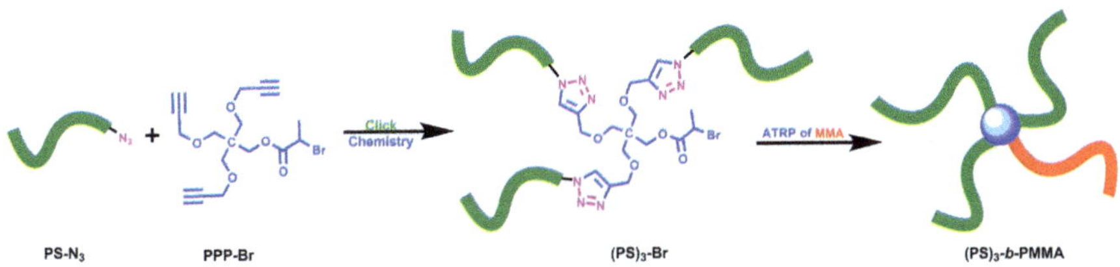


**Figure 4.4** Synthesis of a hetero-arm star $A_3B$-type polymer *via* controlled free radical polymerization and click chemistry.

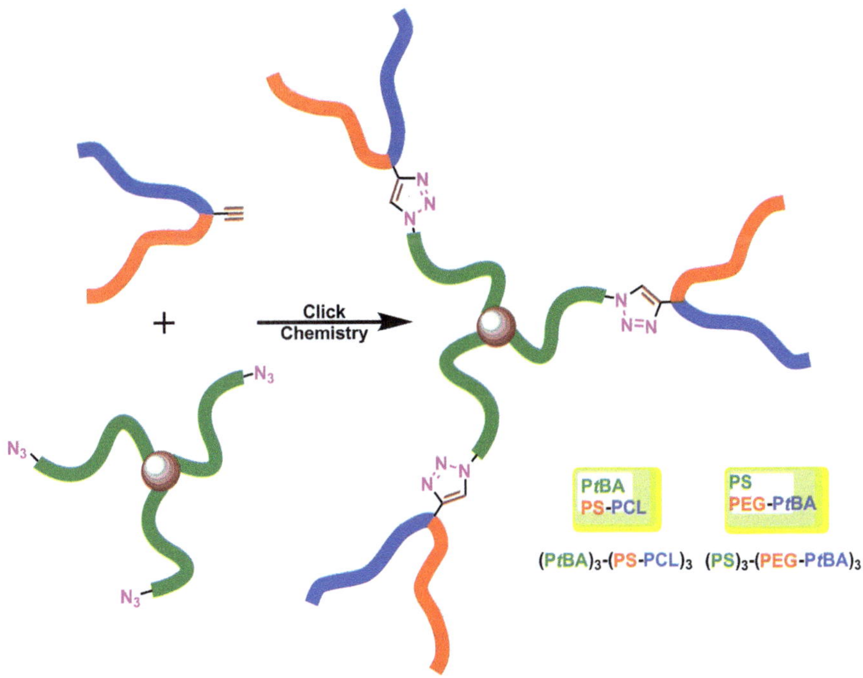


**Figure 4.5** Preparation of dendrimer type miktoarm star terpolymers: (P*t*BA)$_3$-(PS-PCL)$_3$ and (PS)$_3$-(PEG-P*t*BA)$_3$ *via* click reaction.

In a second step, PS-PCL and P*t*BA-PEG block copolymers with an alkyne group at the junction point were obtained by NMP of S, ROP of ε-CL, and ATRP of *t*BA. Finally, PS-PCL and P*t*BA-PEG blocks were linked to the ends of P*t*BA and PS star polymers *via* CuAAC, affording the targeted $(PtBA)_3$-$(PS\text{-}PCL)_3$ and $(PS)_3$-$(PtBA\text{-}PEG)_3$ miktoarm star terpolymers, respectively. In the click reactions, equimolar amounts of azide- and alkyne-functionalized units were used and the obtained products gave monomodal GPC traces with low polydispersity index (PDI) values (1.15 and 1.20), indicating a quantitative click reaction.

An $A_2B_2$ inverse miktoarm star block copolymer was prepared following an elegant approach.[18] Firstly, two-armed and azide end-functionalized linear polystyrene $(N_3\text{-}PS)_2$-core-$(OH)_2$ *via* ATRP of styrene, and subsequent azidification of the bromide terminal units. Then, acetylene end-capped PCL was attached to the termini of the PS blocks *via* CuAAC in the presence of CuBr/PMDETA and the resulting $(PCL\text{-}PS)_2$-core-$(OH)_2$ was used as a macroinitiator in the ROP of ε-CL, affording 4-armed miktoarm star $(PCL\text{-}PS)_2$-core-$(PCL\text{-}OH)_2$. Then, the OH terminal groups of the PCL arms were converted to bromide groups *via* esterification, and the obtained miktoarm star $(PCL\text{-}PS)_2$-core-$(PCL\text{-}Br)_2$ acted as a macroinitiator in the ATRP of S to yield an $A_2B_2$ inverse miktoarm star block copolymer [$(PCL\text{-}PS)_2$-core-$(PCL\text{-}PS)_2$] (Figure 4.6).

Consecutive multi-CuAAC click reactions have also been utilized to produce well-defined miktoarm star polymers. The delicate selection of reagents having alkyne/azide functional groups, along with other suitable functionalities to be converted into new alkyne/azide groups, is of primary importance to prepare star polymers with different arms. The case of a trifunctional compound with one hydroxyl and two alkyne groups is a good example of this method[78] (Figure 4.7). Firstly, the alkyne groups of the compound are reacted with azide end-functionalized polyethylene oxide-*co*-polystyrene $(PEO\text{-}PS\text{-}N_3)$ in the presence of CuBr/PMDETA, and then the hydroxyl unit at the junction point of two PEO-PS blocks is converted into acetylene. Subsequently, the newly formed alkyne unit is reacted with azide-terminated polylactide (PLA), affording a $(PEO\text{-}PS)_2$-PLA miktoarm terpolymer. In another study, a trifunctional compound, with hydroxyl, chloride, and azide groups, was coupled with an alkyne-terminated PEG in the presence of CuBr/PMDETA, yielding

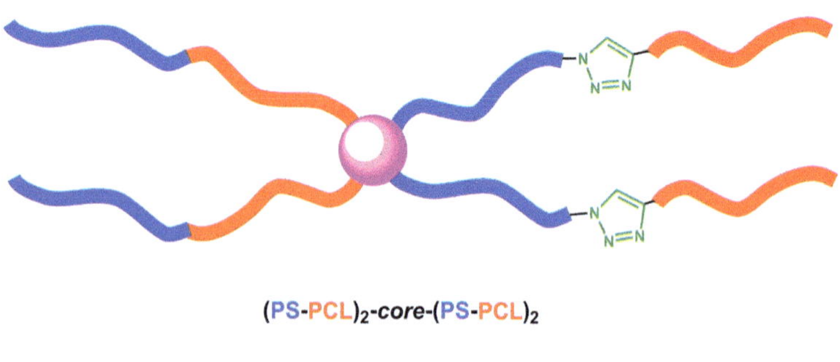


**Figure 4.6** Schematic diagram for the inverse star block copolymer $(PCL\text{-}PS)_2$-core-$(PCL\text{-}PS)_2$.

chloride and hydroxyl functionalized PEG (PEG-Cl-OH). Then, the chloride and hydroxyl functional groups were converted into azide and bromide moieties, respectively, and the obtained polymer (PEG-$N_3$-Br) was used in the ATRP of *tert*-BMA, affording a PEG-*b*-P*t*BMA block copolymer with an azide group at the junction point. PEG-*b*-P*t*BMA was treated with alkyne-terminated poly(2-(diethylamino)ethyl methacrylate) (PDEMA) in a CuAAC reaction in the presence of CuBr/PMDETA in DMF at 60 °C, giving an ABC-type PEG-P*t*BMA-PDEMA miktoarm terpolymer. The excess of the alkyne-terminated PDEMA used in the CuAAC reaction was scavenged by an azide-functionalized Merrifield resin. Finally, the *tert*-butyl side groups of the PEG-P*t*BMA-PDEMA miktoarm star polymer were hydrolyzed using trifluoroacetic acid, affording a PEG-PMAA-PDEMA miktoarm star terpolymer (Figure 4.8).[32]

In an alternative method, the presence of protected alkyne groups in these reactions guaranteed the synthesis of tailor-made miktoarm polymers.[41,47,68,87] A multifunctional compound containing hydroxyl, trimethylsilyl (TMS), and triisopropylsilyl (TIPS)-protected alkyne groups was elegantly utilized in the synthesis of an ABC-type miktoarm star polymer. TMS and TIPS groups were deprotected consecutively in a controlled manner and treated with azide end-functionalized PEG and PS in CuAAC reactions through a coupling-onto method, affording a PEG-*b*-PS block copolymer with a hydroxyl group at the junction point. Finally, a PEG-PS-PCL miktoarm

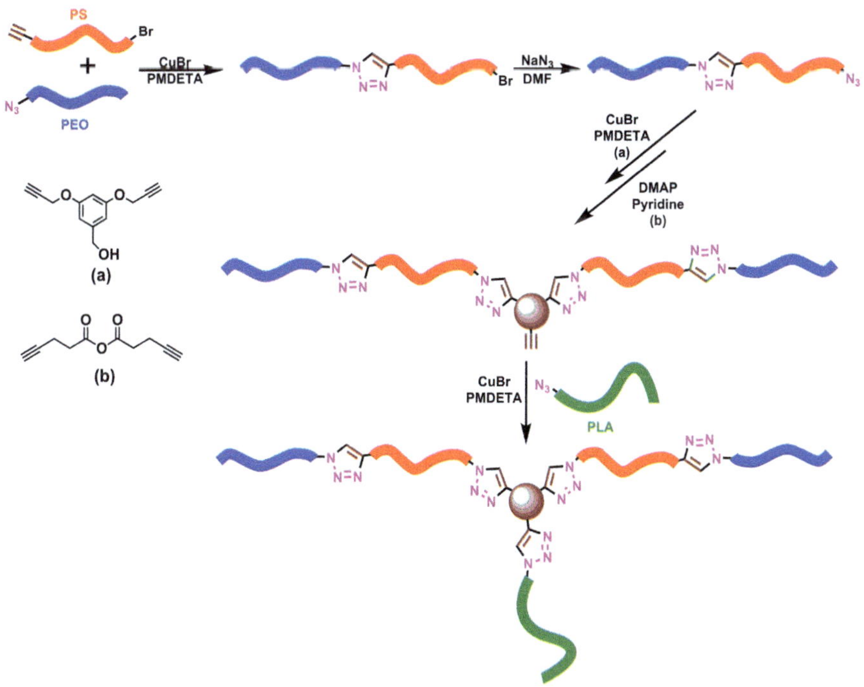


**Figure 4.7** Synthesis of the (PEO-PS)$_2$-PLA terpolymer *via* stepwise click reactions.

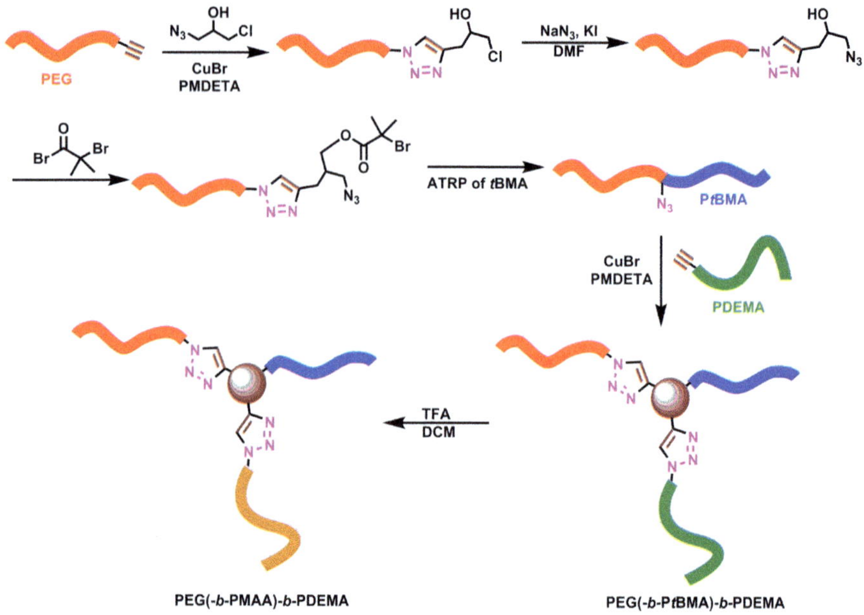


**Figure 4.8** Synthetic route for the preparation of PEG-P*t*BMA-PDEMA and PEG-PMAA-PDEMA ABC miktoarm star terpolymers.

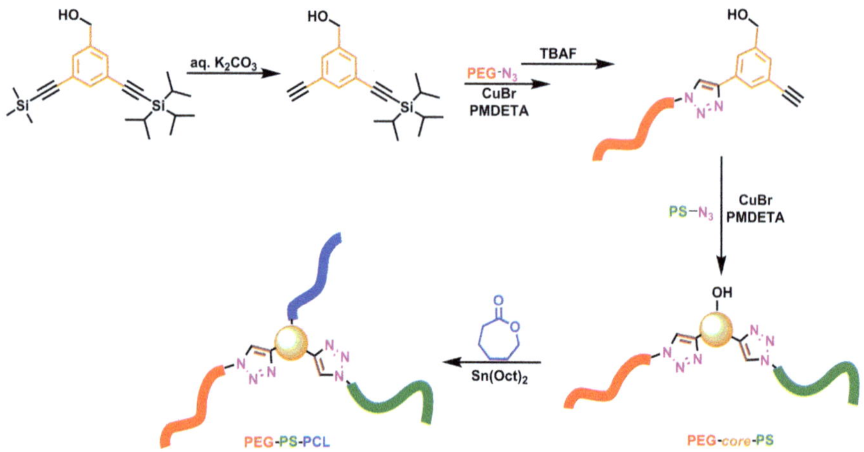


**Figure 4.9** Synthesis of ABC miktoarm star polymers μ(PEG-PS-PCL) *via* consecutive CuAAC click and ring-opening polymerization (ROP) reactions.

star terpolymer was produced *via* a core-first approach using PEG-*b*-PS as the macroinitiator in the ROP of ε-CL (Figure 4.9).[41] In a different example, the aldehyde group of a similar trifunctional compound was coupled to lithium-terminated polystyrene (PS-Li), yielding PS with TIPS and TMS-protected alkyne end-groups. Upon consecutive deprotection of TIPS and TMS,

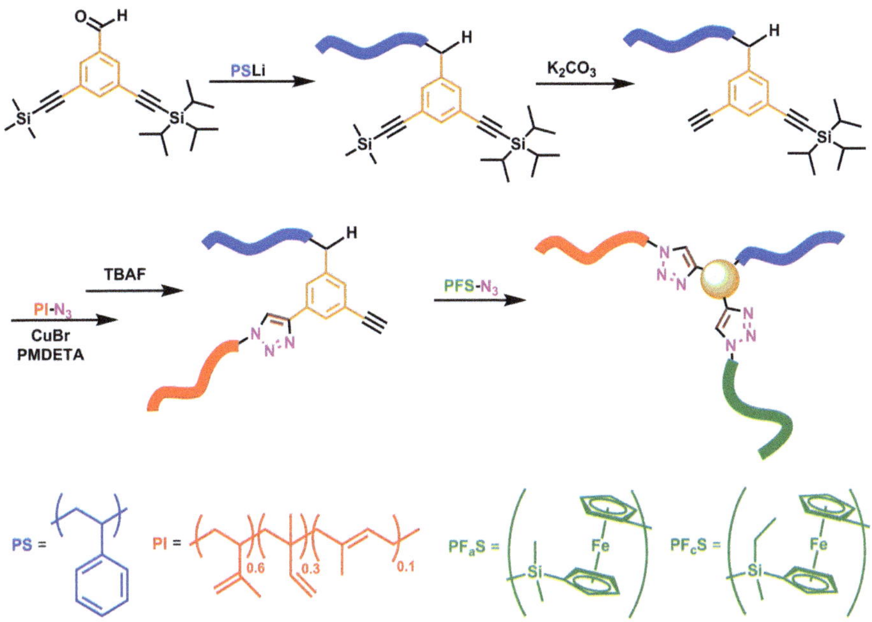


**Figure 4.10** Synthetic route for the preparation of a PS-PFS-PI miktoarm star polymer.

the obtained alkyne units were respectively reacted with azide end-capped polyisoprene (PI) and polyferrocenylsilane (PFS), affording a PS-PFS-PI miktoarm star polymer (Figure 4.10).[68]

Another delicate method requires the use of an excess amount of multi-alkyne/azide functional compounds to couple with an azide/alkyne functionalized macromolecule to prevent multi-point attachment. Then, other alkyne/azide groups can be used in subsequent CuAAC reactions with different polymeric structures.[15,25,35,117] In a typical example, azide end-capped polystyrene (PS-$N_3$) was reacted with 50 equivalents of 3,5-bis(propargyloxy)benzyl alcohol (BPBA) to ensure the coupling of PS-$N_3$ to only one alkyne unit of BPBA. The number average molecular weight ($M_n$) of the obtained polymer from GPC experiments revealed the successful synthesis of homoblock PS with one hydroxyl and one alkyne end-groups (PS-alkynyl-OH). Then, PS-alkynyl-OH was employed as a macroinitiator in the ROP of ε-CL, producing PS-alkynyl-PCL. Finally, azide end-functionalized poly(*N*-isopropyl acrylamide) (PNIPAM-$N_3$) was coupled to PS-alkynyl-PCL *via* a CuAAC reaction in the presence of CuBr/PMDETA, affording a PS-PCL-PNIPAM miktoarm star terpolymer (Figure 4.11).[35] The excess of PNIPAM-$N_3$ was scavenged by an alkyne-functionalized Wang resin.

The orthogonality and tolerability of Cu(I)-catalyzed click reactions (CuAAC) towards many functional groups and reaction environments were explicitly revealed in the one-pot synthesis of miktoarm star polymers using different polymerization methods. One-pot techniques with modular approaches do not require isolation of the intermediates, thus only purification of the targeted

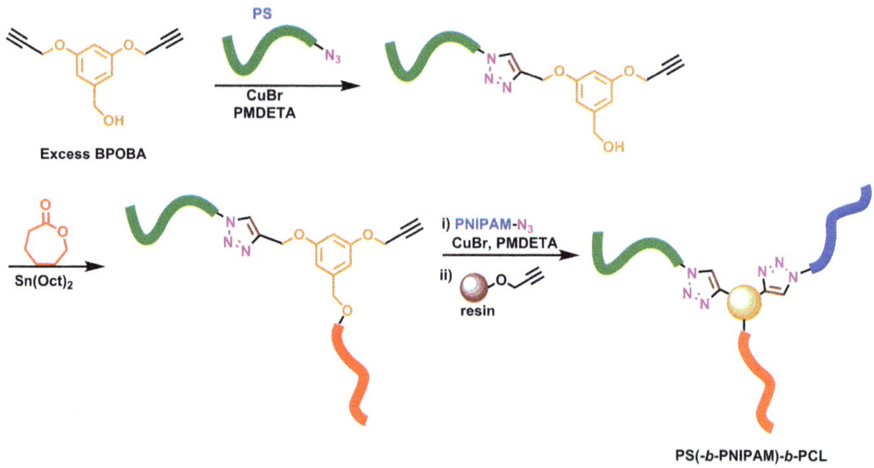


**Figure 4.11** Synthesis of a PS-PCL-PNIPAM amphiphilic ABC miktoarm star terpolymer *via* a combination of ATRP, ROP, and multiple click reactions.

miktoarm stars is performed.[12] The synthesis of well-defined ABC-type PS-PCL-PDMA and PEO-PCL-PDMA miktoarm star polymers was performed in a one-pot manner in the presence of a CuBr/PMDETA/Sn(Oct)$_2$ catalytic mixture *via* simultaneous ATRP of 2-(dimethylamino)ethyl methacrylate (DMA), ROP of ε-CL, and CuAAC reactions of azide end-functionalized PS and PEG with an alkyne functionalized core compound (Figure 4.12).[30]

## 4.3 Diels–Alder Click Reactions

Diels–Alder reactions were invented by Otto Diels and Kurt Alder, who were honored with the Nobel Prize in 1950.[118] These reactions do not require toxic copper salts and involve [4 + 2] cycloaddition reactions between an electron-rich diene (*i.e.*, furan, anthracene, 1,3 cyclopentadiene, and their derivatives) and an electron-poor dienophile (*i.e.*, maleic acid, vinyl ketone, and their derivatives), yielding a stable cyclohexene adduct.[119,120] DA reactions can be performed at room temperature or under relatively mild conditions to afford carbon–carbon bonds or carbon-heteroatom bonds (hetero-Diels–Alder, HAD).[121–123] DA reactions are known to shift backwards (retro DA, rDA) at higher temperatures, generally above 90 °C.[124,125] On the other hand, decomposition reactions of furan-maleimide adducts are observed at temperatures >60 °C in solution.[124]

DA reactions are among the most widely used techniques in organic chemistry laboratories and they are also used for the preparation of macromolecular structures. In the synthesis of miktoarm polymers, DA reactions are employed to attach polymer arms to a core molecule or to provide macromolecular structures or small-molecule compounds with functional groups to be used as initiating sites for controlled polymerization of various monomers. For example, in the synthesis of ABC-type miktoarm star polymers, a trifunctional compound, having an anthracene group, is reacted with

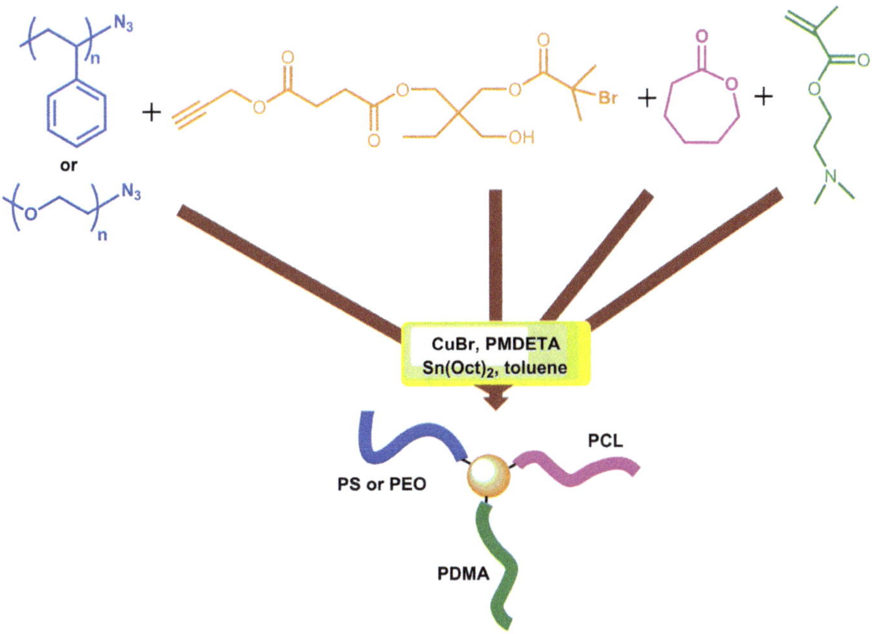


**Figure 4.12** Schematic illustration of the one-pot synthesis of amphiphilic ABC miktoarm star copolymers *via* combination of ATRP, ROP, and click chemistry.

maleimide-terminated PEG in a DA reaction to afford a PEG macroinitiator with a bromide and a unimolecular TEMPO functional group (PEG-MI). The DA reaction is performed in THF at 75 °C for 48 h. Then, PEG-MI is employed in the NMP of S and ATRP of *tert*-butyl acrylate (*t*BA), consecutively, affording the PEG-*t*BA-PS miktoarm star terpolymer (Figure 4.13).[10]

In some cases, maleimide groups are protected with furan or its derivatives to avoid the susceptibility of maleimide to nucleophilic attack.[126] In a typical example,[21] a trifunctional compound with furan-protected maleimide is used as the initiator in the ATRP of *t*-BA to produce P*t*BA with a furan-protected maleimide and a unimolecular TEMPO initiating site (P*t*BA-MI) (Figure 4.14). In the reaction of anthracene end-capped PCL with P*t*BA-MI at 110 °C, furan-protected maleimide units are subjected to a thermal cyclo-reversion reaction and the resulting maleimide groups, in turn, react with the anthracene groups, affording a P*t*BA-*b*-PCL block copolymer. Then, P*t*BA-*b*-PCL is used as a unimolecular macroinitiator in the NMP of S, giving a P*t*BA-PCL-PS miktoarm star polymer. Finally, P*t*BA-PCL-PS is employed in the photopolymerization of MMA in the presence of thioxantoneas as the photosensitizer to synthesize the P*t*BA-PCL-PS-PMMA miktoarm star quaterpolymer.

Several tailor-made miktoarm star polymers have been prepared *via* combination of DA and CuAAC techniques. Multi-miktoarm star block copolymers with PMMA and P*t*BA arms were prepared by consecutive coupling of azide end-functionalized poly(*tert*-butyl acrylate) (P*t*BA-$N_3$) and furan-protected maleimide end-capped poly(methyl methacrylate) (PMMA-MIm) onto PS star

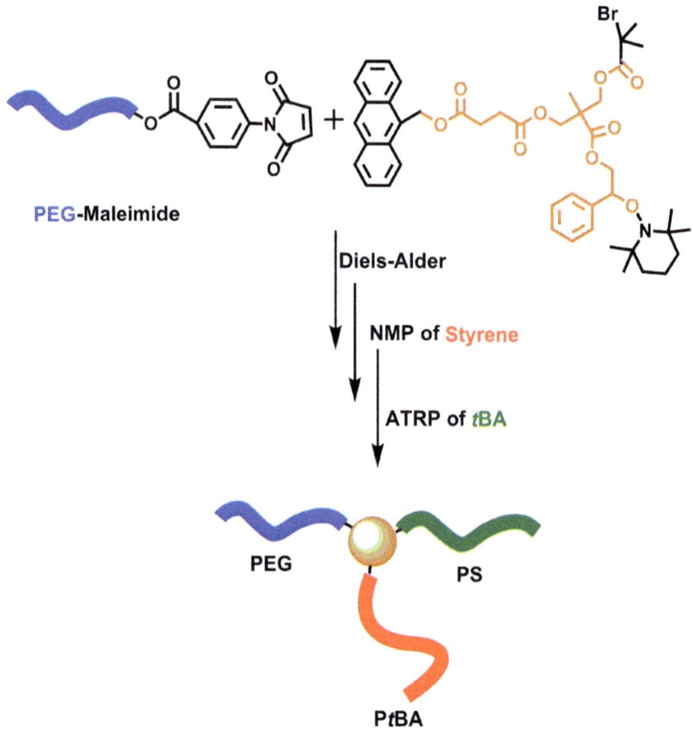


**Figure 4.13** Synthesis of PEG-*t*BA-PS miktoarm star terpolymer *via* combination of DA, stable free radical polymerization (SFRP), and ATRP reactions.

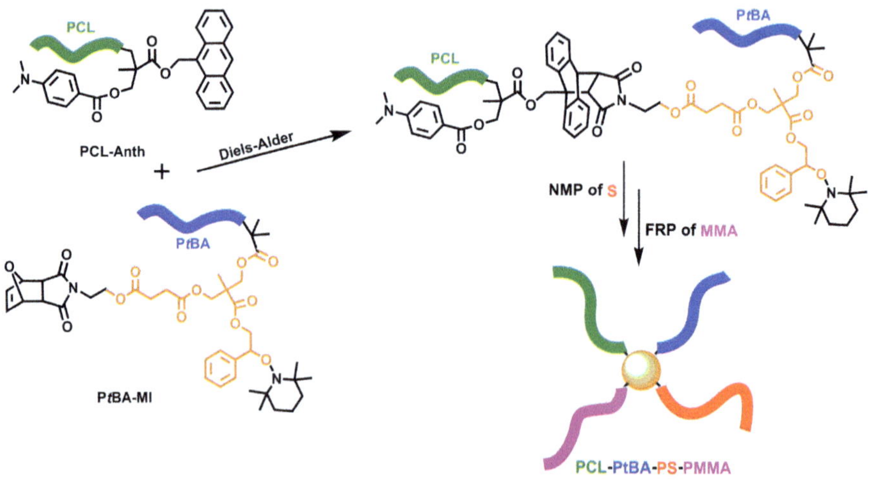


**Figure 4.14** Preparation of a PCL-P*t*BA-PS-PMMA 4-miktoarm star quaterpolymer.

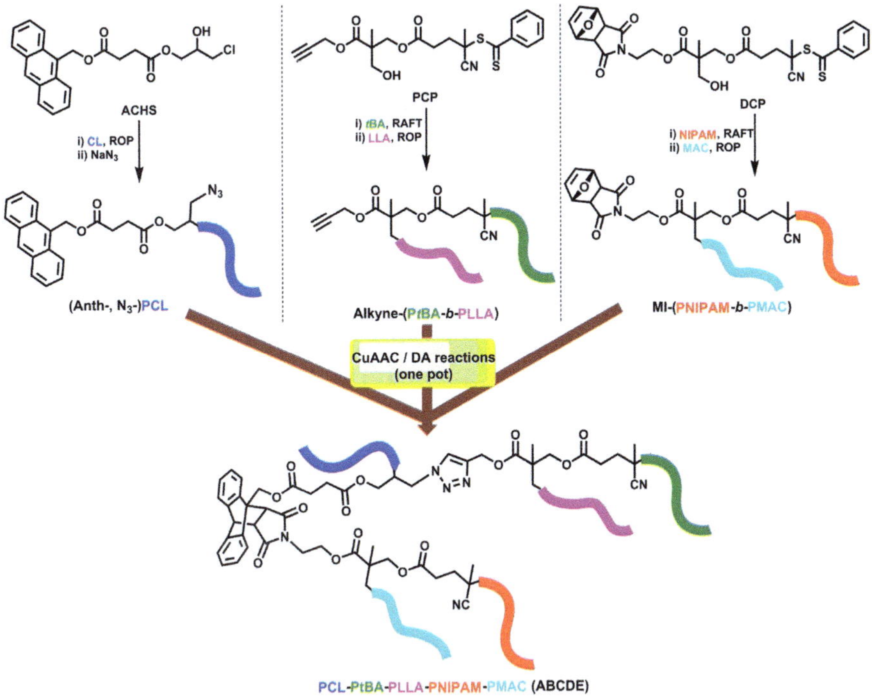


**Figure 4.15** Synthesis of a PCL-P*t*BA-PLLA-PNIPAM-PMAC (ABCDE) miktoarm star copolymer through a one-pot approach, including CuAAC and DA reactions.

polymers bearing alkyne and anthracene terminal groups. Slight excess of azide and maleimide-functionalized polymers were used with respect to the number of arms with alkyne and anthracene groups, respectively. The unreacted polymers were then removed by precipitation.[37] In another study, a well-defined ABCDE star copolymer with PCL (A), P*t*BA (B), poly(L-lactide) (PLLA, C), poly(*N*-isopropylacrylamide) (PNIPAM, D) and poly(5-methyl-5-allyloxycarbonyl-1,3-dioxane-2-one) (PMAC, E) arms was prepared by simultaneous DA and CuAAC reactions (Figure 4.15). The unreacted functionalized polymers were treated with alkyne, azide, anthracene, or maleimide-functionalized silica particles and then removed by filtration through an alumina column.[105]

## 4.4 Thiol-Based Click Reactions

The addition of thiols to double bonds occurs *via* radical pathways and these reactions are widely used in curing reactions, and the synthesis and modification of polymers.[101,127] These reactions can be initiated in the presence of cleavage type (type I) photoinitiators, *i.e.* (2,4,6-trimethylbenzoyl)diphenylphosphine oxide (TMDPO), 2,2-dimethoxy-2-phenyl acetophenone (DMPA), and bis-(2,4,6-trimethylbenzoyl)-phenylphosphine oxide (BAPO); H-abstraction type photo-initiators, *i.e.* benzophenone (BP), thioxanthone (TX), camphorquinone (CQ); or thermal initiators, *i.e.* 2,2′-azobis(isobutyronitrile) (AIBN).[98,128] A well-defined ABC-type miktoarm star polymer was prepared *via* thiol–ene, ROP, and CuAAC reactions. First, thiol end-capped PS was reacted with a trifunctional compound with allyl, hydroxyl, and azide functional groups in the presence of BAPO as the photoinitiator, affording PS with hydroxyl and azide end-groups (PS-OH-$N_3$). Then, the obtained polymer was used as the macroinitiator in the ROP of ε-CL, yielding PS-*b*-PCL with an azide unit at the junction point. Finally, CuAAC between PS-*b*-PCL and alkyne end-capped PEG yielded the PS-PCL-PEG miktoarm star polymer (Figure 4.16).[51] A similar miktoarm star polymer was also prepared *via* three consecutive click reactions, including thiol–ene, CuAAC, and DA reactions (Figure 4.17).[46]

Tailor-made $AB_3$, $A_2B_2$, and $A_3B$ miktoarm star polymers were synthesized by consecutive thiol–ene click ATRP reactions. In the first step, a cyclotetrasiloxane compound with four vinyl groups was consecutively reacted with 2-mercaptoethanol and thiol end-capped PEG in the presence of AIBN,

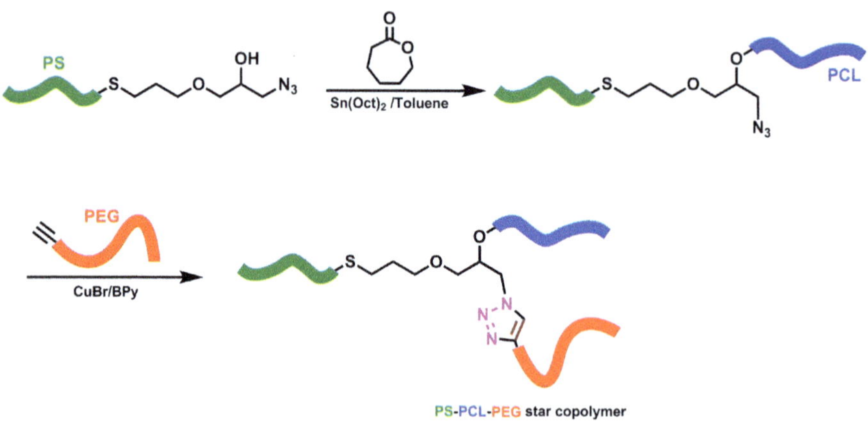


**Figure 4.16** Synthesis of an ABC-type miktoarm star copolymer (PS-PCL-PEG) *via* controlled polymerization techniques in combination with thiol–ene and CuAAC click reactions.

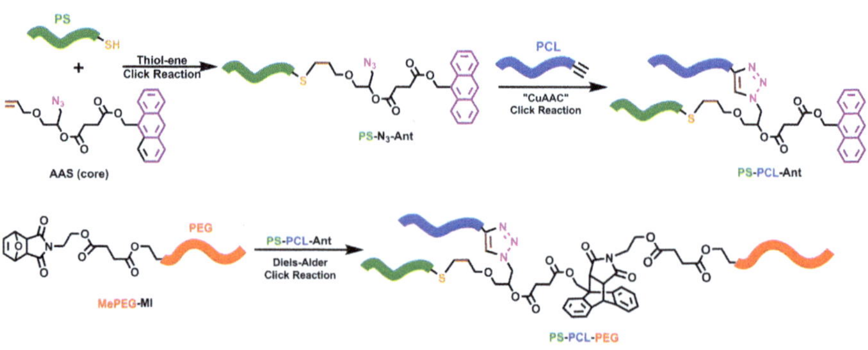


**Figure 4.17** Synthesis of a PS-PCL-PEG star copolymer by three consecutive click reactions.

affording PEG polymers with various arm numbers and hydroxyl-functional groups on the siloxane moiety. Then, the hydroxyl groups were converted into bromide and subsequently used as initiating sites for the ATRP of azobenzene mesogen containing a methacrylate monomer (MA(Az)) to afford 4-miktoarm star copolymers with PEO and PMA(Az) chains (Figure 4.18).[91]

Thiol–ene Michael addition reactions are highly efficient green-chemistry methods and they have been used in various applications, including preparation of small molecular compounds and polymers, *in situ* polymer modifications in biological systems, and surface functionalization. In addition, nucleophile- and base-catalyzed thiol–ene Michael addition mechanisms can be securely used for the conjugation of polymers or polymer end-group modification even under very dilute reaction conditions without leading to significant side products.[129] In the synthesis of ABC miktoarm star polymers, the acrylate end-functionalized group of a polyphosphoester was reacted with 1-thioglycerol to yield a polyphosphoester with two hydroxyl end groups. Then, the ABC miktoarm star polymer was produced through ROP and acyclic diene metathesis (ADMET) polymerizations.[101]

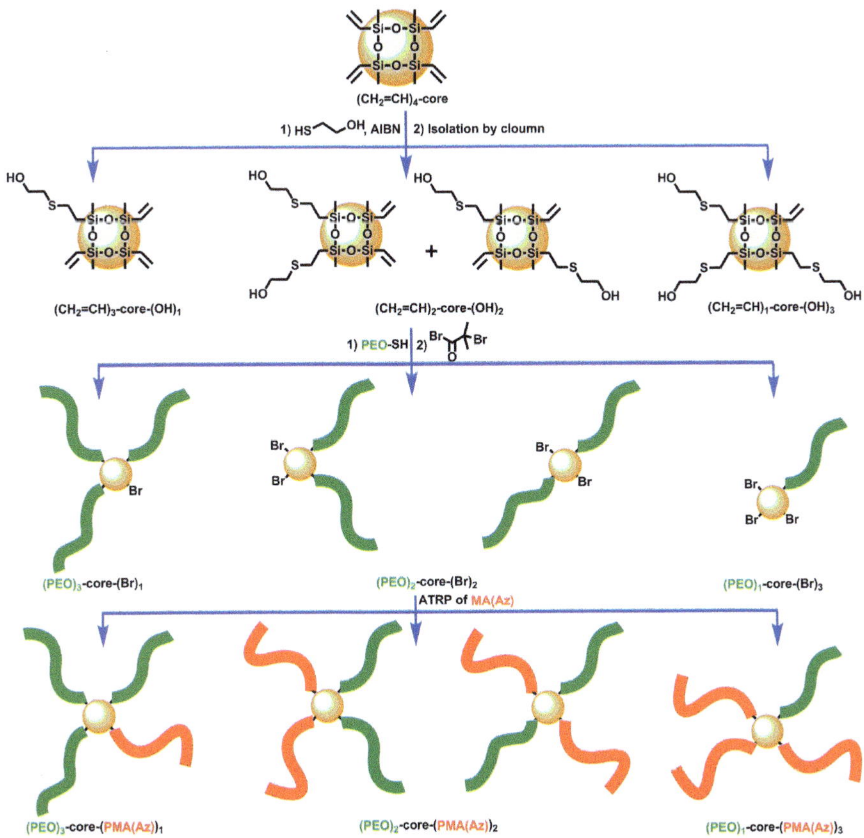


**Figure 4.18** Synthesis of $A_3B$-, $A_2B_2$-, and $AB_3$-type 4-miktoarm star copolymers by combination of thiol–ene click and ATRP reactions.

## 4.5 Atom Transfer Nitroxide Radical Coupling (ATNRC) Click Reactions

Atom transfer nitroxide radical coupling (ATNRC) combines the reversible radical forming ability of atom transfer radical addition (ATRA) reactions from carbon halide bonds and the effective coupling of nitroxide stable radicals with carbon-centered radicals at diffusion controlled rates ($k_t \sim 10^8$ L mol$^{-1}$ s$^{-1}$).[130] Derivatives of 2,2,6,6-tetramethyl-1-piperidinyloxy (TEMPO) are widely used as nitroxy radicals due their chemical stability. A well-defined ABC-type miktoarm star polymer was prepared *via* DA, ATNRC, and CuAAC click reactions. Firstly, a trifunctional compound with alkyne, anthracene, and bromide groups was reacted with furan protected maleimide end-capped PEG in toluene at 110 °C for 24 h, yielding PEG with bromide and alkyne terminal groups. Then, it was ligated with tetramethylpiperidine-1-oxyl-terminated poly(ε-caprolactone) (PCL-TEMPO) and azide-terminated polystyrene (PS-$N_3$) or poly(*N*-butyl oxanorborneneimide) (PONB-$N_3$) in *N*,*N*-dimethylformamide (DMF) at room temperature for 12 h catalyzed by Cu(0)/Cu(I), affording PEG-PCL-PS and PEG-PCL-PONB miktoarm star polymers (Figure 4.19).[55]

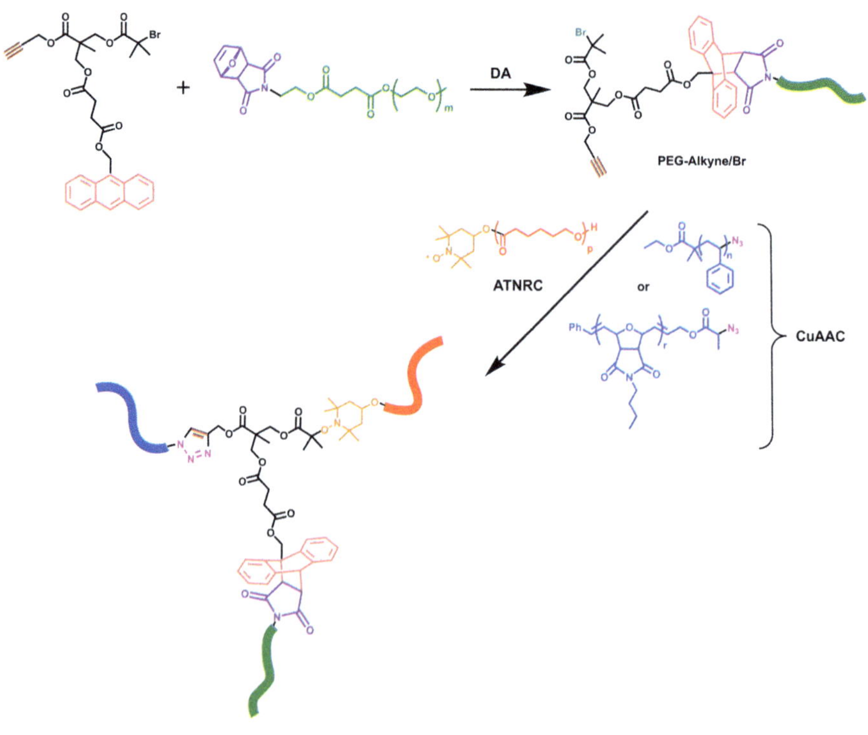


**Figure 4.19** Synthetic pathway for the preparation of PEG-PCL-PS and PEG-PCL-PONB miktoarm star polymers using DA, CuAAC, and ATNRC click reactions.

A tailor-made P*t*BA-PS-PEO miktoarm star polymer was synthesized *via* CuAAC and ATNRC click reactions between azide end-functionalized P*t*BA, alkyne and bromide functionalized polystyrene, and TEMPO end-capped PEG in a one-pot procedure carried out in the presence of CuBr/Cu/PMDETA at 70 °C for 24 h.[22] In another study, three-arm $AB_2$ stars having linear and macrocyclic polymer chains were delicately prepared *via* CuAAC and ATNRC click coupling reactions between a trifunctional compound (with an alkyne and two TEMPO end-functionalized groups) and linear or cyclic polymer chains (with bromide or azide terminal units) (Figure 4.20).[57]

## 4.6 Aldehyde–Aminooxy Click Reactions

Condensation reactions between aldehydes/ketones and α-effect amines (hydrazides or aminooxy reagents) are known as the first bioorthogonal click reactions.[131] They were used for the synthesis and chemical decoration of various biologically important structures, as well as miktoarm star polymers.[43] Double hydrophilic PNIPAM-PEG multi-miktoarm star copolymers were prepared *via* combination of RAFT polymerization of NIPAM, crosslinking of the obtained polymer using 6,6′-(ethane-1,2-diylbis(oxy))bis(3-vinylbenzaldehyde) (EVBA), and aldehyde–aminooxy click reaction between the aldehyde functionalized

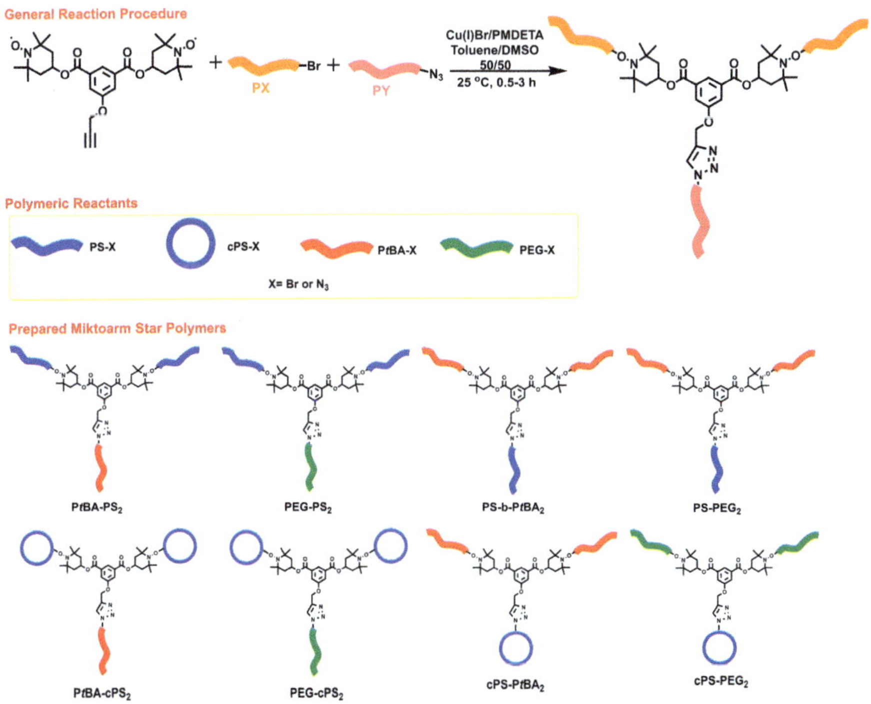


**Figure 4.20** General synthetic strategy for the synthesis of $AB_2$-type 3-arm miktoarm star polymers.

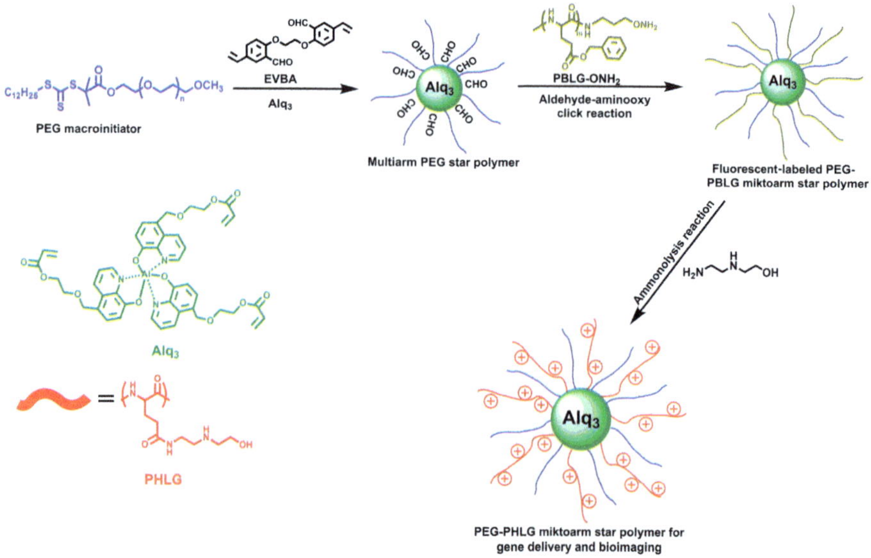


**Figure 4.21** Synthesis of the fluorescent-labeled PEG-poly(EVBA-*co*-$Alq_3$)-PHLG miktoarm star copolymer *via* combination of the RAFT arm-first technique and an aldehyde–aminooxy click reaction.

core and the aminooxy-terminated PEG.[45] A polypeptide-PEG multi-miktoarm star copolymer was prepared elegantly by crosslinking the trithiocarbonate-terminated PEG using EVBA in combination with a fluorescent aluminum tris(8-hydroxyquinoline)-bearing cross-linker (Alq3 crosslinker) to afford a PEG star polymer with a fluorescent crosslinked core. Then, the highly reactive aldehyde functional sites of the obtained core were treated with aminooxy-terminated poly(γ-benzyl-L-glutamate) ($H_2NO$-PBLG), yielding a PBLG–PEG multi-miktoarm star polymer with a fluorescent crosslinked core (Figure 4.21).[103]

## References


1. B. Yang, Y. Zhao, Y. Wei, C. Fu and L. Tao, *Polym. Chem.*, 2015, **6**, 8233–8239.
2. K. Kempe, A. Krieg, C. R. Becer and U. S. Schubert, *Chem. Soc. Rev.*, 2012, **41**, 176–191.
3. O. Altintas, G. Hizal and U. Tunca, *J. Polym. Sci., Part A: Polym. Chem.*, 2008, **46**, 1218–1228.
4. N. Akeroyd and B. Klumperman, *Eur. Polym. J.*, 2011, **47**, 1207–1231.
5. H. C. Kolb, M. G. Finn and K. B. Sharpless, *Angew. Chem., Int. Ed.*, 2001, **40**, 2004–2021.
6. U. Tunca, *Macromol. Rapid Commun.*, 2013, **34**, 38–46.
7. O. Altintas, A. P. Vogt, C. Barner-Kowollik and U. Tunca, *Polym. Chem.*, 2012, **3**, 34–45.

8. O. Altintas, G. Hizal and U. Tunca, *J. Polym. Sci., Part A: Polym. Chem.*, 2006, **44**, 5699–5707.
9. M. R. Whittaker, C. N. Urbani and M. J. Monteiro, *J. Am. Chem. Soc.*, 2006, **128**, 11360–11361.
10. H. Durmaz, F. Karatas, U. Tunca and G. Hizal, *J. Polym. Sci., Part A: Polym. Chem.*, 2006, **44**, 499–509.
11. G. Deng, D. Ma and Z. Xu, *Eur. Polym. J.*, 2007, **43**, 1179–1187.
12. O. Altintas, B. Yankul, G. Hizal and U. Tunca, *J. Polym. Sci., Part A: Polym. Chem.*, 2007, **45**, 3588–3598.
13. J. Zhu, X. Zhu, E. T. Kang and K. G. Neoh, *Polymer*, 2007, **48**, 6992–6999.
14. O. Altintas, A. L. Demirel, G. Hizal and U. Tunca, *J. Polym. Sci., Part A: Polym. Chem.*, 2008, **46**, 5916–5928.
15. C. N. Urbani, C. A. Bell, M. R. Whittaker and M. J. Monteiro, *Macromolecules*, 2008, **41**, 1057–1060.
16. J. Rao, Y. Zhang, J. Zhang and S. Liu, *Biomacromolecules*, 2008, **9**, 2586–2593.
17. G. Wang, X. Luo, C. Liu and J. Huang, *J. Polym. Sci., Part A: Polym. Chem.*, 2008, **46**, 2154–2166.
18. L.-P. Yang, X.-H. Dong and C.-Y. Pan, *J. Polym. Sci., Part A: Polym. Chem.*, 2008, **46**, 7757–7772.
19. L. Yang, H. Zhou, G. Shi, Y. Wang and C.-Y. Pan, *J. Polym. Sci., Part A: Polym. Chem.*, 2008, **46**, 6641–6653.
20. Y.-Y. Yuan, Y.-C. Wang, J.-Z. Du and J. Wang, *Macromolecules*, 2008, **41**, 8620–8625.
21. O. Altintas, G. Hizal and U. Tunca, *Des. Monomers Polym.*, 2009, **12**, 83–98.
22. Q. Fu, G. Wang, W. Lin and J. Huang, *J. Polym. Sci., Part A: Polym. Chem.*, 2009, **47**, 986–990.
23. P.-F. Gou, W.-P. Zhu, N. Xu and Z.-Q. Shen, *J. Polym. Sci., Part A: Polym. Chem.*, 2009, **47**, 6962–6976.
24. Z. Ge, J. Xu, J. Hu, Y. Zhang and S. Liu, *Soft Matter*, 2009, **5**, 3932–3939.
25. C. Li, Z. Ge, H. Liu and S. Liu, *J. Polym. Sci., Part A: Polym. Chem.*, 2009, **47**, 4001–4013.
26. A. Gozgen, A. Dag, H. Durmaz, O. Sirkecioglu, G. Hizal and U. Tunca, *J. Polym. Sci., Part A: Polym. Chem.*, 2009, **47**, 497–504.
27. J. Li, W.-D. He, S.-c. Han, X.-l. Sun, L.-y. Li and B.-y. Zhang, *J. Polym. Sci., Part A: Polym. Chem.*, 2009, **47**, 786–796.
28. A. Vora, K. Singh and D. C. Webster, *Polymer*, 2009, **50**, 2768–2774.
29. J. Li, W.-D. He, N. He, S.-C. Han, X.-L. Sun, L.-Y. Li and B.-Y. Zhang, *J. Polym. Sci., Part A: Polym. Chem.*, 2009, **47**, 1450–1462.
30. Y. Zhang, C. Li and S. Liu, *J. Polym. Sci., Part A: Polym. Chem.*, 2009, **47**, 3066–3077.
31. Y. Zhang, H. Liu, J. Hu, C. Li and S. Liu, *Macromol. Rapid Commun.*, 2009, **30**, 941–947.
32. H. Liu, C. Li, H. Liu and S. Liu, *Langmuir*, 2009, **25**, 4724–4734.
33. W. Zhang, W. Zhang, Z. Zhang, J. Zhu, Q. Pan and X. Zhu, *Polym. Bull.*, 2009, **63**, 467–483.

34. D. E. Lonsdale, M. R. Whittaker and M. J. Monteiro, *J. Polym. Sci., Part A: Polym. Chem.*, 2009, **47**, 6292–6303.
35. Y. Zhang, H. Liu, H. Dong, C. Li and S. Liu, *J. Polym. Sci., Part A: Polym. Chem.*, 2009, **47**, 1636–1650.
36. P.-F. Gou, W.-P. Zhu and Z.-Q. Shen, *J. Polym. Sci., Part A: Polym. Chem.*, 2010, **48**, 5643–5651.
37. A. Dag, H. Durmaz, V. Kirmizi, G. Hizal and U. Tunca, *Polym. Chem.*, 2010, **1**, 621–623.
38. L.-Y. Li, W.-D. He, J. Li, B.-Y. Zhang, T.-T. Pan, X.-L. Sun and Z.-L. Ding, *Biomacromolecules*, 2010, **11**, 1882–1890.
39. I. Javakhishvili and S. Hvilsted, *Polym. Chem.*, 2010, **1**, 1650–1661.
40. P.-F. Gou, W.-P. Zhu and Z.-Q. Shen, *Biomacromolecules*, 2010, **11**, 934–943.
41. K. Khanna, S. Varshney and A. Kakkar, *Macromolecules*, 2010, **43**, 5688–5698.
42. G. M. Soliman, R. Sharma, A. O. Choi, S. K. Varshney, F. M. Winnik, A. K. Kakkar and D. Maysinger, *Biomaterials*, 2010, **31**, 8382–8392.
43. Z.-M. Wu, H. Liang, J. Lu and W.-L. Deng, *J. Polym. Sci., Part A: Polym. Chem.*, 2010, **48**, 3323–3330.
44. E. H. H. Wong, M. H. Stenzel, T. Junkers and C. Barner-Kowollik, *Macromolecules*, 2010, **43**, 3785–3793.
45. Z. Wu, H. Liang and J. Lu, *Macromolecules*, 2010, **43**, 5699–5705.
46. B. Iskin, G. Yilmaz and Y. Yagci, *Polym. Chem.*, 2011, **2**, 2865–2871.
47. A. Sharma, G. M. Soliman, N. Al-Hajaj, R. Sharma, D. Maysinger and A. Kakkar, *Biomacromolecules*, 2012, **13**, 239–252.
48. O. Altintas, U. Tunca and C. Barner-Kowollik, *Polym. Chem.*, 2011, **2**, 1146–1155.
49. P. Pan, M. Fujita, W.-Y. Ooi, K. Sudesh, T. Takarada, A. Goto and M. Maeda, *Polymer*, 2011, **52**, 895–900.
50. Y.-Y. Yuan and J. Wang, *Colloids Surf., B*, 2011, **85**, 81–85.
51. B. Iskin, G. Yilmaz and Y. Yagci, *J. Polym. Sci., Part A: Polym. Chem.*, 2011, **49**, 2417–2422.
52. R. Liu, Z. Li, D. Yuan, C. Meng, Q. Wu and F. Zhu, *Polymer*, 2011, **52**, 356–362.
53. D. E. Lonsdale and M. J. Monteiro, *J. Polym. Sci., Part A: Polym. Chem.*, 2011, **49**, 4603–4612.
54. H. Durmaz, A. Dag, U. Tunca and G. Hizal, *J. Polym. Sci., Part A: Polym. Chem.*, 2012, **50**, 2406–2414.
55. U. S. Gunay, H. Durmaz, E. Gungor, A. Dag, G. Hizal and U. Tunca, *J. Polym. Sci., Part A: Polym. Chem.*, 2012, **50**, 729–735.
56. A. Hanisch, H. Schmalz and A. H. E. Müller, *Macromolecules*, 2012, **45**, 8300–8309.
57. J. Kulis, Z. Jia and M. J. Monteiro, *Macromolecules*, 2012, **45**, 5956–5966.
58. X. Huan, D. Wang, R. Dong, C. Tu, B. Zhu, D. Yan and X. Zhu, *Macromolecules*, 2012, **45**, 5941–5947.

59. C.-g. Mu, X.-d. Fan, W. Tian, Y. Bai and X. Zhou, *Polym. Chem.*, 2012, **3**, 1137–1149.
60. B. V. K. J. Schmidt, M. Hetzer, H. Ritter and C. Barner-Kowollik, *Polym. Chem.*, 2012, **3**, 3064–3067.
61. P. Pan, M. Fujita, W.-Y. Ooi, K. Sudesh, T. Takarada, A. Goto and M. Maeda, *Langmuir*, 2012, **28**, 14347–14356.
62. J. Song and B.-K. Cho, *ACS Macro Lett.*, 2012, **1**, 1375–1378.
63. C. Ye, G. Zhao, M. Zhang, J. Du and Y. Zhao, *Macromolecules*, 2012, **45**, 7429–7439.
64. H. Yin, H. C. Kang, K. M. Huh and Y. H. Bae, *J. Mater. Chem.*, 2012, **22**, 19168–19178.
65. K. Aissou, A. Nunns, I. Manners and C. A. Ross, *Small*, 2013, **9**, 4077–4084.
66. K. Aissou, H. K. Choi, A. Nunns, I. Manners and C. A. Ross, *Nano Lett.*, 2013, **13**, 835–839.
67. A. Nunns, C. A. Ross and I. Manners, *Macromolecules*, 2013, **46**, 2628–2635.
68. A. Nunns, G. R. Whittell, M. A. Winnik and I. Manners, *Macromolecules*, 2014, **47**, 8420–8428.
69. T. Isono, Y. Kondo, I. Otsuka, Y. Nishiyama, R. Borsali, T. Kakuchi and T. Satoh, *Macromolecules*, 2013, **46**, 8509–8518.
70. T. Isono, I. Otsuka, Y. Kondo, S. Halila, S. Fort, C. Rochas, T. Satoh, R. Borsali and T. Kakuchi, *Macromolecules*, 2013, **46**, 1461–1469.
71. T. Isono, I. Otsuka, D. Suemasa, C. Rochas, T. Satoh, R. Borsali and T. Kakuchi, *Macromolecules*, 2013, **46**, 8932–8940.
72. C. Gao, Y. Wang, P. Gou, X. Cai, X. Li, W. Zhu and Z. Shen, *J. Polym. Sci., Part A: Polym. Chem.*, 2013, **51**, 2824–2833.
73. F. Hatakeyama, T. Yamamoto and Y. Tezuka, *ACS Macro Lett.*, 2013, **2**, 427–431.
74. T. Liu, Y.-f. Zhang and S.-y. Liu, *Chin. J. Polym. Sci.*, 2013, **31**, 924–937.
75. L. Li, X.-W. Kan, X.-X. Deng, C.-C. Song, F.-S. Du and Z.-C. Li, *J. Polym. Sci., Part A: Polym. Chem.*, 2013, **51**, 865–873.
76. J. Park, H. C. Moon and J. K. Kim, *J. Polym. Sci., Part A: Polym. Chem.*, 2013, **51**, 2225–2232.
77. W. Sun, X. He, C. Gao, X. Liao, M. Xie, S. Lin and D. Yan, *Polym. Chem.*, 2013, **4**, 1939–1949.
78. J. Song, E. Lee and B.-K. Cho, *J. Polym. Sci., Part A: Polym. Chem.*, 2013, **51**, 446–456.
79. W. Yuan, X. Liu, H. Zou, J. Li, H. Yuan and J. Ren, *Macromol. Chem. Phys.*, 2013, **214**, 1580–1589.
80. F. Bahadori, A. Dag, H. Durmaz, N. Cakir, H. Onyuksel, U. Tunca, G. Topcu and G. Hizal, *Polymers*, 2014, **6**, 214.
81. Y. Chu, H. Yu, Y. Zhang, G. Zhang, Y. Ma, R. Zhuo and X. Jiang, *J. Polym. Sci., Part A: Polym. Chem.*, 2014, **52**, 3346–3355.
82. E. Doganci, M. Gorur, C. Uyanik and F. Yilmaz, *J. Polym. Sci., Part A: Polym. Chem.*, 2014, **52**, 3390–3399.

83. T. Satoh, N. Nishikawa, D. Kawato, D. Suemasa, S. Jung, Y. Y. Kim, M. Ree and T. Kakuchi, *Polym. Chem.*, 2014, **5**, 588–599.
84. Y. Y. Kim, S. Jung, C. Kim, B. J. Ree, D. Kawato, N. Nishikawa, D. Suemasa, T. Isono, T. Kakuchi, T. Satoh and M. Ree, *Macromolecules*, 2014, **47**, 7510–7524.
85. R. Phinjaroenphan, Y. Y. Kim, B. J. Ree, T. Isono, J. Lee, S. Rugmai, H. Kim, S. Maensiri, T. Kakuchi, T. Satoh and M. Ree, *Macromolecules*, 2015, **48**, 5816–5833.
86. K. Miao, H. Liu and Y. Zhao, *Polym. Chem.*, 2014, **5**, 3335–3345.
87. G. M. Soliman, R. Redon, A. Sharma, D. Mejía, D. Maysinger and A. Kakkar, *Macromol. Biosci.*, 2014, **14**, 1312–1324.
88. H. Yin, H. C. Kang, K. M. Huh and Y. H. Bae, *Colloids Surf., B*, 2014, **116**, 128–137.
89. B. V. K. J. Schmidt and C. Barner-Kowollik, *Polym. Chem.*, 2014, **5**, 2461–2472.
90. E. Doganci, M. A. Tasdelen and F. Yilmaz, *Macromol. Chem. Phys.*, 2015, **216**, 1823–1830.
91. M. Asadul Hoque, H. Komiyama, H. Nishiyama, K. Nagai, T. Kawauchi and T. Iyoda, *J. Polym. Sci., Part A: Polym. Chem.*, 2016, **54**, 1175–1188.
92. R. Alizadeh and M. Ghaemy, *Polymer*, 2015, **66**, 179–191.
93. B. Huang, M. Chen, S. Zhou and L. Wu, *Polym. Chem.*, 2015, **6**, 3913–3917.
94. L. Ma, Y. Lin, H. Tan, J. Zheng, F. Liu, G. Zhang, J. Liu and T. Tang, *RSC Adv.*, 2015, **5**, 34466–34474.
95. I. Otsuka, Y. Zhang, T. Isono, C. Rochas, T. Kakuchi, T. Satoh and R. Borsali, *Macromolecules*, 2015, **48**, 1509–1517.
96. A. Moquin, A. Sharma, Y. Cui, A. Lau, D. Maysinger and A. Kakkar, *Macromol. Biosci.*, 2015, **15**, 1744–1754.
97. J. Park, H. C. Moon, C. Choi and J. K. Kim, *Macromolecules*, 2015, **48**, 3523–3530.
98. S. S. Patil, B. V. Tawade and P. P. Wadgaonkar, *J. Polym. Sci., Part A: Polym. Chem.*, 2016, **54**, 844–860.
99. H. Yao, W. Tian, Y. Liu, Y. Bai, D. Liu, T. Liu, M. Qi, M. Wang and Y. Liu, *RSC Adv.*, 2015, **5**, 34557–34565.
100. K. Saito, T. Isono, H.-S. Sun, T. Kakuchi, W.-C. Chen and T. Satoh, *Polym. Chem.*, 2015, **6**, 6959–6972.
101. J. Qiu, J. Zhang, F. Yu, J. Wei and L. Ding, *J. Polym. Sci., Part A: Polym. Chem.*, 2016, **54**, 692–701.
102. J. Gaitzsch, V. Chudasama, E. Morecroft, L. Messager and G. Battaglia, *ACS Macro Lett.*, 2016, **5**, 351–354.
103. J. Huang, H. Liang, D. Cheng and J. Lu, *Polym. Chem.*, 2016, **7**, 1792–1802.
104. Y. Satoh, K. Miyachi, H. Matsuno, T. Isono, K. Tajima, T. Kakuchi and T. Satoh, *Macromolecules*, 2016, **49**, 499–509.
105. H. Liu, W. Pan, M. Tong and Y. Zhao, *Polym. Chem.*, 2016, **7**, 1603–1611.
106. J. Zhu, Y. Liu, L. Xiao and P. Zhou, *Macromol. Chem. Phys.*, 2016, **217**, 773–782.
107. R. Huisgen, *Angew. Chem., Int. Ed. Engl.*, 1963, **2**, 565–598.

108. C. W. Tornøe, C. Christensen and M. Meldal, *J. Org. Chem.*, 2002, **67**, 3057–3064.
109. V. V. Rostovtsev, L. G. Green, V. V. Fokin and K. B. Sharpless, *Angew. Chem., Int. Ed.*, 2002, **41**, 2596–2599.
110. P. Avti, D. Maysinger and A. Kakkar, *Molecules*, 2013, **18**, 9531.
111. K. Khanna, S. Varshney and A. Kakkar, *Polym. Chem.*, 2010, **1**, 1171–1185.
112. F. Himo, T. Lovell, R. Hilgraf, V. V. Rostovtsev, L. Noodleman, K. B. Sharpless and V. V. Fokin, *J. Am. Chem. Soc.*, 2005, **127**, 210–216.
113. T. Romero, A. Caballero, A. Tárraga and P. Molina, *Org. Lett.*, 2009, **11**, 3466–3469.
114. Q.-Y. Cao, T. Pradhan, S. Kim and J. S. Kim, *Org. Lett.*, 2011, **13**, 4386–4389.
115. C. J. Duxbury, D. Cummins and A. Heise, *J. Polym. Sci., Part A: Polym. Chem.*, 2009, **47**, 3795–3802.
116. E. Gungor, G. Hizal and U. Tunca, *J. Polym. Sci., Part A: Polym. Chem.*, 2008, **46**, 6703–6711.
117. D. E. Lonsdale and M. J. Monteiro, *Chem. Commun.*, 2010, **46**, 7945–7947.
118. O. Diels and K. Alder, *Justus Liebigs Ann. Chem.*, 1928, **460**, 98–122.
119. J. Sauer and R. Sustmann, *Angew. Chem., Int. Ed. Engl.*, 1980, **19**, 779–807.
120. K. C. Nicolaou, S. A. Snyder, T. Montagnon and G. Vassilikogiannakis, *Angew. Chem., Int. Ed.*, 2002, **41**, 1668–1698.
121. H. Durmaz, B. Colakoglu, U. Tunca and G. Hizal, *J. Polym. Sci., Part A: Polym. Chem.*, 2006, **44**, 1667–1675.
122. B. Gacal, H. Durmaz, M. A. Tasdelen, G. Hizal, U. Tunca, Y. Yagci and A. L. Demirel, *Macromolecules*, 2006, **39**, 5330–5336.
123. A. J. Inglis, S. Sinnwell, M. H. Stenzel and C. Barner-Kowollik, *Angew. Chem., Int. Ed.*, 2009, **48**, 2411–2414.
124. M. L. Szalai, D. V. McGrath, D. R. Wheeler, T. Zifer and J. R. McElhanon, *Macromolecules*, 2007, **40**, 818–823.
125. J. R. McElhanon and D. R. Wheeler, *Org. Lett.*, 2001, **3**, 2681–2683.
126. Z. Lu, R. Weber and R. J. Twieg, *Tetrahedron Lett.*, 2006, **47**, 7213–7217.
127. C. R. Becer, R. Hoogenboom and U. S. Schubert, *Angew. Chem., Int. Ed.*, 2009, **48**, 4900–4908.
128. M. Uygun, M. A. Tasdelen and Y. Yagci, *Macromol. Chem. Phys.*, 2010, **211**, 103–110.
129. D. P. Nair, M. Podgórski, S. Chatani, T. Gong, W. Xi, C. R. Fenoli and C. N. Bowman, *Chem. Mater.*, 2014, **26**, 724–744.
130. D. Yang, C. Feng and J. Hu, *Polym. Chem.*, 2013, **4**, 2384–2394.
131. J. M. Baskin and C. R. Bertozzi, *QSAR Comb. Sci.*, 2007, **26**, 1211–1219.

CHAPTER 5

# Micellar and Emulsion-Assisted Drug Delivery: Comparison of Miktoarm Star Polymers and Block Copolymers

ULRIKE WAIS[a,c], JIANHUA LIU[b], TAO HE*[b] AND HAIFEI ZHANG*[a]

[a]Department of Chemistry, University of Liverpool, Oxford Street, Liverpool, L69 7ZD, UK; [b]School of Chemistry and Chemical Engineering, Hefei University of Technology, Hefei, China; [c]Institute of Chemical and Engineering Science, 1 Pesek Road, Jurong Island 627833, Singapore
*E-mail: zhanghf@liverpool.ac.uk, taohe@hfut.edu.cn

## 5.1 Introduction

At the turn of the 19th century, Nobel Prize winner Paul Ehrlich coined the term 'magic bullet' for a therapeutic reagent that never misses its target, after allegedly visiting an opera where bullets never missed, no matter the aim of the shooter. Being a medical doctor, he thought of antibodies but, unknowingly to him, the notion of nanoparticles and targeted drug delivery was born.[1]

Targeted drug delivery can be divided into two general control mechanisms, *i.e.*, temporal and distribution control.[2] Drugs exhibiting temporal control mechanisms can be released from a carrier at specific times in

Polymer Chemistry Series No. 25
Miktoarm Star Polymers: From Basics of Branched Architecture to Synthesis, Self-assembly and Applications
Edited by Ashok Kakkar

Published by the Royal Society of Chemistry, www.rsc.org

short bursts or over an extended period of time in small, continual dosages. This is especially beneficial for fast metabolised drugs. Controlled release of these drugs would enable the matching of elimination and release and thus keep the dosage within the therapeutic window.[3–5] The second mechanism is distribution control, which describes the concept of a drug carrier being designed to release its loading precisely at the site of action. As a result, attacks on healthy tissue or other major damage on healthy organs can be prevented.[6–8] As such, distribution control is especially interesting in cancer research, where traditional chemotherapy damages the healthy bone marrow so much that many patients need to stop the treatment before completion. Distribution control can, at the same time, be used to overcome natural barriers in the metabolism (*e.g.* the blood–brain barrier).[9] The aforementioned mechanisms are not mutually exclusive and can be incorporated into one carrier system for efficient and non-harmful drug transport. Polymer nanoparticle carrier systems provide an ideal platform to this end, in particular nanoparticles prepared from block polymers and star-shaped polymers. Both of their distinctive structural features give rise to a myriad of possible designs and applications. The nature of these makes it possible to design and synthesise stimulus-responsive drug nanocarriers. Depending on the external environment, these polymeric nanoparticles can assemble or disassemble into micelles and encapsulate or release drugs. Depending on the functional monomers chosen, it is possible to synthesise polymers sensitive to changes in temperature, pH, salt concentration, or radiation.[10–13] As already mentioned above, this kind of specific release mechanism is especially interesting for cancer treatment, since cancer cells and the malignant surrounding tissue exhibit abnormal environments compared to that around healthy tissue (Figure 5.1).[14] Polymeric nanoparticles, in particular, have the added benefit of being able to exploit the enhanced permeability and retention (EPR) effect. The EPR effect is a special property exhibited by tumour tissue. Enhanced growth of malignant tissue leads to the formation of new blood vessels (angiogenesis), which show structural deficiencies, *e.g.*, poor alignment, lack of muscle tissue and wide fenestration, as well as a lack of efficient lymphatic drainage. All this leads to a higher agglomeration of nanoparticles in tumour cells than in healthy cells.[15,16]

## 5.2 Stimulus-Responsive Micellisation and Drug Delivery

### 5.2.1 Synthesis of Block Copolymer and Miktoarm Star Polymer

Amphiphilic block copolymers with various structures such as linear, star, or graft (Figure 5.2) can be used to form vesicles or micelles for drug delivery. This is only possible if the polymer exhibits a tightly controlled constitution, *e.g.* the type and composition of monomers. Established polymerisation

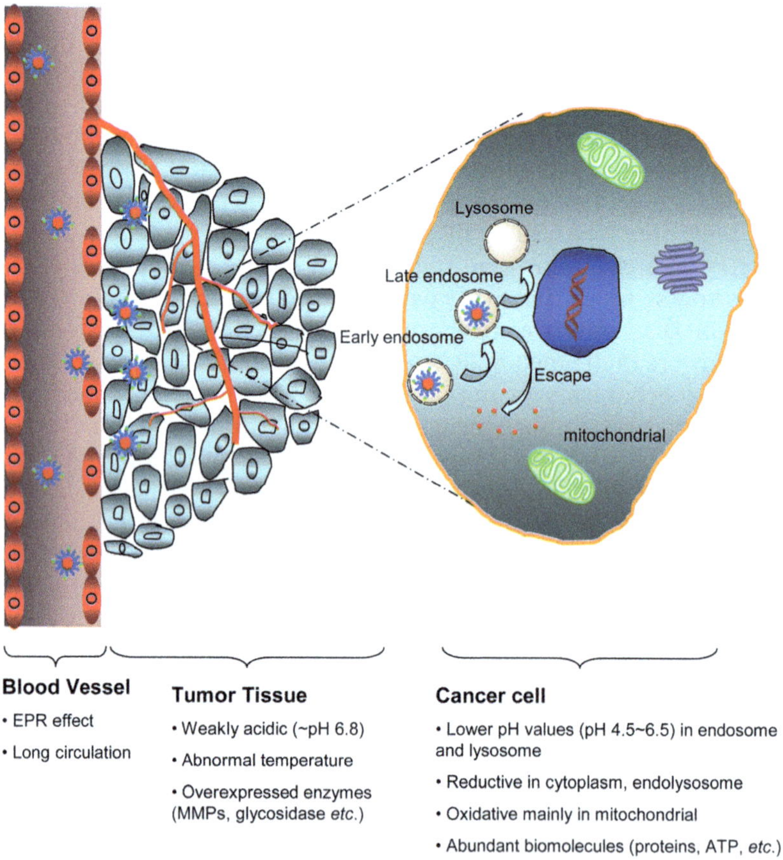


**Figure 5.1** Schematic representation of the abnormal environment found in malignant tissue and the cells used for active and passive targeting. (Reproduced from ref. 14 with permission from the Royal Society of Chemistry)

techniques, such as conventional radical polymerisations, often lead to statistical polymers. Living or controlled chain-growth polymerisation, on the other hand, paired with highly purified materials and hence minimal termination or transfer reactions, gives well-controlled polymeric structures. Different techniques, such as anionic polymerisation,[17] nitroxide-mediated radical polymerisation,[18,19] atom transfer radical polymerisation (ATRP), reversible addition-fragmentation (RAFT) or even olefin metathesis polymerisation,[20,21] can be used depending on the functionality of the monomer. The two most commonly used living radical polymerisation techniques are ATRP and RAFT. ATRP was first reported in 1995.[22] This technique uses a copper (or other metal) complex as a catalyst for the design of polymers with narrow molecular weights and tight structural control, while the mild reaction conditions allows for the absence of protection groups for most functionalities. The RAFT technique was first reported in 1998.[23] This method uses a

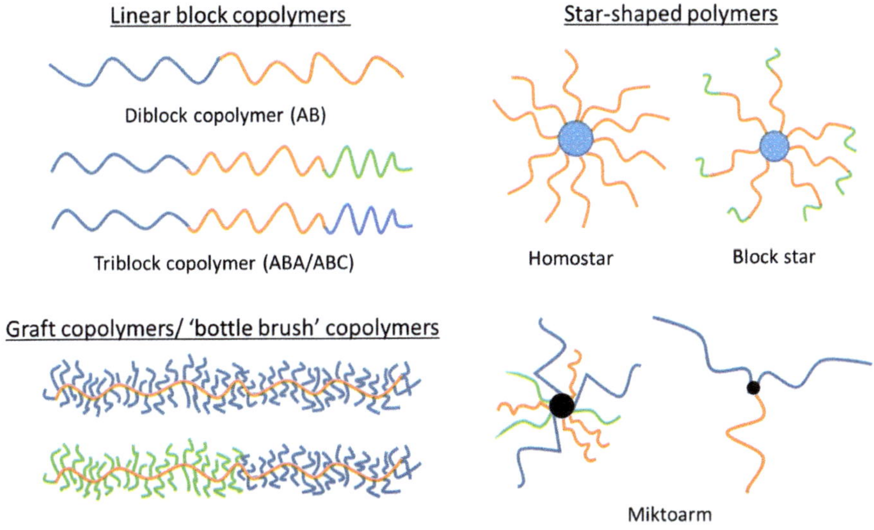


**Figure 5.2** Schematic representation of different types of block and graft polymers.

thiol-based so-called RAFT agent to yield polymers with a defined molecular weight, low polydispersity index, and known end-functionality.

Excellent reaction control and tolerance against most functional groups opens up the possibility of highly advanced structural architectures. The wide variety of reaction media broadens even further the scope of starting materials that can be utilised.[24] Often, polymerisation techniques are combined to prepare a specific polymer, since certain techniques work better than others for different monomers.[25,26] Linear or branched di- or triblock copolymers can be synthesised using two general approaches: coupling of two or more chains *via* linkers or sequential addition. The latter technique is used more often than the former due to the difficulties in end-functionalising certain polymer chains for coupling reactions, which always result in low yield of the targeted structures and the disadvantage of chain growth and degradation under the harsh conditions of coupling reactions. In the sequential addition, the key points are the order and time of monomer feeds, since the living chain of the first block has to initiate the second monomer polymerisation. Conversion of the first monomer needs to be very high; otherwise no control over the structure can be achieved. Core crosslinkers are often added as polymer groups on the hydrophobic chain and polymerised thermally or with ultraviolet (UV) light irradiation after micellisation. Careful control of the length and crosslinking is required to avoid impeaching on the space needed for drugs.[27]

Sequential addition can also be used to form graft polymers, though in this case macromonomers of branched and backbone monomers are added to form the backbone *in situ*, the so-called 'grafting through'. Other techniques include the separate synthesis of backbone and graft arms and subsequent post-polymeric reactions ('grafting onto') and the synthesis of a backbone with active sides to initiate arm polymerisation, which leads to better control of branch numbers but harder characterisation ('grafting from').

Star-homo or block polymers can be synthesised using the 'core first' method, where multifunctional initiators propagate the polymerisation of arm A onto the core. Arm B is subsequently either added as a pre-existing block or the polymerisation is initiated by the living chain of A. However, in order to form uniform arms, all sites need to be equally reactive and have the same initiation rate, which has to be higher than the propagation rate, with steric hindrance and repulsions kept to a minimum. The other option is the 'arms first' technique, where the arms are polymerised as already described for linear di- and triblock polymers. The 'arm' polymers can be subsequently added *via* coupling reactions to the core. However most linking reactions need long reaction times. Another obstacle in the synthesis of star block polymers is the big influence the topology of the block (inner and outer position) has on the properties in solution. Hence, a block that may be easy to connect to an inner linker may not be the one that is suitable in the outer part.[28] All these commonly used synthetic methods have the drawbacks of long reaction times with multiple steps and constant purification processes in between reactions to achieve the targeted polymers, as well as difficulties in accurate characterisation. Miktoarm polymers are a class of star polymers where at least two chemically different arms emit from a common core, as such they are also sometimes called heterostar polymers.[29] Miktoarm polymers can be synthesised using multifunctional initiator sites, which are selective for particular classes of monomers or polymerisations,[30] or *via* a multifunctional linking strategy.[31] This makes for easier syntheses and less chances of side reactions on already formed chains, as well as opening up a greater variety of different combinations of monomers.[32]

### 5.2.2 Micellisation and Drug Delivery

To avoid renal clearance and sieving by pulmonary capillaries, particles need to be between 0.01 and 3.0 μm, with particles larger than 0.2 μm being cleared more rapidly. Particles under 0.4 μm can furthermore make use of the EPR effect.[33] Polymeric nanoparticles usually exhibit sizes in the range of 0.05 to 0.2 μm and as such are excellent for drug delivery. The first polymeric delivery system was reported in 1976 and used crosslinked polyacrylamide,[34] which was shortly replaced by the biodegradable and biocompatible polymers polylactide (PLA) and poly(lactide-*co*-glycolide) (PLGA).[35,36] Although better uptake *in vitro* could be observed, the problem of short *in vivo* circulation times remained. This could be solved through the design of so-called 'stealth' particles, where the surface groups show non-specific and low interaction with proteins. Poly(ethylene glycol) (PEG) or polymer-containing PEG blocks are widely used to coat stealth particles. Stealth particles exhibit increased blood circulation and reduced uptake and capitation by the mononuclear phagocyte system (MPS).[37,38] This effect is known for highly ionic or hydrophobic groups[39] and prevents recognition by the reticuloendothelial system (RES),[40] as well as enzymatic degradation.[27]

Amphiphilic polymers can self-assemble into micelles. The driving force behind the self-assembly is the different solubility of the hydrophilic and hydrophobic blocks in water.[41] The main parameters for the characterisation of micelles are the aggregation number and micellar size, as well as the shape. These are influenced by the temperature, solvent, polymer concentration and polymer characteristics, such as the molecular weight, block architecture, *etc.*[42] The aggregation number, *i.e.*, the number of molecules needed to form a stable micelle, is closely relevant to the critical micelle concentration (CMC), a concentration 'below which virtually no micelles are detected and the limit above which virtually all additional surfactant molecules form micelles'.[43] For linear block copolymers, the aggregation number and CMC is relatively high. Hence, the formed micelles will easily dissociate because of the dilution in the blood stream. To prevent this from happening, cross-linked branches are introduced to enhance the micelle stability below the CMC.[27] Direct comparison with star block polymers and miktoarm polymers showed that the aggregation number decreases dramatically for star shaped polymers, while maintaining the same hydrodynamic radii due to the greater difficulty of arms to be accommodated into a micelle.[44] The hydrodynamic radius decreases as well for the micelles formed from star block polymers compared to those of linear di-block polymers of the same molecular weight, since the hydrophobic and hydrophilic parts in a star architecture can be broken into smaller arms as well as displaying higher flexibility (Figure 5.3).[42,45]

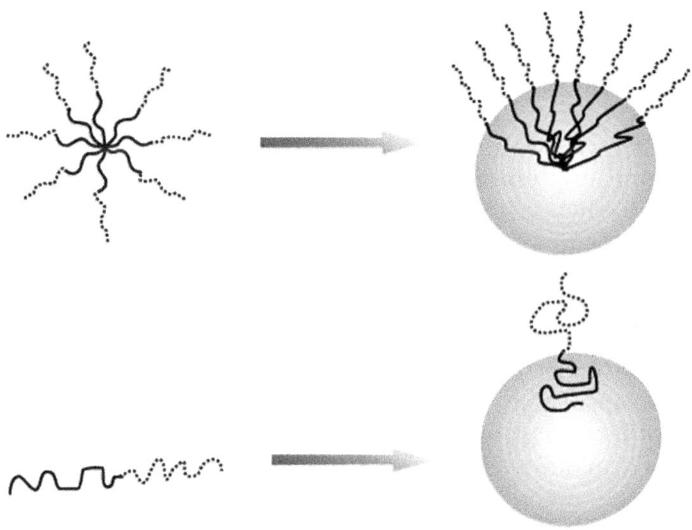

**Figure 5.3** Schematic representation of micellisation of a star block polymer and a linear diblock polymer. (Reprinted with permission from ref. 42 [G. Mountrichas, M. Mpiri and S. Pispas, *Macromolecules*, 2005, **38**, 940–947]. Copyright 2005 American Chemical Society.)

Miktoarms in particular are able to self-assemble and transform into very interesting structures by just adjusting the length of one or more arms, *e.g.* transformation of a hamburger-like structure to a worm-like structure,[29] and can further form multicompartment micelles due to the possibility of forming structures of three mutually immiscible arms connected *via* a central point.[46] They are also known to mimic the natural architecture of phospholipid bilayers better than their linear counterparts.[47] Homostar polymers, on the other hand, seldomly tend to form micelles due to their tendency to be either hydrophilic or hydrophobic. However, homostars possessing longer arms and slightly bigger cores may entrap molecules in their corona, forming unimolecular micelles, with a hydrophobic core and a hydrophilic corona.[48] A comparison of polymer architectures and factors controlling drug delivery outcomes is given in Table 5.1.[49]

A phenomena well known for block copolymers is 'schizophrenic micellisation', which describes the transformation *via* molecular dissolution of a micelle in water to an inverted micelle, *e.g.*, the hydrophobic block becomes hydrophilic (Figure 5.4).[50,51] The transformation is initiated by an outside stimulus, such as the temperature, electrolyte concentration, or pH, and is a mechanism by which a drug cargo can be released at the target side.[52,53] Other mechanisms involve the irreversible cleavage of polymer chains to destroy the micelle structure. Such mechanisms can often be found in enzymatic degradation, but can also be temperature or pH dependent.[54,55]

### 5.2.2.1 *pH-Responsive Delivery*

Systematic infection sites are well known for exhibiting acidic extracellular microenvironments. Hence, most tumour sites show a lower pH due to the production of lactic acid and hydrolysis of ATP in hypoxic regions of the tumour, which can be utilised for targeted drug delivery.[10] The incorporation of pH-sensitive polymers, such as poly-2-vinylpyridine (P2VP), poly[(2-diethylamino)ethyl methacrylate] (PDEAEMA), poly(methylmethacrylate) (PMMA), poly[2-(diethylamino)ethyl methacrylate] (PDEA), poly-2-dimethylaminoethyl acrylate (PDMAEA) or poly(γ-methyl-ε-caprolactone) (PMCL), can lead to site specific drug release *via* swelling, disassembly, or cleavage (Table 5.2).[14] Drug release due to swelling and disassembly is mostly found in micelle carriers or gels where a drug–polymer conjugated core is formed (physical encapsulation) due to hydrophobic interactions.[49] Protonation of a functional block, *e.g.*, with amine groups, leads to either straightening of the block or arm due to electrostatic repulsion[56,57] and subsequent swelling, or a shift in the solubility of the block, or disassembly of the micelle due to pH-sensitive cleavage, followed by disassembly of the micelle (Figure 5.5).[58–62] Chemically encapsulated drugs, through chemical bonding of a drug to a block, can be released by cleavage of pH-sensitive linkers.[63,64] Owing to the slow hydrolysis rates of biodegradable linkers and the scarcity of reactive functional groups on drugs, physical encapsulation is more widely used.[42,49]

**Table 5.1** Controlling factors for desired drug delivery outcomes for different types of polymer architectures (Adapted with permission from ref. 49 [L. Qiu and Y. Bae, *Pharm. Res.*, 2006, **23**, 1–30]. Copyright 2006 Springer Science + Busines Media, Inc).

| Architecture | Drug carrier | Property | Controlling factor |
|---|---|---|---|
| Block polymer | Micelle | Shape, CMC, size | Proportion of blocks A and B (C)<br>Electrical charge |
| | | Encapsulation | Hydrophobic interactions between hydrophobic block and drug |
| | | Bio-distribution | Molecular weight proportion of blocks<br>Surface charge/hydrophilicity<br>Surface groups/ligands |
| | | Drug release | Hydrophobic interactions between hydrophobic block and drug |
| | Polymersome | Shape | Proportion of A to B (C) block and/or interaction to added polymers |
| | | Membrane thickness | Length of hydrophobic chain |
| | Hydrogel | Sol–gel transition temperature and critical concentration | Molecular weight<br>Proportion of A to B block |
| Graft polymer | Micelle | CMC | Graft ratio |
| | Hydrogel | Lower critical solution temperature | Graft ratio<br>Molecular weight |
| Star polymer | Micelle | CMC (unimolecular *vs.* aggregate) | Arm number<br>Arm length (proportion of block A to B)<br>Topology |
| | | Drug encapsulation | Dimension of hydrophobic core<br>Arm number and topology |
| | Hydrogel | Lower critical solution temperature and gel strength | Arm number |
| Miktoarm star polymer | Micelle | Shape, CMC, size | Proportion and solubility of A and B (C) arms<br>Length of A and B (C) arms<br>Electrical charge |
| | | Bio-distribution | Molecular weight<br>Proportion of arms<br>Surface hydrophilicity<br>Surface groups or ligands and charge |
| | | Drug encapsulation | Hydrophobic interactions between hydrophobic block and drug |
| | Polymersome | Shape (*e.g.* spherical, multicompartment) | Solubility of arm A and B (C) in water and in each other |
| | | Membrane thickness | Length of hydrophobic chain |

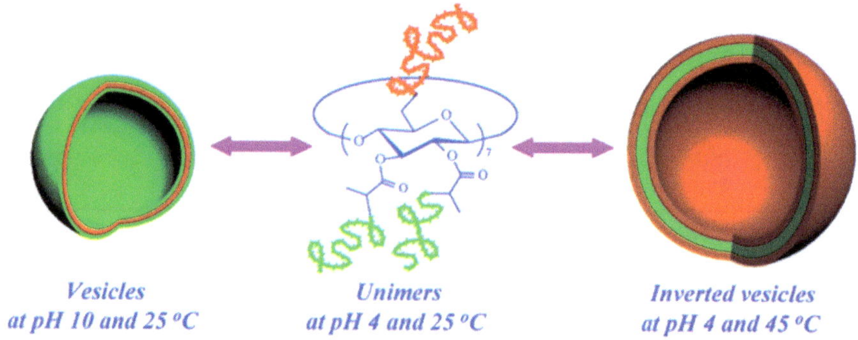


**Figure 5.4** Schematic representation of 'schizophrenic' micellisation of an $A_nB_m$ miktoarm star polymer. (Reproduced from ref. 51 with permission from the Royal Society of Chemistry.)

**Table 5.2** Examples of pH-responsive polymers with a p$K_a$ or cleavable bonds near or within the pH range of normal and tumour tissues (Adapted from ref. 14 with permission from the Royal Society of Chemistry).

| Structure | Name | p$K_a$ |
|---|---|---|
| 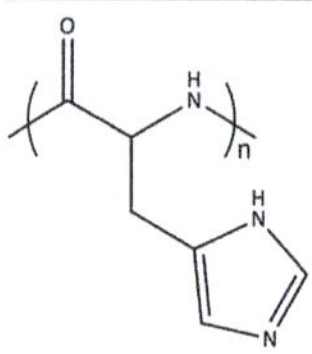 | Poly(histidine) | ~6.5 |
| 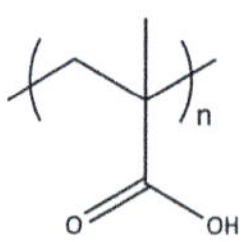 | Poly(methacrylic acid) | ~5.6 |
| 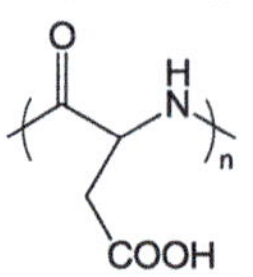 | Poly(aspartate) | ~4.9 |
| 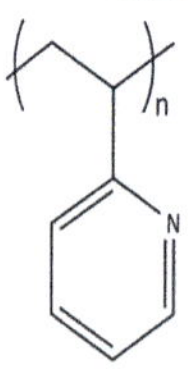 | Poly(2-vinylpyridine) | 5.0 |
| 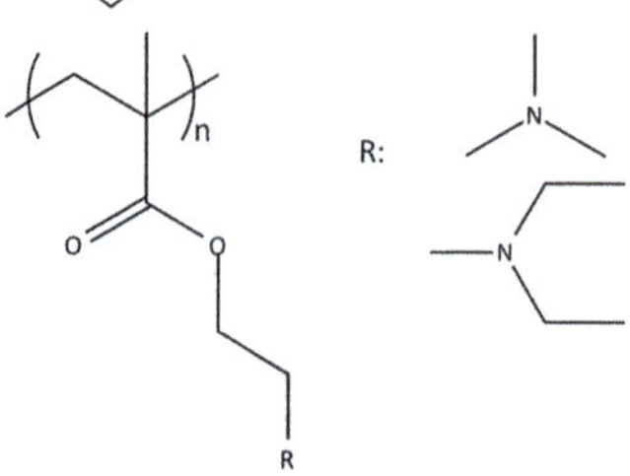 | Poly[2-(dimethylamino) ethyl methacrylate] | 7.4 |
| | Poly[2-(diethylamino) ethyl methacrylate] | ~7.3 |
| 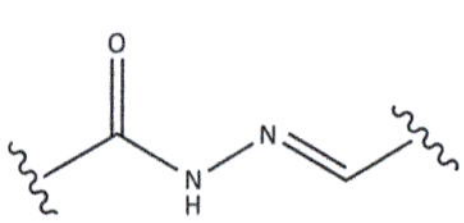 | Hydrazone | Cleavable bond by acidic environment in endosomes and lysosomes |

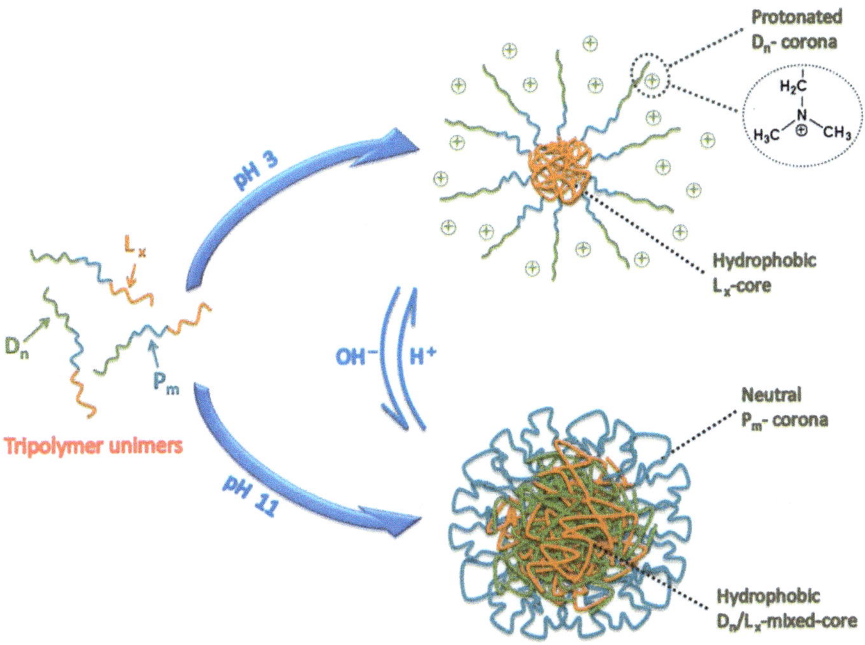


**Figure 5.5** Schematic representation of the pH-dependent self-assembly of an amphiphilic triblock copolymer. (Reprinted with permission from ref. 62 [J. Huang, J. Xu, K. Chen, T. Wang, C. Cui, X. Wei, R. Zhang, L. Li and X. Guo, *Ind. Eng. Chem. Res.*, 2015, **54**, 1564–1575]. Copyright 2015 American Chemical Society.)

Miktoarm polymers of the ABC type are able to self-assemble into a wide variety of interesting structures, while linear triblock polymers tend to assemble into Janus micelles.[56] Higher order structures are generally achieved by the addition of a second compound, *e.g.*, a second diblock that interacts electrostatically with the existing micelle.[58,65,66] Liu *et al.* reported the pH-dependent structural changes of an ABC miktoarm polymer, μ-[polystyrene][poly(ethylene oxide)][poly(2-(dimethyl amino) ethyl acrylate)] (μ-SODA). At pH 2.6, spherical micelles with a mixed corona of PEO and protonated PDMAEA could be observed, while an increase in the pH led to the formation of multicompartment worm-like structures. Alternating sub-compartments of different lengths of insoluble, deprotonated PDMAEA and PS were found due to the immiscibility of PS and PDMAEA, which were structural reminiscent of cell organelles.[67] In this study, structural changes were achieved by proton-buffering components and subsequent solubility changes. Saito and colleagues showed in a study using μ-[poly(ethylethylene)][poly(ethylene oxide)][poly(γ-methyl-ε capolactone)] (μ-EOC) that structural transitions could be achieved by pH-sensitive cleavage of one arm of the ABC miktoarm polymer. At neutral pH, worm-like multicompartment structures with sub-domains of PMCL and PEE of up to 1 μm length and 36 nm width could be observed. At pH 12, the PMCL chain was cleaved off close to the junction point, leading to the formation of an EO diblock and PMCL homopolymer. Subsequently, the wormlike structure transitioned to spherical micelles with single wormlike structures protruding outwards over a period of 14 days, due to a slower cleavage mechanism.[61]

Micelle-to-micelle transitions of structures exhibiting defined nanodomains could be an exciting platform for time-dependent and targeted release of more than one drug at the same site. However, miktoarm polymers forming predominantly spherical core–shell micelles exhibit as well favourable characteristics in comparison to their well-established linear counterparts. A study carried out by Van Bustele and co-workers demonstrated that an ABC miktoarm polymer of poly(ε-caprolactone) (PCL), poly(ethylene oxide)(PEO) and poly(2-vinypyridine)(P2VP) could form stable micelles over the whole pH range, while the corresponding diblock of P2VP and PEO could only form micelles at pH ≥ 5. This was attributed to the higher hydrophobicity of the PCL arm. The micellar size was found to be a function of the pH value. At low pH, complete protonation of P2VP led to stretching of the arms and electrostatic repulsion while, at high pH, the deprotonated P2VP was insoluble and micelle contraction was observed.[56] Similar studies done on a PEG((-*b*-PMMA)-*b*-PDEA) zwitterionic miktoarm polymer showed the formation of micelles with a neutral PDEA core and a hybrid PEG/PMAA corona in alkaline media. Decreasing the pH led to micelles with poly-ion cores due to charge compensation between partially protonated PMAA and PDEA (pH 5–7) and to micelles of hydrogen bonded complex cores and protonated PDEA coronas (pH ≤ 4). Lowering the pH led to an increase in size although, due to co-stabilisation of PEG and ionised PMAA, the particles were still smaller compared to those with linear polymers.[60] This system nicely describes the ability of a miktoarm terpolymer to self-assemble in water *via* three different driving forces, *i.e.*, interpolyelectrolyte complexation, hydrogen-bonding and hydrophobic interactions.[28]

A direct comparison of a binary mixture of a diblock polymer of PEO-PDEA and PMAA showed that, in alkaline media, diblock micelles and the PMAA homopolymer do not interact. Lowering the pH led to the interaction of PMAA with partially protonated PDEA and the formation of slightly bigger micelles (pH 6–8.5). When the pH was decreased further, aggregates of hydrogen bonded PEO/PMAA cores with cationic PDEA coronas could be observed.[58] This destabilisation with the decreasing pH could also be observed on a similar triblock polymer of PEO-PDEA-PSEMA. Micelle-to-micelle transition with the increasing pH of 'flower'-like micelles *via* large aggregates, both exhibiting high polydispersity, to well defined core–shell micelles in alkaline medium could be observed.[59] Compared to miktoarm polymers of the similar type, linear block polymers exhibited bigger particle sizes of 30–120 nm (in comparison to 9–90 nm), as well as lower stability (disassembly instead of swelling) over a wide range of pH.

As already mentioned, physical encapsulation of drugs and the subsequent release by swelling or disassembly may be preferred over chemical encapsulation. However di- and triblock polymers have high CMC values in comparison to miktoarm polymers and are prone to dissociate in the blood stream; hence, there is the need to either crosslink the core or shell and chemically encapsulate the drug.[57,68,69] For example, a pH-sensitive block polymer may contain a hydrazone bond, which is cleaved in an acidic environment to release the conjugated drug. Although superior accumulation in targeted cells could be observed, full drug release took 45 to 184 h.[12,63] Physical encapsulation in pH-sensitive polymersomes composed of block polymer was also found to offer stability.[70]

Star block polymers can chemically and physically encapsulate drug molecules, depending on the size and blocks as well as their topology. Hence, a potentially pH-responsive star block terpolymer of hexa(ethylene glycol) methacrylate (HEGMA), MMA and 2-(dimethylamino)ethyl methacrylate (DMAEMA)-*core* displayed unimolecular micelles for DMAEMA-*b*-HEGMA-*b*-MMA-*core*, while aggregates were observed for HEGMA-*b*-DMAEMA -*b*-MMA-*core* topologies.[71] Prabaharan *et al.* synthesised a folate conjugated Boltorn® H40-poly(L-aspertate-doxorubicin)-*b*-poly(ethylene glycol) block star polymer that existed as unimolecular micelles in aqueous medium, thereby overcoming the thermodynamic instabilities of micelles. Release of doxorubicin was triggered by pH-sensitive cleavage and drug release of 90% was completed in 45 h.[64] The use of PDEAEMA in an amphiphilic triblock star copolymer led to the formation of a three layered doxorubicin loaded spherical micelle in water. At pH 5, the micelle swelled and the drug was released due to the protonation of PDEAEMA. Yet, the release of 80% of the encapsulated drug still took up to 90 h.[72] Direct comparison with an amphiphilic miktoarm polymer comprised of the same polymer blocks (PCL, PDEAEMA and PPEGMA) showed CMC values as low as 0.003–0.0035 mg $L^{-1}$, compared to the 2.2–4.0 mg $L^{-1}$ values of the block star copolymer. Micelle swelling between pH values of 8 to 4 due to protonation of the PDEAEMA arm was observed as well. However, the highly charged micelles due to the more mobile protonated arms than in the block structure prevented aggregation and could extend the blood circulation time and facilitate the interaction of the micelles with the negatively charged cell membranes. Increasing the amount of PDEAEMA led to an entrapment efficiency of up to 71% and, subsequently, 85% of the drug was released after 48 h.[57]

Micelle-to-micelle transitions were used by Yin *et al.* to encapsulate and release a hydrophilic dye from a polymersome of PEG-$(\text{poly}(\text{L-hystidine}))_2$. At pH values below 6.8, the imidazole became ionised and less hydrophobic. The loss of stability led to a transition of the structure from spherical to cylindrical back to spherical and, finally, to unimers. Almost complete release was achieved after 4 h at pH 6.8.[47] Homostar polymers and brush graft polymers are rarely used to form micelles and encapsulate drugs. They are mostly used in the formation of hydrogels, that is, highly absorbent polymeric networks. Depending on the constitution of the graft- and star polymers, pH-responsive swelling and subsequent on–off release profiles can be obtained.[73,74]

#### 5.2.2.2 Temperature-Responsive Delivery

The treatment of cancer with heat by exploiting the higher temperature sensitivity of tumours mostly brought on by secondary infections has been an established technique for thousands of years. However, insufficient equipment and lack of understanding of hyperthermia led to a decrease in the use of hyperthermia therapy in the last century. During the last 20 years, research on equipment and systematic processes has revived the concept of hyperthermia therapy, which can be defined as the elevation of a specific, or non-specific, part of the body to temperatures of 39–45 °C for a certain period of time.[75] This concept can be applied to polymeric drug delivery *via* the use of thermoresponsive polymers such as poloxamers, triblock polymer poly(ethylene oxide)-*b*-poly(propylene oxide)-*b*-poly(ethylene oxide) (PEO-PPO-PEO), as well as poly(*N*-isopropylacrylamide) (PNIPAM). Polymers with a temperature-responsive block exhibit a specific lower critical solution temperature (LCST), where the stimulus-responsive block changes its solubility due to conformational changes.[76,77] The LCST is dependent on the concentration and block length, as well as the other copolymers used. Depending on the other copolymers, temperature dependent schizophrenic micellisation can often be achieved. Cai *et al.* gave an example of a non-linear polymer exhibiting schizophrenic micellisation behaviour. Y-shaped miktoarm polymers of commercially available Jeffamines, monoamine capped poly(alkylene oxides), and poly(sulfobetaine methacrylate) (JM2005-$(SBMA_{20})_2$) showed micelles of about 85 nm with a PSBMA core below 12 °C, unimers at 12–30 °C, and well-defined spherical micelles of 21 nm with a PSBMA shell above 30 °C. This observation could be explained by different LCSTs. PSBMA is insoluble below 20 °C due to interchain/intrachain electrostatic interactions, while PPO is insoluble above 30 °C. In the temperature range of 20–30 °C, both polymers are soluble. Hence, molecular dissolution occurs.[78] Direct comparison with similar diblock polymers revealed a tendency to form smaller particles.[79] Schizophrenic micellisation and reversible transitions dependent of a specific temperature can be used to obtain switchable drug carriers.[80] Li and colleagues showed this on a Y-shaped PNIPAM-*b*-$(PLL)_2$ (PLL = poly(L-lysine)) shell crosslinked miktoarm polymer. A solution of PNIPAM-*b*-$(PLL)_2$ was heated above the LCST. The core–shell micelles with a PNIPAM core and subsequently crosslinked PLL shell could form stable polymersomes over a wide range of temperatures and pH. Crosslinking prevented immediate drug release in favour of cumulative release at temperatures and pH values where both blocks are hydrophilic and molecular dissolution could be observed. Heating above the LCST led to shrinkage of the polymersomes and the drug was subsequently ‘squeezed’ out.[81] Many examples of thermoresponsive linear di- and triblock polymers can be found in the literature for drug delivery[82–84] and gene delivery,[10,85,86] as well as some brush-type polymers for drug delivery.[87] In recent years, most hyperthermia therapies have used magnetic iron or iron oxide nanoparticles, which can be tracked by magnetic resonance imaging (MRI) and heated at very specific systematic sites through the application of a magnetic field. Increasing the solubility of these particles can be achieved by block polymer surface coating.[88–90] The incorporation of these iron oxide nanoparticles in thermoresponsive drug-loaded block polymer micelles provides an ideal platform for a trackable and specific delivery system. Elegantly, the applied magnetic field leads to a localised temperature increase, which facilitates the tumour cell death and temperature sensitive release of the encapsulated drugs.[81–92] Miktoarm polymer analogues are, to the best of the author's knowledge, not known although smaller aggregation numbers, lower CMCs and smaller particle sizes would make these polymers ideal candidates for thermoresponsive imaging-guided drug delivery.

#### *5.2.2.3 Redox-Responsive Delivery*

The significant difference in the redox potential between the oxidising extracellular and the reducing intracellular environment is the basis on which redox stimulus-responsive drug or gene carriers function.[10] In the oxidising extracellular environment, disulphide bonds are maintained outside the cell, while the high content of gluthathione (GSH) inside the cell, particularly in the cytosol and nucleus, prevents the formation of disulphide bonds and reduces them to thiols.[14,93] Malignant tissue, *e.g.*, tumours, are known to overexpress GSH and can exhibit up to 10 times higher concentrations compared to normal tissue.[14] Hence, the most favoured functional groups for redox-responsive drug delivery are disulphide bonds. Redox stimulated drug and gene delivery can be found in excellent reviews[94,95] and research papers.[96–99] Accumulation of nanoparticles into malignant tissues is mostly achieved by passive targeting, *e.g.* the EPR effect, if no other active targeting moieties are used.

Due to their relatively low CMC and small sizes, miktorarm star polymers could be favourable platforms for redox-responsive drug delivery. Unfortunately, not many examples of single responsive miktoarm polymers have been reported. A recent report by Zhou *et al.* showed that a Y-shaped miktoarm polymer of PEG and PMMA bonded together *via* a disulphide bond (mPEG-SS-$PMMA_2$) could entrap and release Methotrexate (MTX) in a sustained manner compared to the burst release of free MTX (Figure 5.6).[100] The slow release was mainly attributed to the reductive cleavage and subsequent shedding of PEG, as well as the hydrophobic interaction of the PMMA core and MTX. It was further possible to effectively reduce the viability of HeLa cells.[100] The same enhanced drug delivery could be shown on four arm star block polymers of PEG and PCL bridged by disulphide bonds. Higher uptake and greater tumour reduction by released DOX, compared to free DOX, could be observed, which could be enhanced even further by the incorporation of ligands, although the CMC values were still higher than those found for miktoarm polymers.[101] A study by Bapat and colleagues comparing disulphide core crosslinked star block polymers and their corresponding miktoarm polymers of poly(styrene-*alt*-maleic anhydride)-poly(*N*-isopropylacrylamide) found that it was possible to reversibly break the

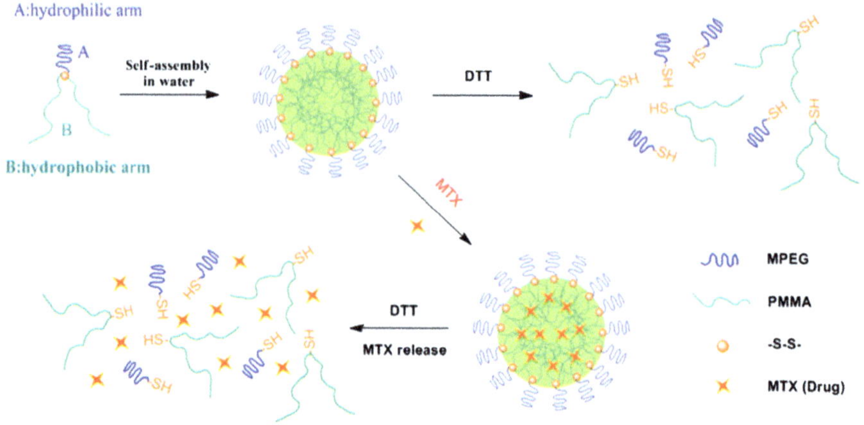


**Figure 5.6** Schematic representation of the self-assembly and dithiothreitol (DTT) mediated disassembly of a redox responsive miktoarm polymer. (Reprinted with permission from ref. 100. [Q.-H. Zhou, J. Lin, L.-D. Li and L. Shang, *Colloid. Polym. Sci.*, 2015, **293**, 2291–2300]. Copyright 2015 Springer-Verlag Berlin Heidelberg.)

disulphide bond by reduction and reform it by oxidisation in air. Hence, it was possible to open and close the formed structures in a continuous cycle. It was interestingly further shown that star block polymers with short polystyrene blocks tended to transform into crosslinked gels, which, after addition of a reducing agent, formed a sol. Miktoarm polymers, on the other hand, formed super aggregates in water, *e.g.* swelling from 31 nm in THF to 97 nm in water, due to the incorporation of two hydrophilic arms.[102]

### 5.2.2.4 *Multi-Responsive Delivery*

The multi-responsive behaviour of polymers is the possibility of applying more than one external stimulus to facilitate a structural change. Higher treatment efficiency and better targeting can be achieved by utilising more than one stimulus to release a cargo from a polymer delivery system.[103] These stimuli can be temperature, pH, redox or UV-light. The combination of pH- and temperature-responsive polymers is one of the most used techniques to target tumour tissue, due to the acidic microenvironment as well as the elevated temperature in the tissue.[14] These polymers also exhibit 'schizophrenic' micellisation behaviour, which can be used as an elegant drug release mechanism. Ge and colleagues described the synthesis and subsequent pH and temperature dependent schizophrenic behaviour of a PNIPAM-*b*-$(PDEA)_4$ star polymer. Above the LCST and at pH 4, micelles with a PNIPAM core could be formed. Both polymers were soluble and unimolecularly dispersed in solution at 20 °C and pH 4, while micelles with a PDEA core formed at 20 °C and pH 10.[104] The same group also showed the synthesis and schizophrenic behaviour of a Janus-type $A_7B_{14}$ heteroarm star polymer of A = PNIPAM and

B = DEA with a rigid toroidal β-CD (CD = cyclodextrin) core.[51] In water at pH 10 and 25 °C, they observed hollow vesicles of PDEA incorporated into the hydrophobic membrane walls, which dissolved as unimolecular polymers by decreasing the pH to 4. An increase of the pH to 10 led to inverse, larger hollow vesicles of PNIPAM incorporated into the membrane walls (Figure 5.4).

The formation of vesicles from linear diblock copolymers generally needs delicate design of the polymer as well as hydrophilic/hydrophobic balance. Reversible inversion and schizophrenic behaviour for vesicles are often achieved only by stabilisation with secondary structures, such as an added polymer and the formation of a binary mixture[58] or peptides.[105] Hence, these findings for star-shaped double hydrophilic block copolymers were quite unexpected. It should be noted that miktoarm star polymers exhibit smaller sized micelles, as well as lower CMC values, compared to their diblock analogues, while still maintaining the same 'schizophrenic' behaviour.[106,107] A similar graft polymer of a chitosan backbone, PDMAEMA and PNIPAM grafts (CS(-*g*-PDMAEMA)-*g*-PNIPAM) showed the same 'schizophrenic' formation of micelles, although reverse micellisation could be observed at pH 5–6, due to the lower $pK_a$ value of 5.5 of CS.[108] Interestingly, increasing the pH above 7 led to the formation of a third 'onion'-like structure, with a CS core surrounded by a layer of hydrophobic PDMAEMA with hydrophilic PNIPAM arms on the surface. Applying the same principle to miktoarm polymers could lead to interesting multicompartment structures, which could release their cargo at different pH values.

Liu *et al.* synthesised an $AB_3$ miktoarm polymer with a short PMMA arm and three hydrophilic p(NIPAAm-*co*-DMEAEMA) arms and examined it as a drug delivery system for MTX.[103] They found that the formed core–shell particles had a higher entrapment efficiency (EE) than conventional micelles and that the LCST increased with the decreasing pH, due to changes in the hydrophilic and hydrophobic PDMEAME ratio incorporated into PNIPAAm. At physiological pH 7.4, the LCST was 36.6 °C and faster drug release could be observed. This system could be used for orally transported drugs to preserve the drug loading in highly acidic environments (gastric) and release it in environments with physiological pH (intestinal).

Another combination for dual-responsiveness is thermo- and photosensitive polymers. To introduce photosensitive properties, azobenzene moieties are often incorporated into polymers, since they can undergo *cis* to *trans* isomerisation under UV radiation, which can ultimately lead to a shape change and increased dipole moment. Blasco and co-workers designed an $AB_3$ miktoarm polymer of an azopolymer and three identical PDEAA arms.[109] At temperatures above the LCST, they observed micelle collapse, which was reversible by cooling down. Radiation of the micelles with UV light for 10 min induced the isomerisation of *trans* to *cis* and, subsequently, an increase in the polarity and a decrease in size, accompanied by the presence of less defined micellar aggregates (Figure 5.7). This process was reversible as well. Encapsulated Nile Red could be released in a controlled way by photo-induced micelle deformation.

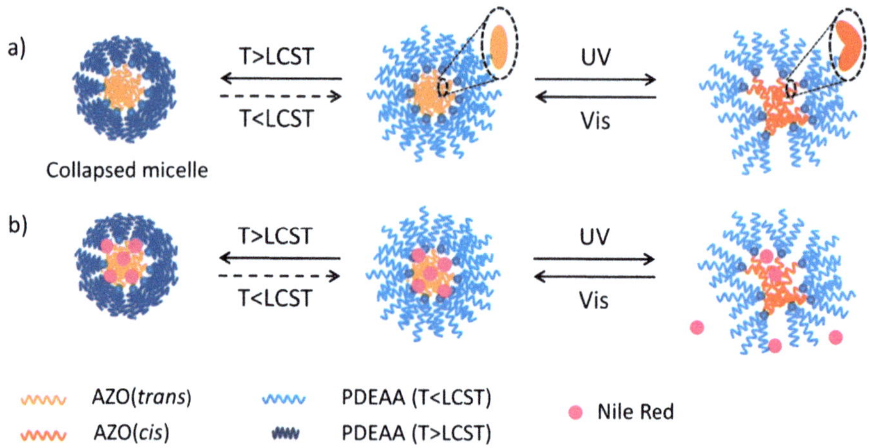


**Figure 5.7** Schematic representation of a thermo- and photo-responsive miktoarm star polymer, showing (a) temperature- and UV-dependent structural changes and (b) responsive release of Nile Red. (Reproduced from ref. 109 with permission from the Royal Society of Chemistry.)

Comparison with linear thermo- and photo-responsive block polymers shows the importance of the position of the azobenzene group. Feng *et al.* incorporated the azobenzene unit into the second block of a diblock polymer. Their PEO-*b*-P($AZO_{10}$-*co*-$NIPAM_{10}$) diblock polymer showed temperature-sensitive release of Nile Red, due to shrinkage, but no micelle disruption by UV radiation.[110] However, when Jochum *et al.* synthesised a PEO-*b*-PNIPAM diblock polymer with 5% azobenzene side groups in the PNIPAM block, they observed not just the incorporation of a hydrophobic dye at higher temperatures, but also the partial reversible disruption of micelles under UV light.[111] In the examples given above, the azobenzene function was always able to freely move due to it sticking out to the side of the polymer main chain. Boissiere *et al.* tried another approach and incorporated azobenzene groups as linear segments into PNIPAM.[112] They obtained flower-like micelles in cold water with the azobenzene groups as the core. Although the azo segments had no influence on the LCST, the multiple loops and the possibility of multiple folding led to the micelles being still dispersed in water above the LCST, albeit with a bigger hydrodynamic radii.

Up to this point, all examples given were dual responsive polymers. Miao *et al.* synthesised two $AB_2C_2$ miktoarm polymers with disulphide bearing linker elements for redox-responsive behaviour.[113] The star polymer $PEG(PCL)_2$-$(PNIPAM)_2$ carried a thermo-responsive group, while $PEG(PCL)_2(PAA)_2$ showed pH-responsive behaviour. Hence, they could be used as dual responsive star polymers, while a 1 : 1 mixture of these polymers showed subsequently triple responsive behaviour, due to aggregation of polymer 1 and 2, while avoiding the difficult synthesis of quaterpolymers. At higher temperatures, acidic pH and in the presence of dithiothreitol (DTT), *i.e.*, the microenvironment of tumour tissue, they could observe a sharp increase in the cumulative release of encapsulated DOX, due to the changes of hydrogen bonding and electrostatic repulsion between drug, copolymer, and the microenvironment. The triple stimuli allowed for better on demand adjustment of the release kinetics within a wide range *via* external stimuli. These triple responsive behaviour is also known in linear block polymers,[114] but in contrast to the aforementioned star polymer, they cannot be used as dual responsive polymers.

## 5.3 Emulsions

Emulsions are defined by the International Union of Pure and Applied Chemistry (IUPAC) as "a fluid colloidal system in which liquid droplets and/or liquid crystals are dispersed in a liquid. The droplets often exceed the usual limit of size for colloids. An emulsion is denoted by O/W if the continuous phase is an aqueous solution and by W/O if the continuous phase is an organic liquid (an 'oil')". More generally speaking, an emulsion can be described as one liquid dispersed in the form of droplets in an immiscible second liquid called the continuous phase.[115,116] As already described above, depending on the continuous phase, emulsions can be either O/W or W/O. While this is true for most emulsions, there is also the possibility to form multiple emulsions, water-in-oil-in-water (W/O/W) or oil-in-water-in-oil (O/W/O) emulsions.[115,117] Emulsions are kinetically unstable and prone to phase separation to reach an energetically favourable state *via* flocculation, creaming, coalescence or Ostwald ripening.[118,119] To prevent phase separation, stabilisers such as surfactants or polymers are added. Due to their amphiphilic nature, surfactants and certain polymers can reduce the interfacial tension between two immiscible phases. If the stabilising agent is a solid particle, the emulsion is called a 'Pickering emulsion'.[120] Pickering emulsions are known to be especially stable due to the steric and electrostatic repulsion arising from the particles adsorbed on oil–water interfaces.[121]

Block copolymers and miktoarm polymers can be designed as amphiphilic stabilisers for emulsions, due to their specific structural characteristics. The wide range of polymeric blocks to choose from leads to an extensive assortment of possible stabilisers with particular features such as stimulus-responsive phase inversion or de-emulsification and drug encapsulation for oral or topical applications.[122]

### 5.3.1 Emulsions Stabilised by Linear/Branched Di- and Triblock Copolymers

Classical surfactants are comprised of a hydrophilic and a hydrophobic part. This pattern can be imitated in di- and triblock copolymers by choosing suitable hydrophilic and hydrophobic polymers. As with surfactants, block polymers spontaneously form micelles in solution, whereby the insoluble part of the polymer forms a core and is shielded from the surrounding solvent by the soluble part of the polymer (corona). This is further enhanced by repulsive forces between both blocks.[123] The stabilisation of emulsions is furthermore favoured by the CMC, as well as the lower molecular mobility exhibited by block copolymers, which leads to a decrease in molecule desorption from particle surfaces.[124] Huang *et al.* recently published the synthesis and subsequent emulsification properties of a poly[*N*,*N*-(dimethylamino) ethyl methacrylate-*b*-poly(ethylene glycol) methyl ether methacrylate-*b*-lauryl methacrylate] (PDMA-*b*-PPMA-*b*-PLMA) triblock copolymer. Due to the non-ionic hydrophilic poly(ethylene glycol) side group and the hydrophobic PLMA block, 1 : 1 wt% O/W emulsions of dodecane and crude oil could be formed. By varying the amount of PLMA, the interfacial tension of dodecane droplets in water could be tuned. Longer hydrophilic block segments led to a sharp decrease in the interfacial tension, since the polymer was more firmly adsorbed on the interface and a higher degree of droplet coverage could be achieved. This in turn meant that a lower concentration was needed to form stable emulsions, though if the segment was too long, steric hindrance and electrostatic repulsions increased, and destabilisation was observed. The synthesised polymer also showed pH-responsive behaviour due to the PDMA block. At higher pH values, the interfacial tension decreased, as expected. Interfacial tension increased at pH values under 5, where PDMA was protonated and cationic, leading to electrostatic repulsions and loss of adsorption as well as hydrophobicity. However, at concentrations above the CMC, the opposite behaviour was observed for long chains of PLMA. It was suggested that protonation of PDMA led to additional salt screening and electrostatic repulsion of the adsorbed polymer.[62]

Another example of a pH-sensitive tri-block copolymer is poly(ethylene-oxide)-*block*-glycerol monomethacrylate-*block*-2-(diethylamino) ethyl methacrylate (PEO-PGMA-PDEA) which, after esterification with succinic anhydride, gave a zwitterionic triblock polymer. In aqueous medium, an inner core of PEO and poly-(succinic anhydride glycerol monomethacrylate) (PSAGMA) was formed surrounded by a PDEA corona. At pH 8, the carboxylic acid functions of PSAGMA formed inter-micellar crosslinks and stable O/W emulsions could be formed. By lowering the pH to 3, PDEA was protonated and became hydrophilic. However, by increasing the pH again, the de-emulsification process was found to be reversible.[125]

Another stimulus interesting for drug delivery is the temperature. The first group to show phase inversion of a polystyrene stabilised 'Pickering' emulsion dependent of the temperature was Bink and co-workers with a poly[2-(diethylamino)ethyl methacrylate-*block*-methyl methacrylate] (PDMA-*b*-PMMA) diblock polymer. Emulsions were prepared at pH 8.1 to avoid destabilisation due to protonation of the hydrophilic PDMA block. It was observed that 1 : 1 wt% *n*-hexadecane/water emulsions formed at room temperature remained as stable O/W emulsions even when heated up to 70 °C. However, when the emulsions were formed at different temperatures, O/W emulsions were observed below 52 °C and W/O emulsions above 65 °C. Between 52 and 65 °C, O/W, W/O and multiple emulsions coexisted, a common phenomenon observed during inversion.[126] They attributed this change to the wettability of the block polymer from predominantly hydrophilic (O/W) to hydrophobic (W/O) at higher temperatures, due to the breaking up of hydrogen bonds surrounding the PDMA block.[127]

Besnard *et al.* combined these two stimuli to form temperature- and pH-sensitive multiple emulsions for the release and encapsulation of hydrophilic drugs. A diblock copolymer containing a polystyrene (PS) block and a block of randomly distributed units of styrene and 2-(dimethylamino) ethyl methacrylate ($PS_x$-*b*-P($S_y$-*co*-$DMAEMA_z$)) was synthesised. DMAEMA was chosen for its sensitivity towards pH changes and temperature changes. PDMAEMA has a low LCST of around 40 °C in water. Above the LCST, PDMAEMA becomes more hydrophobic, due to the breakage of hydrogen bonds. With the increasing pH, they could detect an inversion from O/W to W/O emulsions. At pH 5.8, just under the inversion zone, they found not just coexisting O/W, W/O, and W/O/W, but also more complex systems such as O/W/O/W. The same could be observed by increasing the temperature from 5 to 45 °C and maintaining the pH value at 6, where the polymer is partially protonated and, hence, the only trigger is the change in temperature. Changes in pH are an easier and coarser trigger for emulsion inversion since coulomb interactions are stronger than hydrogen bonds, as polymers are able to stabilise better emulsions, be it classical O/W or W/O or even multiple emulsions, when one block is charged.[128] These observations were then used to show that a hydrophilic dye could be captured in a W/O emulsion, and released by pH changes due to phase inversion, then recaptured by reforming a W/O emulsion.[129]

Oral administered nanoemulsions, *e.g.*, emulsions with droplet sizes under 100 nm, stabilised by di- and triblock copolymers were produced and used as computer tomography (CT) contrast agents. It was possible to stabilise emulsions of iodinated oil and water with droplet sizes of around 75–100 nm, thus preventing rapid renal excretion and high cellular distribution as observed in particles under 5.5 nm, while at the same time being small enough not to cause lung embolisms.[130,131] Emulsions formed by diblock polymers furthermore showed superior stability in comparison to triblock and lipid stabilised emulsions, due to their high surface coverage and deeper penetration of the oil droplets. *In vivo* testing showed an increase in Houndsfield units (HU), and a slower uptake and later clearance of the contrast agent (uptake into smaller organs and clearance without any damage).[132]

Besides direct oral or parental administration, polymer stabilised emulsions can be used to form polymersomes for controlled release. After forming an emulsion and evaporating the organic solvent, polymersomes with an incorporated drug were formed in aqueous suspension or as dried materials. Such polymersomes could be prepared from a PEO-poly(β-benzyl-l-aspartate) (PEO-PBLA) block copolymer with incorporated doxorubicin. With a size of 37 nm and slow release (80% doxorubicin still in the polymersome after 100 h), they are ideal to exploit the EPR effect and show long-time release at the target site.[133] In another study, hydrophilic CdSe was incorporated into a diblock poly(*n*-butyl acrylate)-poly(acryl acid) (PBA-PAA) polymersome. Remarkably, the polymersomes were formed *via* a one-step procedure using a microfluidic device to form W/O/W emulsions, the only restriction being that the utilised polymer had to exist as unimers, otherwise it could aggregate and destabilise the inner oil droplets. The procedure led to full size control, as well as complete separation of the inner and outer aqueous layer. Wall thickness could be tuned *via* the initial polymer concentration and were stable up to several days. Formation of emulsions with more than one inner water droplet led to a foam-like structure upon drying, and polymersomes consisting of many smaller compartments could also be obtained.[134]

Another possibility to enhance the drug delivery and uptake is the formation of nanosized drug particles, which are inherently sensitive against agglomeration and difficult to form. One approach to form drug nanoparticles is by emulsion freeze drying.[135,136] An O/W emulsion of a branched block copolymer containing a poorly water-soluble drug was formed and then rapidly frozen. After solvent removal by lyophilisation, drug nanoparticles were formed *in situ* in the porous polymer scaffold, which readily produced an aqueous drug nanosuspension by simply dissolving the scaffold in water, and remained unchanged in storage for up to one year.[137]

### 5.3.2 Emulsions Stabilised by Star-Shaped Homo/Block Polymers and Bottle-Brush Copolymers

In addition to miktoarm polymers, which will be discussed later, other star-shaped constitutions are also available from a (non-)polymeric core and polymeric side arms. If the arms are made up of one type of monomers they are called homostars while, when two or three blocks of different polymers are used as arms, they are known as block stars. In contrast to linear or branched block copolymers, these do not self-assemble in a medium by folding into micelles, but already exhibit polar and non-polar inner and outer spheres. By straightening and turning all sidearms to one side, the same general constitution of a classical surfactant can be achieved, which in turn makes it possible for star-shaped polymers to stabilise emulsion droplets with sizes between a few microns up to tens of microns.[138] The required flexibility to form surfactant-like structures is dependent on the tight control of the side-arm number and core size. Because of this, and the relative novelty of these structures, not a lot of research into emulsions stabilised by star-shaped polymers has been reported.

The flexible behaviour of side arms was shown on porphyrin-[poly(methyloxazoline)-*b*-poly(ethyloxazoline)]$_4$ (Por-[PMOZ-*b*-PEOX]$_4$), a star block polymer with a porphyrin core for colour labelling and four side arms of PMOZ-*b*-PEOX. Direct comparison of the corresponding homostars to the star copolymers showed superior stabilisation of a 1:10 chloroform to water emulsion. This was attributed to the two zones of weakly hydrophobic PEOX and strong hydrophilic PMOZ found in the star block. It was possible to form O/W and W/O emulsions using the same polymer, which was observed through the colour change of the porphyrin core when facing towards the continuous phase or being shielded by the side arms.

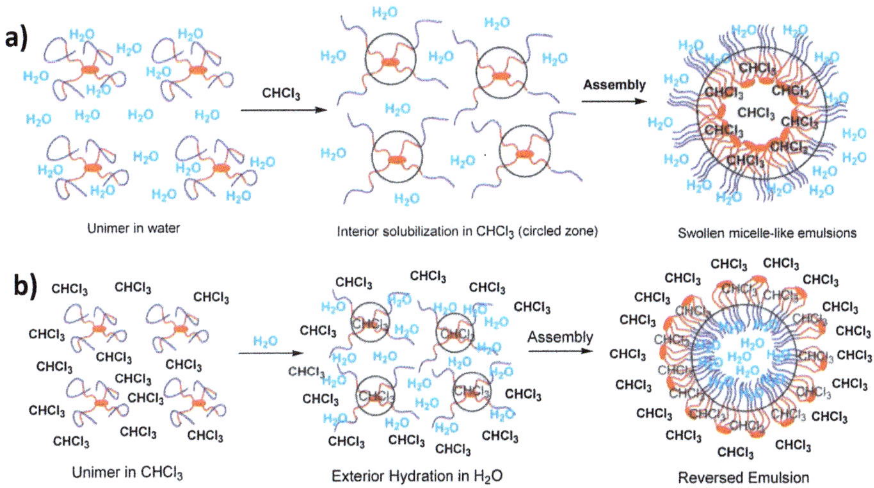


**Figure 5.8** Schematic representation of O/W (a) and W/O (b) emulsions stabilised by the same star block copolymer. (Reproduced from ref. 139 with permission from the Royal Society of Chemistry.)

The ability to stabilise emulsions was explained as following: in water (chloroform), the interior segment of a unimer is solvated by chloroform (water), which in turn results in the straightening of the side arms. Attractive forces between solvated interior segments lead to the spontaneous formation of micelles with incorporated chloroform (water) of sizes between 0.5 and 1.1 μm (Figure 5.8).[139] Spheres with diameters of 1–2 μm suspended in water could be obtained after evaporation of chloroform, which could be used for drug delivery.[139]

Other possible polymeric structures known to be able to stabilise emulsions are bottle-brush polymers. These comb-type graft copolymers consist of a main chain with one or more different linear side arms along the length of the chain.[140] Due to this structure, a better surface coverage and smaller droplet sizes can be obtained in most cases.[141,142] Stabilisation and amphiphilicity are possible *via* different routes depending on the type of polymer. Double brushes, for example, are synthesised using a linker molecule and two different graft arms. If one arm is hydrophilic and the other hydrophobic, the polymer will have a hydrophilic and a hydrophobic side after linkage, due to electrostatic repulsion of the two graft arms during assembly. One side of the polymer will consequently face the continuous phase, resulting in the polymer covering the droplet lengthwise. Chain flexibility enables the formation of smaller droplets and thus nano- and microemulsions.[141] Mathur *et al.* synthesised a comb-type graft copolymer with a poly(methacrylic acid) backbone and short grafts of poly(ethylene glycol). Under neutral and basic conditions, the polymer was completely hydrophilic. Decreasing the pH led to the formation of hydrogen bonds between the ether group of the PEG graft arms and the acid group of the MAA chain. As a result, only

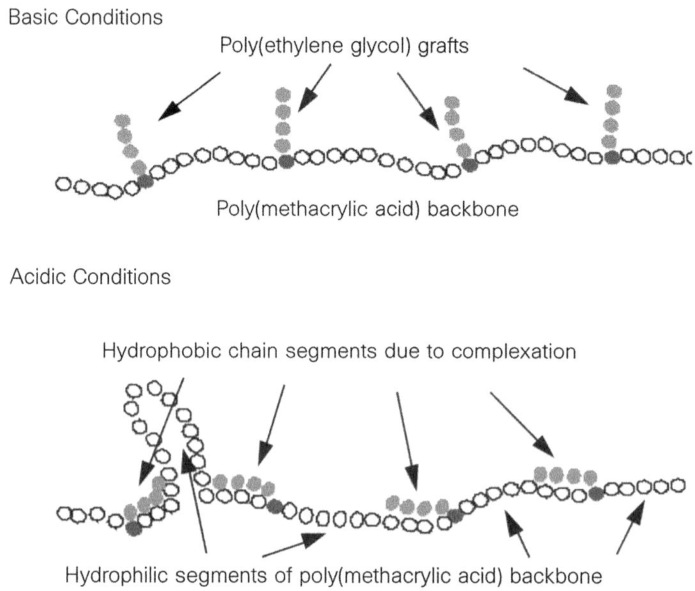


**Figure 5.9** Schematic representation of a graft copolymer with a poly(methacrylic acid) backbone and grafts of poly(ethylene glycol) under basic and acidic (complex promoting) conditions. (Reprinted with permission from ref. 143 [A. M. Mathur, B. Drescher, A. B. Scranton and J. Klier, *Nature*, 1998, **392**, 367–370]. Copyright 1998 Nature Publishing Group.)

alkyl groups were facing the solvent interface and thus hydrophobic regions could be created (Figure 5.9). Increasing and decreasing the pH showed reversible emulsion breaking and formation. Increasing the MAA-to-EG ratio led to emulsion breakage at low pH values. Due to the high flexibility and high steric requirements of these complex regions, droplet sizes were around 6 μm.[143]

Another option for inducing amphiphilic behaviour and emulsion stabilisation is the synthesis of tertiary graft copolymers. Emulsion stabilisation is possible in three ways. The first two possibilities are that two graft arms may be hydrophobic while the other arm is hydrophilic and *vice versa*. The third possibility is that one chain is water-soluble, the other is not, and the third arm is insoluble in both water and organic solvents. This can lead to the formation of a barrier of insoluble graft arms between water and the solvent and hence, encapsulated emulsion droplets or capsules can be obtained.[144] In a study with poly(glycidyl methacrylate)-*g*-[polystyrene-*r*-poly(acrylic acid)-*r*-poly(ethylene glycol) methyl ether], a tertiary graft polymer, nanoemulsions of benzyl benzoate/water and doxorubicin were formed and found to be stable for up to one year. By tuning the ratio between MPEG and PAA, pH-responsive release at pH 5 (the physiological pH of most cancerous cells) could be observed. These polymer complexes were non-toxic for healthy cells, but the emulsion droplets could enter cancer cells due to the EPR effect and the polymer complex ruptured, releasing doxorubicin.[142]

### 5.3.3 Emulsions Stabilised by Miktoarm Polymers

Miktoarms show unique self-assembly characteristics, which may be exploited to form emulsions. Ba and colleagues first demonstrated that polystyrene-polyoxirane ($PS(PEO)_2$) and polystyrene-poly*tert*-butylstyrene ($PS(PTBS)_2$) could stabilise O/W and W/O emulsions, while their linear diblock counterparts were only able to form stable O/W microemulsions.[145] A direct comparison of star block, homostar, and miktoarm polymers as emulsion stabilising agents showed that miktoarm polymers were better at forming emulsions. In those emulsions stabilised by miktoarms, no excess oil, higher capacity for the organic phase, and lower viscosity were observed. All tested star-shaped polymers were soluble in the organic phase, while their linear counterparts were only soluble in water. This generally led to the surfactant being more readily available.[146,147]

An emulsion may be formed with the stabiliser being initially present in the discontinuous phase, implying that the stabiliser molecules must have moved at one point to the phase interface. This was shown in the example of a PEO-PBA-poly(DVB) miktoarm star polymer. Although dispersed in the oil phase, due to the high affinity of PEO and PBA to xylene, a W/O unexpectedly formed after vortex mixing, which was attributed to the high mobility of the miktoarm star polymers. This may be the case for homostar or star block polymers; however, due to their arms being only hydrophilic or hydrophobic, the phase in which they are soluble will also be the continuous phase and no migration will be observed, although phase inversion is possible by addition of the discontinuous phase solvent.[124,139]

Star-shaped polymers can generally form emulsions with smaller droplet sizes in the 100 to 200 nm range, whereas their linear and branched counterparts usually form droplets that are a few microns in size. Miktoarm polymers are able to form even smaller droplets, due to their higher flexibility and the possibility to finely tune the polarity. Figure 5.10 shows a

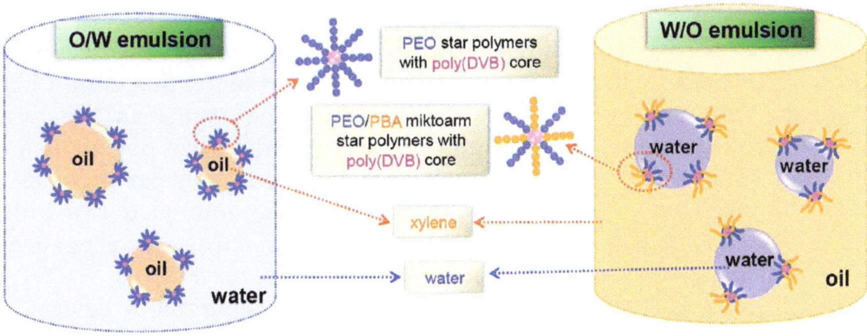


**Figure 5.10** Schematic representation of emulsions stabilised by homostar and miktoarm polymers. (Reprinted with permission from ref. 124 [W. Li, Y. Yu, M. Lamson, M. S. Silverstein, R. D. Tilton and K. Matyjaszewski, *Macromolecules*, 2012, **45**, 9419–9426.]. Copyright 2012 American Chemical Society.)

schematic representation of emulsions stabilised by homostar and miktoarm polymers.[124] A study comparing homostar and miktoarm polymers consisting of PDMA and PMEA in different ratios showed that emulsions formed with homostar polymers were either not stable for prolonged periods of time, or could not even be formed without excess oil or water phases.[148] This was thought to be because miktoarm star-shaped polymers are able to form aggregates and form 'Pickering' emulsions. Increasing the ratio of the less polar PMEA to the highly polar PDMA led to bigger droplet sizes. However, even at high ratios, only droplets with average size 54 nm could be found. Miktoarm polymers with a ratio of PDMA to PMEA close to two, exhibited stable O/W emulsions only when the oil content reached more than 60 wt%. Further decrease of the polarity by increasing the amount of PMEA to a ratio close to one led to phase inversion depending on the oil content. At lower oil phase weight percentages, stable O/W emulsions were observed, as expected. Surprisingly, at oil content of 40 wt%, a stable multiple emulsion was observed with coexisting O/W, W/O and W/O/W emulsions. At higher oil contents, irregular shaped W/O droplets were predominantly observed. Excess of PMEA gave stable O/W emulsions for oil contents of 10 to 20 wt% with regular, spherical shaped droplets. Any further increase of oil led to excess oil phases and unstable multiple emulsions.[148] This showed that, by finely tuning the ratios of differently polar arms on miktoarm polymers, not only different emulsion types could be achieved but also different droplet sizes, which was not possible with homostar or star block polymers.

Up to this point, all of the discussed miktoarm star polymers were of the star-shaped variety with a central core surrounded by different arms. One category of this structural pattern is Y-shaped miktoarm star polymers, where a core or linker connects two arms (A) with one structurally different arm (B). Compared to miktoarm polymers with a higher amount of arms, these Y-shaped polymers are smaller, more mobile, and more akin to the hydrophilic/hydrophobic head and tail constitution of surfactant molecules, as well as to amphiphilic block copolymers. A direct comparison of a dual responsive poly(styrene)-di-poly(*N*,*N*-dimethyl aminoethyl methacrylate) (PS-$(PDMAEMA)_2$) with its linear diblock counterpart found that almost half the concentration was needed to stabilise the same emulsions. This was explained by the better self-assembly of Y-shaped miktoarm polymers into micelles with a more compact core.[44] Each miktoarm polymer occupies a larger surface area at the core–corona interface to accommodate two arms. As such, a smaller core size and thus droplet size, as well as a lower aggregation number and therefore a lower concentration, are needed.

Y-shaped miktoarm polymers can, as their linear counterparts, be used to form polymersomes for drug delivery. Maglio *et al.* reported the formation of nanocapsules of roughly 100 nm with a loading efficiency around 50% (Figure 5.11).[149] A favourable release profile of an initial burst, followed by sustained slow release, was demonstrated. Addition of a surfactant further slowed the release down, due to lower drug mobility.[149] These examples show

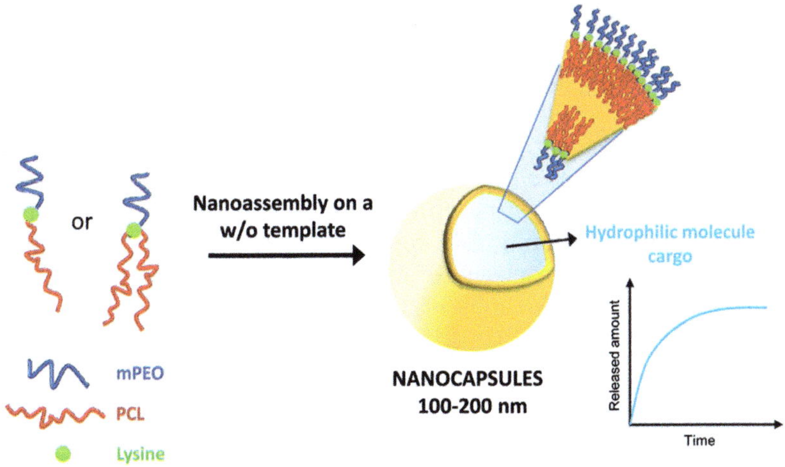


**Figure 5.11** Schematic encapsulation of a hydrophilic cargo into a nanocapsule with lipid-like structure *via* a W/O emulsion using a Lys-mPEO$(PCL)_2$ miktoarm star block polymer. (Reprinted with permission from ref. 149 [G. Maglio, F. Nicodemi, C. Conte, R. Palumbo, P. Tirino, E. Panza, A. Ianaro, F. Ungaro and F. Quaglia, *Biomacromolecules*, 2011, **12**, 4221–4229.]. Copyright 2011 American Chemical Society.)

that smaller vesicles can be achieved more easily and the potential for higher versatility towards the encapsulation of hydrophilic and hydrophobic drugs *via* emulsion formation in comparison to linear diblock copolymers, homo- and blockstar, as well as bottle-brush polymers, is feasible, although more research needs to be done in this field.

## 5.4 Conclusions and Perspective

In the last few years, increasing efforts have gone into the improvement of targeted drug delivery to malignant tissue, ideally without harming healthy tissue. Amphiphilic linear block polymers as drug carrier systems are well-researched platforms for delivery, with some polymer drug conjugates being in different phases of clinical trials. The use of polymer blocks sensitive to external stimuli increases the chance of target-specific delivery. Tight reaction control and the need for constant purification processes after each step, however, make linear di- and triblock polymers hard to obtain on a large scale. Relatively high CMC values and the subsequent need for crosslinking, which in turn can lead to low encapsulation, higher hydrodynamic radii, and dissimilarity to biological structures, further limit their application. Bottle-brush polymers, homostar and star block polymers exhibit lower CMC values and smaller micelle sizes, and they do not tend to disassemble at high dilution (*e.g.* in the blood stream) and are thus better candidates for passive targeting. Even so, these polymers are mainly used in the formation of hydrogels with highly interlinked structures.

The unique structure of miktoarm polymers results from the easy synthesis of arms with different characteristics, as well as low CMC values, small particle sizes, interesting and easily obtained architectures (*e.g.* multicompartments, hamburger or worm-like structures) and high drug encapsulation. The favourable micellisation behaviour brought on by the use of stimulus-responsive polymers can also be easily obtained for miktoarm star polymers as for their linear counterparts, while still maintaining all other advantages. The ability to easily obtain multicompartmental vesicles or micelles is especially interesting for the delivery of multiple drugs to the same target side. The possibility to combine different stimulus-responsive polymers could lead to accurate and enhanced drug delivery in response to changes in extracellular/intracellular microenvironments or at different parts of the cell. Another exciting and emerging area is the use of miktoarm polymers for theranostics, the combination of therapeutic and diagnostic agents.

Micellar/nanoparticle delivery systems are mainly used for intravenous administration, which is associated with discomfort for many patients. Hence, drug delivery systems for oral administration are desirable. One such platform is the formation and use of emulsions. Linear block polymers are a well-established tool for emulsion formation and for drug delivery *via* emulsions because of their non-toxicity and biocompatibility in comparison to surfactants. Thus, the herein mentioned examples are just a few of the research being done not just in drug delivery, but also in cosmetics or materials sciences. Graft polymers and homo- and block-star polymers exhibit more favourable CMC values, as well as the possibility to form smaller droplets. However, these types of polymers are less used for the formation and application of emulsions as their linear counterparts. Miktoarm polymers, on the other hand, seem to be a promising platform to obtain nano- and microemulsions, due to their ability to form very small droplets/particles as a result of their enhanced self-assembly capability.

The research of miktoarm polymers for drug delivery *via* micellar carriers, polymersomes or emulsions is still in its early stages; however, promising results and favourable characteristics have already been obtained compared to their well-established linear and non-linear counterparts. More research is needed to get a clearer picture, where each polymeric carrier can afford the best treatment.

## List of Abbreviations

**μ-EOC** μ-[Poly(ethylethylene)][poly(ethylene oxide)][poly(γ-methyl-ε-caprolactone)]
**μ-SODA** μ-[Polystyrene][poly(ethylene oxide)][poly(2-(dimethyl amino) ethyl acrylate)]
**ATP** Adenosine triphosphate
**ATRP** Atom transfer radical polymerisation
**CD** Cyclodextrin
**CMC** Critical micelle concentration
**CS** Chitosan
**CT** Computer tomography

| | |
|---|---|
| **DEA** | (2-Diethylamino)ethyl methacrylate |
| **DEAEMA** | (2-Diethylamino)ethyl methacrylate |
| **DMA** | (Dimethylamino) ethyl |
| **DMAEA** | (2-Diethylamino)ethyl acrylate |
| **DOX** | Doxorubicin |
| **DTT** | Dithiothreitol |
| **DVB** | Divinylbenzene |
| **EE** | Entrapment efficiency |
| **EOX** | Ethyloxazoline |
| **EPR** | Enhanced permeability and rentetion |
| **GMA** | Glycerol monomethacrylate |
| **GSH** | Gluthathione |
| **HEGMA** | Hexa(ethylene glycol) methacrylate |
| **HU** | Houndsfield units |
| **LCST** | Lower critical solution temperature |
| **LMA** | Lauryl methacrylate |
| **MAA** | Methacrylic acid |
| **MCL** | γ-Methyl-ε-caprolactone |
| **MMA** | Methylmethacrylate |
| **MOZ** | Methyloxazoline |
| **MPS** | Mononuclear phagocyte system |
| **MRI** | Magnetic resonance imaging |
| **MTX** | Methotrexate |
| **NIPAM** | *N*-Isopropylacrylamide |
| **O/W** | Oil-in-water |
| **O/W/O** | Oil-in-water-in-oil |
| **P2VP** | Poly-2-vinylpyridine |
| **PAA** | Poly(acryl acid) |
| **PBA** | Poly(*n*-butyl acrylate) |
| **PCL** | Poly(ε-caprolactone) |
| **PEE** | Poly(ethylethylene) |
| **PEG** | Poly(ethylene glycol) |
| **PEGMA** | PEG methyl acrylate |
| **PEO** | Poly(ethylene oxide) |
| **PLA** | Polylactide |
| **PLGA** | Poly(lactide-*co*-glycolide) |
| **PLL** | Poly(L-lysine) |
| **Por** | Porphyrin |
| **PPO** | Poly(phenyloxide) |
| **PS** | Polystyrene |
| **RAFT** | Reversible addition-fragmentation |
| **RES** | Reticuloendothelial system |
| **SBMA** | Sulfobetaine methacrylate |
| **TBS** | Tertiobutylstyrene |
| **UV** | Ultraviolet |
| **W/O** | Water-in-oil |
| **W/O/W** | Water-in-oil-in-water |

## References


1. J. Kreuter, *Int. J. Pharm.*, 2007, **331**, 1–10.
2. K. E. Uhrich, S. M. Cannizzaro, R. S. Langer and K. M. Shakesheff, *Chem. Rev.*, 1999, **99**, 3181–3198.
3. X. Zhang, L. Meng, Q. Lu, Z. Fei and P. J. Dyson, *Biomaterials*, 2009, **30**, 6041–6047.
4. R. Jeetah, A. Bhaw-Luximon and D. Jhurry, *J. Biomed. Nanotechnol.*, 2014, **10**, 1810–1840.
5. S. Yu, W. Shen, L. Yu, Y. Hou, J. Han and H. M. Richards, *J. Pain*, 2014, **15**, 835–844.
6. A. Katsnelson, *Nat. Biotechnol.*, 2014, **32**, 306.
7. D. Böhme and A. G. Beck-Sickinger, *J. Pept. Sci.*, 2015, **21**, 186–200.
8. H. Xu, H. Ma, P. Yang, X. Zhang, X. Wu, W. Yin, H. Wang and D. Xu, *Colloids Surf., B*, 2015, **136**, 729–734.
9. R. K. Oberoi, K. E. Parrish, T. T. Sio, R. K. Mittapalli, W. F. Elmquist and J. N. Sarkaria, *Neuro-Oncology.*, 2016, **18**, 27–36.
10. S. Ganta, H. Devalapally, A. Shahiwala and M. Amiji, *J. Controlled Release*, 2008, **126**, 187–204.
11. K. S. Soppimath, D. C. W. Tan and Y. Y. Yang, *Adv. Mater.*, 2005, **17**, 318–323.
12. J.-Z. Du, X.-J. Du, C.-Q. Mao and J. Wang, *J. Am. Chem. Soc.*, 2011, **133**, 17560–17563.
13. Y. Li, B. S. Lokitz and C. L. McCormick, *Angew. Chem., Int. Ed.*, 2006, **45**, 5792–5795.
14. Z. Ge and S. Liu, *Chem. Soc. Rev.*, 2013, **42**, 7289–7325.
15. Y. Matsumura and H. Maeda, *Cancer Res.*, 1986, **46**, 6387–6392.
16. J. Fang, H. Nakamura and H. Maeda, *Adv. Drug Delivery Rev.*, 2011, **63**, 136–151.
17. N. Hadjichristidis, M. Pitsikalis, S. Pispas and H. Iatrou, *Chem. Rev.*, 2001, **101**, 3747–3792.
18. I. Li, B. Howell, R. Koster and D. Priddy, *Macromolecules*, 1996, **29**, 8554–8555.
19. M. Messerschmidt, L. Häußler, B. Voit, T. Krause and W.-D. Habicher, *Macromol. Symp.*, 2004, **210**, 111–120.
20. H. S. Bazzi and H. F. Sleiman, *Macromolecules*, 2002, **35**, 9617–9620.
21. M. Sandholzer, S. Bichler, F. Stelzer and C. Slugovc, *J. Polym. Sci., Part A: Polym. Chem.*, 2008, **46**, 2402–2413.
22. J.-S. Wang and K. Matyjaszewski, *J. Am. Chem. Soc.*, 1995, **117**, 5614–5615.
23. J. Chiefari, Y. K. Chong, F. Ercole, J. Krstina, J. Jeffery, T. P. T. Le, R. T. A. Mayadunne, G. F. Meijs, C. L. Moad, G. Moad, E. Rizzardo and S. H. Thang, *Macromolecules*, 1998, **31**, 5559–5562.
24. D. J. Keddie, *Chem. Soc. Rev.*, 2014, **43**, 496–505.
25. X. Jiang, Y. Li, G. Lu and X. Huang, *Polym. Chem.*, 2013, **4**, 1402–1411.
26. M. J. Nasrullah, A. Vora and D. C. Webster, *Macromol. Chem. Phys.*, 2011, **212**, 539–549.

27. A. Rösler, G. W. M. Vandermeulen and H.-A. Klok, *Adv. Drug Delivery Rev.*, 2012, **64**, 270–279.
28. Z. Iatridi and C. Tsitsilianis, *Polymers*, 2011, **3**, 1911.
29. K. Khanna, S. Varshney and A. Kakkar, *Polym. Chem.*, 2010, **1**, 1171–1185.
30. T. He, D. Li, X. Sheng and B. Zhao, *Macromolecules*, 2004, **37**, 3128–3135.
31. J. Rao, Y. Zhang, J. Zhang and S. Liu, *Biomacromolecules*, 2008, **9**, 2586–2593.
32. N. Hadjichristidis, S. Pispas and G. Floudas, *Block Copolymers: Synthetic Strategies, Physical Properties, and Applications*, John Wiley & Sons, 2003.
33. M. Beck-Broichsitter, J. Nicolas and P. Couvreur, *Eur. J. Pharm. Biopharm.*, 2015, **97**(Part B), 304–317.
34. G. Birrenbach and P. P. Speiser, *J. Pharm. Sci.*, 1976, **65**, 1763–1766.
35. R. Gurny, N. A. Peppas, D. D. Harrington and G. S. Banker, *Drug Dev. Ind. Pharm.*, 1981, 7, 1–25.
36. J. M. Anderson and M. S. Shive, *Adv. Drug Delivery Rev.*, 1997, **28**, 5–24.
37. G. S. Kwon and T. Okano, *Adv. Drug Delivery Rev.*, 1996, **21**, 107–116.
38. G. S. Kwon and K. Kataoka, *Adv. Drug Delivery Rev.*, 1995, **16**, 295–309.
39. C.-L. Lo, C.-K. Huang, K.-M. Lin and G.-H. Hsiue, *Biomaterials*, 2007, **28**, 1225–1235.
40. G.-H. Hsiue, C.-H. Wang, C.-L. Lo, C.-H. Wang, J.-P. Li and J.-L. Yang, *Int. J. Pharm.*, 2006, **317**, 69–75.
41. W. Li, J. Li, J. Gao, B. Li, Y. Xia, Y. Meng, Y. Yu, H. Chen, J. Dai, H. Wang and Y. Guo, *Biomaterials*, 2011, **32**, 3832–3844.
42. G. Mountrichas, M. Mpiri and S. Pispas, *Macromolecules*, 2005, **38**, 940–947.
43. A. Wilkinson, *IUPAC. Compendium of Chemical Terminology*, Blackwell Scientific Publications, Oxford, 2 edn, 1997.
44. S. Pispas, N. Hadjichristidis, I. Potemkin and A. Khokhlov, *Macromolecules*, 2000, **33**, 1741–1746.
45. H. Li, D. Yang, Y. Gao, H. Li and J. Xu, *RSC Adv.*, 2015, **5**, 96377–96386.
46. Z. Li, E. Kesselman, Y. Talmon, M. A. Hillmyer and T. P. Lodge, *Science*, 2004, **306**, 98–101.
47. H. Yin, H. C. Kang, K. M. Huh and Y. H. Bae, *J. Mater. Chem.*, 2012, **22**, 19168–19178.
48. M. A. R. Meier, J.-F. Gohy, C.-A. Fustin and U. S. Schubert, *J. Am. Chem. Soc.*, 2004, **126**, 11517–11521.
49. L. Qiu and Y. Bae, *Pharm. Res.*, 2006, **23**, 1–30.
50. S. Liu, N. C. Billingham and S. P. Armes, *Angew. Chem., Int. Ed.*, 2001, **40**, 2328–2331.
51. Z. Ge, J. Xu, J. Hu, Y. Zhang and S. Liu, *Soft Matter*, 2009, **5**, 3932–3939.
52. S. Liu and S. P. Armes, *Angew. Chem., Int. Ed.*, 2002, **41**, 1413–1416.
53. J. Rao, Z. Luo, Z. Ge, H. Liu and S. Liu, *Biomacromolecules*, 2007, **8**, 3871–3878.
54. C. E. Callmann, C. V. Barback, M. P. Thompson, D. J. Hall, R. F. Mattrey and N. C. Gianneschi, *Adv. Mater.*, 2015, **27**, 4611–4615.

55. Z. Gan, T. F. Jim, M. Li, Z. Yuer, S. Wang and C. Wu, *Macromolecules*, 1999, **32**, 590–594.
56. K. V. Butsele, C. A. Fustin, J. F. Gohy, R. Jérôme and C. Jérôme, *Langmuir*, 2009, **25**, 107–111.
57. W. Lin, S. Nie, Q. Zhong, Y. Yang, C. Cai, J. Wang and L. Zhang, *J. Mater. Chem. B*, 2014, **2**, 4008–4020.
58. J. V. M. Weaver, S. P. Armes and S. Liu, *Macromolecules*, 2003, **36**, 9994–9998.
59. Y. Cai and S. P. Armes, *Macromolecules*, 2004, **37**, 7116–7122.
60. H. Liu, C. Li, H. Liu and S. Liu, *Langmuir*, 2009, **25**, 4724–4734.
61. N. Saito, C. Liu, T. P. Lodge and M. A. Hillmyer, *ACS Nano*, 2010, **4**, 1907–1912.
62. J. Huang, J. Xu, K. Chen, T. Wang, C. Cui, X. Wei, R. Zhang, L. Li and X. Guo, *Ind. Eng. Chem. Res.*, 2015, **54**, 1564–1575.
63. Y. Bae, S. Fukushima, A. Harada and K. Kataoka, *Angew. Chem., Int. Ed.*, 2003, **42**, 4640–4643.
64. M. Prabaharan, J. J. Grailer, S. Pilla, D. A. Steeber and S. Gong, *Biomaterials*, 2009, **30**, 5757–5766.
65. J.-F. Lutz, S. Geffroy, H. von Berlepsch, C. Bottcher, S. Garnier and A. Laschewsky, *Soft Matter*, 2007, **3**, 694–698.
66. W. Zhang, L. Shi, Z. J. Miao, K. Wu and Y. An, *Macromol. Chem. Phys.*, 2005, **206**, 2354–2361.
67. C. Liu, M. A. Hillmyer and T. P. Lodge, *Langmuir*, 2009, **25**, 13718–13725.
68. G. Gaucher, M.-H. Dufresne, V. P. Sant, N. Kang, D. Maysinger and J.-C. Leroux, *J. Controlled Release*, 2005, **109**, 169–188.
69. J. Lee, E. C. Cho and K. Cho, *J. Controlled Release*, 2004, **94**, 323–335.
70. Y. Du, W. Chen, M. Zheng, F. Meng and Z. Zhong, *Biomaterials*, 2012, **33**, 7291–7299.
71. N. Stavrouli, A. I. Triftaridou, C. S. Patrickios and C. Tsitsilianis, *Macromol. Rapid Commun.*, 2007, **28**, 560–566.
72. Y. Q. Yang, B. Zhao, Z. D. Li, W. J. Lin, C. Y. Zhang, X. D. Guo, J. F. Wang and L. J. Zhang, *Acta Biomater.*, 2013, **9**, 7679–7690.
73. A. M. Lowman, M. Morishita, M. Kajita, T. Nagai and N. A. Peppas, *J. Pharm. Sci.*, 1999, **88**, 933–937.
74. W. Wu, J. Liu, S. Cao, H. Tan, J. Li, F. Xu and X. Zhang, *Int. J. Pharm.*, 2011, **416**, 104–109.
75. N. R. Datta, S. G. Ordóñez, U. S. Gaipl, M. M. Paulides, H. Crezee, J. Gellermann, D. Marder, E. Puric and S. Bodis, *Cancer Treat. Rev.*, 2015, **41**, 742–753.
76. H. G. Schild, *Prog. Polym. Sci.*, 1992, **17**, 163–249.
77. L. Klouda, *Eur. J. Pharm. Biopharm.*, 2015, **97**(Part B), 338–349.
78. Y. Cai, Y. Tang and S. P. Armes, *Macromolecules*, 2004, **37**, 9728–9737.
79. J. V. M. Weaver, S. P. Armes and V. Butun, *Chem. Commun.*, 2002, 2122–2123, DOI: 10.1039/B207251N.
80. J. E. Chung, M. Yokoyama, M. Yamato, T. Aoyagi, Y. Sakurai and T. Okano, *J. Controlled Release*, 1999, **62**, 115–127.

81. L.-Y. Li, W.-D. He, J. Li, B.-Y. Zhang, T.-T. Pan, X.-L. Sun and Z.-L. Ding, *Biomacromolecules*, 2010, **11**, 1882–1890.
82. S. Qin, Y. Geng, D. E. Discher and S. Yang, *Drug Delivery*, 2006, **6**, 7.
83. M. Nakayama, T. Okano, T. Miyazaki, F. Kohori, K. Sakai and M. Yokoyama, *J. Controlled Release*, 2006, **115**, 46–56.
84. C. Cheng, H. Wei, B.-X. Shi, H. Cheng, C. Li, Z.-W. Gu, S.-X. Cheng, X.-Z. Zhang and R.-X. Zhuo, *Biomaterials*, 2008, **29**, 497–505.
85. E. Piskin, *Expert Rev. Med. Devices*, 2005, **2**, 501–509.
86. E. Pişskin, S. Dinçer and M. Türk, *J. Biomater. Sci., Polym. Ed.*, 2004, **15**, 1181–1202.
87. Y. Hu, V. Darcos, S. Monge and S. Li, *Int. J. Pharm.*, 2015, **491**, 152–161.
88. N. A. D. Burke, H. D. H. Stöver and F. P. Dawson, *Chem. Mater.*, 2002, **14**, 4752–4761.
89. L. E. Euliss, S. G. Grancharov, S. O'Brien, T. J. Deming, G. D. Stucky, C. B. Murray and G. A. Held, *Nano Lett.*, 2003, **3**, 1489–1493.
90. B.-S. Kim, J.-M. Qiu, J.-P. Wang and T. A. Taton, *Nano Lett.*, 2005, **5**, 1987–1991.
91. M. Talelli, C. J. F. Rijcken, T. Lammers, P. R. Seevinck, G. Storm, C. F. van Nostrum and W. E. Hennink, *Langmuir*, 2009, **25**, 2060–2067.
92. S. Louguet, B. Rousseau, R. Epherre, N. Guidolin, G. Goglio, S. Mornet, E. Duguet, S. Lecommandoux and C. Schatz, *Polym. Chem.*, 2012, **3**, 1408–1417.
93. G. Saito, J. A. Swanson and K.-D. Lee, *Adv. Drug Delivery Rev.*, 2003, **55**, 199–215.
94. F. Meng, W. E. Hennink and Z. Zhong, *Biomaterials*, 2009, **30**, 2180–2198.
95. R. Cheng, F. Feng, F. Meng, C. Deng, J. Feijen and Z. Zhong, *J. Controlled Release*, 2011, **152**, 2–12.
96. Y. Xu, F. Meng, R. Cheng and Z. Zhong, *Macromol. Biosci.*, 2009, **9**, 1254–1261.
97. H. Sun, B. Guo, R. Cheng, F. Meng, H. Liu and Z. Zhong, *Biomaterials*, 2009, **30**, 6358–6366.
98. S. Cerritelli, D. Velluto and J. A. Hubbell, *Biomacromolecules*, 2007, **8**, 1966–1972.
99. S. Takae, K. Miyata, M. Oba, T. Ishii, N. Nishiyama, K. Itaka, Y. Yamasaki, H. Koyama and K. Kataoka, *J. Am. Chem. Soc.*, 2008, **130**, 6001–6009.
100. Q.-H. Zhou, J. Lin, L.-D. Li and L. Shang, *Colloid Polym. Sci.*, 2015, **293**, 2291–2300.
101. C. Shi, X. Guo, Q. Qu, Z. Tang, Y. Wang and S. Zhou, *Biomaterials*, 2014, **35**, 8711–8722.
102. A. P. Bapat, J. G. Ray, D. A. Savin and B. S. Sumerlin, *Macromolecules*, 2013, **46**, 2188–2198.
103. Y. Liu, X. Cao, M. Luo, Z. Le and W. Xu, *J. Colloid Interface Sci.*, 2009, **329**, 244–252.
104. Z. Ge, Y. Cai, J. Yin, Z. Zhu, J. Rao and S. Liu, *Langmuir*, 2007, **23**, 1114–1122.

105. J. Rodríguez-Hernández and S. Lecommandoux, *J. Am. Chem. Soc.*, 2005, **127**, 2026–2027.
106. Y. Zhang, T. Wu and S. Liu, *Macromol. Chem. Phys.*, 2007, **208**, 2492–2501.
107. Y. Zhang, H. Liu, J. Hu, C. Li and S. Liu, *Macromol. Rapid Commun.*, 2009, **30**, 941–947.
108. H. Bao, L. Li, L. H. Gan, Y. Ping, J. Li and P. Ravi, *Macromolecules*, 2010, **43**, 5679–5687.
109. E. Blasco, B. V. K. J. Schmidt, C. Barner-Kowollik, M. Pinol and L. Oriol, *Polym. Chem.*, 2013, **4**, 4506–4514.
110. Z. Feng, L. Lin, Z. Yan and Y. Yu, *Macromol. Rapid Commun.*, 2010, **31**, 640–644.
111. F. D. Jochum and P. Theato, *Chem. Commun.*, 2010, **46**, 6717–6719.
112. O. Boissiere, D. Han, L. Tremblay and Y. Zhao, *Soft Matter*, 2011, **7**, 9410–9415.
113. K. Miao, H. Liu and Y. Zhao, *Polym. Chem.*, 2014, **5**, 3335–3345.
114. P. Schattling, F. D. Jochum and P. Theato, *Polym. Chem.*, 2014, **5**, 25–36.
115. S. Raynal, J. L. Grossiord, M. Seiller and D. Clausse, *J. Controlled Release*, 1993, **26**, 129–140.
116. H. Zhang and A. I. Cooper, *Soft Matter*, 2005, **1**, 107–113.
117. M. Gallarate, M. E. Carlotti, M. Trotta and S. Bovo, *Int. J. Pharm.*, 1999, **188**, 233–241.
118. D. J. McClements, *Food Emulsions: Principles, Practices, and Techniques*, CRC Press, 2nd edn, 2004.
119. C. Washington, *Adv. Drug Delivery Rev.*, 1996, **20**, 131–145.
120. R. Aveyard, B. P. Binks and J. H. Clint, *Adv. Colloid Interface Sci.*, 2003, **100–102**, 503–546.
121. B. P. Binks, *Curr. Opin. Colloid Interface Sci.*, 2002, **7**, 21–41.
122. U. Wais, A. W. Jackson, T. He and H. Zhang, *Nanoscale*, 2016, **8**, 1746–1769.
123. *Solvents and Self-Organization of Polymers*, ed. S. E. Webber, P. Munk and Z. Tuzar, Springer Science & Business Media, Netherlands, 2012.
124. W. Li, Y. Yu, M. Lamson, M. S. Silverstein, R. D. Tilton and K. Matyjaszewski, *Macromolecules*, 2012, **45**, 9419–9426.
125. S. Fujii, Y. Cai, J. V. M. Weaver and S. P. Armes, *J. Am. Chem. Soc.*, 2005, **127**, 7304–7305.
126. K. Shinoda and H. Saito, *J. Colloid Interface Sci.*, 1969, **30**, 258–263.
127. B. P. Binks, R. Murakami, S. P. Armes and S. Fujii, *Angew. Chem.*, 2005, **117**, 4873–4876.
128. G. Romet-Lemonne, J. Daillant, P. Guenoun, J. Yang, D. W. Holley and J. W. Mays, *J. Chem. Phys.*, 2005, **122**, 064703.
129. L. Besnard, F. Marchal, J. F. Paredes, J. Daillant, N. Pantoustier, P. Perrin and P. Guenoun, *Adv. Mater.*, 2013, **25**, 2844–2848.
130. R. F. Mattrey and D. A. Aguirre, *Acad. Radiol.*, 2003, **10**, 1450–1460.
131. P. Dawson, D. O. Cosgrove and R. G. Grainger, in *Textbook of Contrast Media*, ed. P. Dawson, D. O. Cosgrove and R. G. Grainger, CRC Press, 1999.

132. A. de Vries, E. Custers, J. Lub, S. van den Bosch, K. Nicolay and H. Grüll, *Biomaterials*, 2010, **31**, 6537–6544.
133. G. Kwon, M. Naito, M. Yokoyama, T. Okano, Y. Sakurai and K. Kataoka, *J. Controlled Release*, 1997, **48**, 195–201.
134. E. Lorenceau, A. S. Utada, D. R. Link, G. Cristobal, M. Joanicot and D. A. Weitz, *Langmuir*, 2005, **21**, 9183–9186.
135. H. Zhang, D. Wang, R. Butler, N. L. Campbell, J. Long, B. Tan, D. J. Duncalf, A. J. Foster, A. Hopkinson and D. Taylor, *Nat. Nanotechnol.*, 2008, **3**, 506–511.
136. N. Grant and H. Zhang, *J. Colloid Interface Sci.*, 2011, **356**, 573–578.
137. U. Wais, A. W. Jackson, Y. Zuo, Y. Xiang, T. He and H. Zhang, *J. Controlled Release*, 2016, **222**, 141–150.
138. Q. Qiu, G. Liu and Z. An, *Chem. Commun.*, 2011, **47**, 12685–12687.
139. R.-H. Jin, *J. Mater. Chem.*, 2004, **14**, 320–327.
140. S. Milner, *Science*, 1991, **251**, 905–914.
141. Y. Li, J. Zou, B. P. Das, M. Tsianou and C. Cheng, *Macromolecules*, 2012, **45**, 4623–4629.
142. F. Liu, S. Lin, Z. Zhang, J. Hu, G. Liu, Y. Tu, Y. Yang, H. Zou, Y. Mo and L. Miao, *Biomacromolecules*, 2014, **15**, 968–977.
143. A. M. Mathur, B. Drescher, A. B. Scranton and J. Klier, *Nature*, 1998, **392**, 367–370.
144. F. Liu, J. Hu, G. Liu, C. Hou, S. Lin, H. Zou, G. Zhang, J. Sun, H. Luo and Y. Tu, *Macromolecules*, 2013, **46**, 2646–2657.
145. G. Huynh Ba, K. Jérôme and P. Teyssié, *Colloid Polym. Sci.*, 1979, **257**, 1294–1296.
146. S. C. Hadjiyannakou, A. I. Triftaridou and C. S. Patrickios, *Polymer*, 2005, **46**, 2433–2442.
147. M. E. Matteucci, M. A. Hotze, K. P. Johnston and R. O. Williams, *Langmuir*, 2006, **22**, 8951–8959.
148. X. Shi, M. Miao and Z. An, *Polym. Chem.*, 2013, **4**, 1950–1959.
149. G. Maglio, F. Nicodemi, C. Conte, R. Palumbo, P. Tirino, E. Panza, A. Ianaro, F. Ungaro and F. Quaglia, *Biomacromolecules*, 2011, **12**, 4221–4229.

CHAPTER 6

# Synthetic Articulation of Miktoarm Polymers for Applications in Biology

ANJALI SHARMA[a,b], RISHI SHARMA[b], AHMED ABOUELMAGD[a] AND ASHOK KAKKAR*[a]

[a]Department of Chemistry, McGill University, 801 Sherbrooke St. West, Quebec, Montreal, H3A 0B8, Canada; [b]Center for Nanomedicine/Wilmer Eye Institute, Department of Ophthalmology, Johns Hopkins School of Medicine, Baltimore, Maryland 21287, USA
*E-mail: ashok.kakkar@mcgill.ca

## 6.1 Introduction

The design and synthesis of complex materials with tailored architectures and predetermined compositions has always been an intense area of research in polymer science, since it offers a platform to address challenges in fields ranging from nanomedicine to catalysis and electronics.[1–11] New, improved and well-defined polymeric materials are constantly being fabricated using modern synthetic tools and combination methodologies.[12–14] Miktoarm star polymers have recently garnered a great deal of attention owing to their unique physicochemical solution, bulk, and interphase properties.[13,15–21] These are composed of several polymeric chains tethered to a central core, and this architecture leads to much lower intrinsic viscosities and smaller

Polymer Chemistry Series No. 25
Miktoarm Star Polymers: From Basics of Branched Architecture to Synthesis, Self-assembly and Applications
Edited by Ashok Kakkar

Published by the Royal Society of Chemistry, www.rsc.org

radii of gyration, due to a highly compact structure compared to their linear analogs of similar molecular weight. Miktoarm stars are composed of a central core, to which various types of polymeric chains with different polymeric composition or molecular weight are appended (Figure 6.1). These hetero-arm or asymmetric polymers are becoming increasingly important for applications in biology due to their programmable structure and unique self-assembly behavior.[17,22–28] The precisely controlled structure of miktoarm star polymers provides opportunities to develop complex biomaterials by incorporating different arms with desirable features, for example imaging, targeting, and/or solubility, *etc.* The number of research activities toward improved and facile syntheses has steadily increased since the pioneering work done by Hadjichristidis *et al.*[29–41]

In asymmetric star polymers, polymeric arms of different chemical compositions and variations in hydrophobicity/hydrophilicity are tethered to a central core. Such amphiphilic miktoarm polymers self-assemble in aqueous solution to yield nano-ordered supramolecular micellar structures, with a core composed of hydrophobic polymeric segments and a corona made up of hydrophilic chains (Figure 6.2). While the outer corona serves to stabilize the overall supramolecular assembly and provides an opportunity to attach various functional ligands on the surface of the micelles, the internal core acts as a reservoir to hold hydrophobic guests, *e.g.*, drug molecules, with increased aqueous solubility and controlled release.

Amphiphilic miktoarm star co-polymers can also self-assemble to form polymeric vesicles known as polymersomes depending on the composition and length of the hydrophilic and hydrophobic segments.[31,42,43] Polymersomes have the combined advantages of both liposomes and self-assembled block co-polymer structures. Polymersomes are composed of a hydrophilic

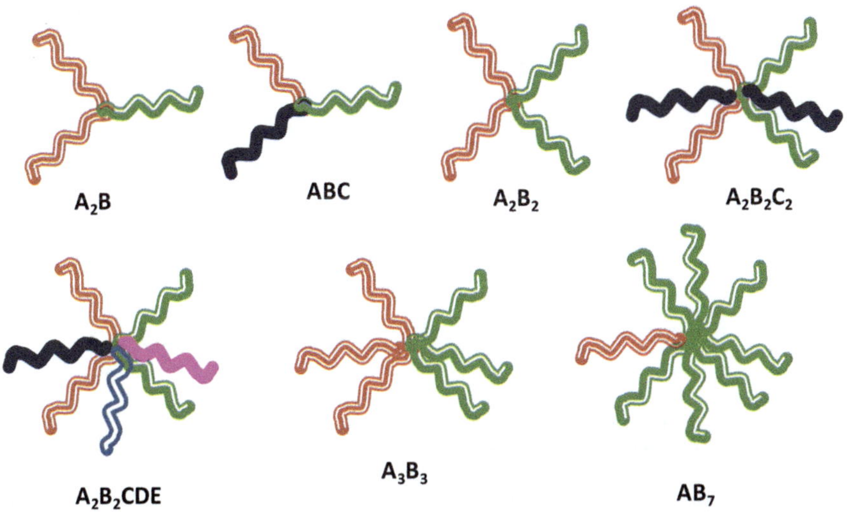


**Figure 6.1** Schematic representation of various miktoarm star polymers.

interior and an outer hydrophobic membrane made up of entangled polymeric chains adjusted in such a way that hydrophilic polymers surround hydrophobic segments. The bioactive molecules can be encapsulated into both the hydrophilic interior and the outer hydrophobic bi-layer membrane (Figure 6.3). Vesicles can provide stability to the drug molecules, increase their blood residence time, enhance their cellular uptake, and impart sustained and controlled release properties.[44–47] Thus, polymersomes serve as an ideal platform to develop synergistic drug delivery systems.

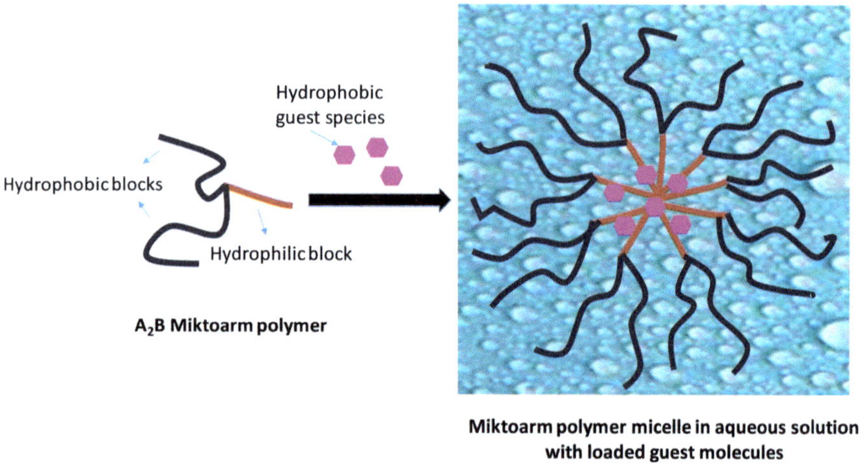


**Figure 6.2** Supramolecular self-assembly of $A_2B$ miktoarm star copolymers in aqueous solution.

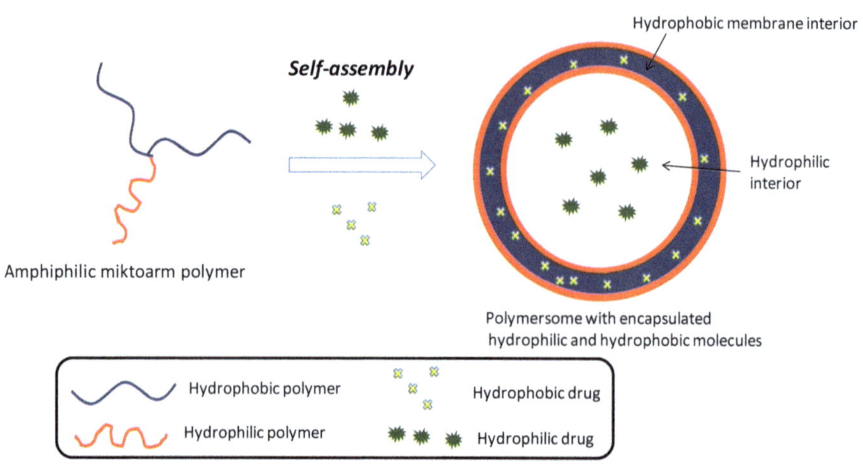


**Figure 6.3** Polymersome formation composed of amphiphilic $AB_2$ miktoarm stars and encapsulation of both hydrophilic and hydrophobic drug molecules.

## 6.2 Miktoarm Polymers *vs.* Traditional Block Co-Polymers and Hyperbranched Structures for Biological Applications

Miktoarm polymers offer dual advantages from both the branched shaped topology and block co-polymer structure, which makes them potential candidates to develop novel supramolecular assemblies for drug delivery applications.[48–51] The overall architecture provides attractive and unique physical/chemical and supramolecular properties to miktoarm stars that are not evident in their linear counterparts.[20,21,52–54] Miktoarm polymers involve less arm entanglements as compared to the linear analogs of the same molecular weight, leading to much less viscous solutions. Although miktoarm stars are branched polymers, they differ uniquely in their properties as compared to hyperbranched structures. The hyperbranched polymers[55] or dendrimers[56–58] are made up of a highly branched repeating structural pattern, creating a dense outer shell with an sterically hindered surface. On the other hand, miktoarm stars are composed of linear polymers and, due to the absence of internal branching, the overall density around the core decreases from the core toward the surface, leading to a less hindered outer periphery. Moreover, amphiphilic miktoarm stars made up of polymers of different chemical compositions can exhibit molecular level phase separation and self-assemble into supramolecular structures in the nanometer size range. In addition, due to the flexibility and freedom in the design of mioktoarm stars, these polymers can be synthesized in a tailor-made fashion to fit the needs of biological applications. The presence of multiple arms in miktoarm stars leads to a versatile platform to physically incorporate or covalently link several different combinations of therapeutic molecules, fluorescent dyes, or targeting moieties. It is even possible to design intrinsically fluorescent traceable miktoarm stars by using a fluorescent dye as the core of the polymeric micelles.[50] Miktoarm polymers can be specifically designed to develop stimulus-responsive supramolecular structures, which can fall apart at the targeted site to release their cargo.[28,59–63] It has been well documented that aggregated supramolecular morphologies produced by miktoarm polymers in solution provide higher drug loading, narrower size distribution, and much better delivery efficiencies than similar assemblies of linear block co-polymers.[64,65]

## 6.3 Challenges Toward the Synthesis of Miktoarm Polymers

The design and development of novel synthetic methodologies for the precise construction of well-defined mixed-arm star polymers continue to attract increasing interest due to the considerable challenges associated with their synthesis. The synthesis of miktoarm polymers is much more demanding compared to that of homo-arm star polymers or block co-polymers due to the requirement of precise incorporation of multiple polymeric arms. It involves synthetic routes with multiple steps, and requires isolation and characterization after introduction of each polymeric chain. Several strategies have been developed to successfully incorporate multiple polymer segments in a single star shaped macromolecular architecture, and the introduction of click chemistry in the synthesis of macromolecules has also helped advance the field.[14,41] A summary of some of the most commonly used synthetic methods to construct miktoarm polymers is discussed below.

## 6.3.1 Methods Based on Living Anionic Polymerization

### *6.3.1.1 Silyl Chloride Based Method*[66–69]

The silyl chloride method involves sequential deactivation of different living anionic polymeric chains on multivalent silyl chloride compounds. This method for the synthesis of miktoarm stars is one of the earliest established methods and was developed by Hadjichristidis and coworkers.[66,70] A wide variety of complex star polymers ($A_nB$,[71–73] $A_nB_n$,[68,74] ABC,[70] and ABCD[68]) were synthesized using this method, which involves the use of living anionic polymerization around a chlorosilane compound serving as the core of the star polymer. The different reactivity of the various living anionic polymers for each chlorosilyl group in multifunctional chlorosilane compound holds the key for the success of this method. Hadjichristidis and coworkers developed an ABC miktoarm-ter-polymer containing polystyrene (PS), polyisoprene (PI), and poly(1,3-butadiene) (PBd) (Scheme 6.1).[70]

The order of attachment of living anionic polymer arms is highly crucial for the success of this methodology and depends on the reactivity of each living polymer chain. Since poly(styryllithium) (PSLi) is more sterically hindered, it will have less reactivity toward $MeSiCl_3$ than poly(isoprenyllithium) (PILi), therefore the order follows the most sterically hindered living polymer chain (PSLi) to react first, and the least sterically hindered chain (PBdLi) last.

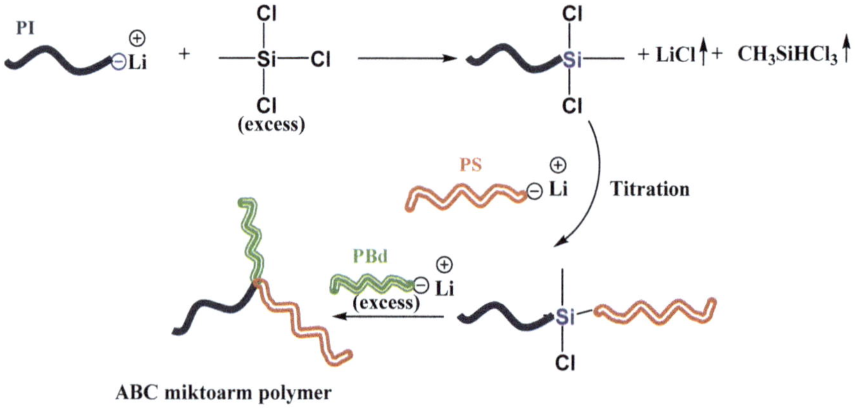


**Scheme 6.1** Synthesis of an ABC miktoarm polymer using the chlorosilane method.

The synthesis is initiated by reacting living PI chains with large excess of methyltricholorosilane to develop end-capped PI functionalized dichlorosilanes. The excess of silyl compound is then removed by evaporation, followed by reaction of the most sterically hindered living PS chain in a titration manner. The end point of titration is closely monitored to ensure the complete formation of the intermediate product (PS)(PI)Si($CH_3$)Cl. Finally, the ABC miktoarm polymer is obtained by reacting the product with a small excess of living PBd chains. It is important to add the chlorosilane compound in excess to ensure the formation of the mono-substituted polymeric product. Using a similar technique, an ABCD miktoarm-quarter-polymer containing four different polymeric segments, PS, PI, PBd, and poly(4-methylstyrene) (P4MeS), was successfully synthesized by Hadjichristidis and coworkers.[68] Once again, the order of the different polymeric chains controlled the reaction sequence. The branches were incorporated in a step by step manner involving two titration steps, with PS chosen to be reacted first and PBdLi last. Several other miktoarm polymers have been synthesized using the chlorosilane method, including, $A_2B$, $A_3B$, $A_nB$ ($n > 3$) and $A_nB_n$ ($n > 2$).[68,71,72,74,75]

### 6.3.1.2 *Divinylbenzene (DVB) Based Method*

The DVB method was developed by Funke and coworkers,[76,77] and it is specifically employed to construct $A_nB_n$ type miktoarm star polymers.[78] This method involves three steps: (i) the synthesis of living chains of polymer A, (ii) initiation of polymerization of a small amount of DVB by the living chains, giving rise to a living star polymer containing multiple active sites at its core, and (iii) finally, the polymerization of B on these active sites resulting in the formation of $A_nB_n$ miktoarm stars. The DVB method presents several problems, including improper control of the structural parameters, inability to characterize independently the polymer B arms, *etc.*, and thus is not very successful for the synthesis of well-defined miktoarm stars.

### 6.3.1.3 *Diphenylethylene (DPE) Derivative Based Method*

The DPE method starts with the preparation of a DPE-end capped linear polymer chain, which is then reacted with another living polymer arm to obtain a diblock copolymer with an anionic focal point. This macro-anion is further used to polymerize another monomer to obtain the desired miktoarm star polymer. Several groups have reported the use of the DPE method to develop a wide variety of miktoarm stars ($A_2B_2$,[79,80] ABC,[81–83] and ABCD[84,85]). Fujimoto and coworkers[86] reported the synthesis of PS, poly(dimethylsiloxane) (PDMS), and poly(*tert*-butyl methacrylate) (P*t*BuMA) tri-arm miktoarm stars using the DPE method (Scheme 6.2). The synthesis starts with the incorporation of a DPE chain-end to PDMS. End-reactive PDMS is obtained by anionic polymerization of hexamethylcyclotrisiloxane using the lithium salt of *p*-(dimethylhydroxy)silyl-α-phenylstyrene as the initiator. The subsequent addition reaction of living PS chains with chain-end DPE-functionalized PDMS leads

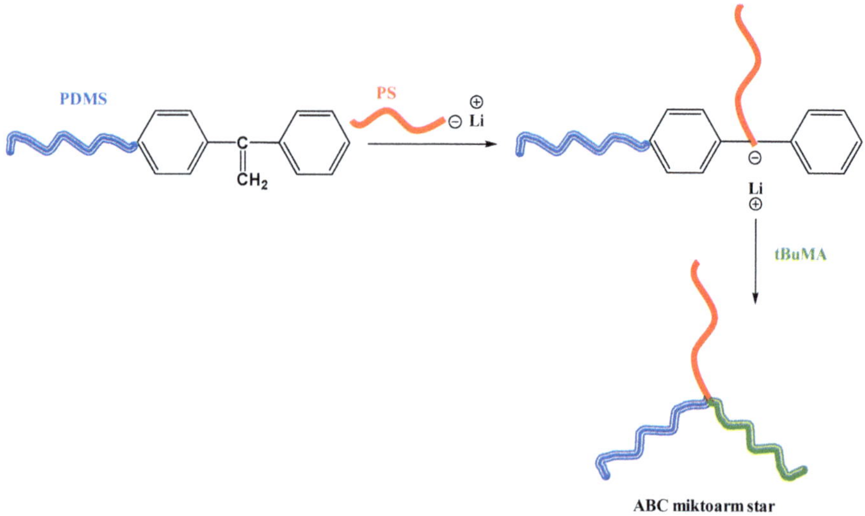


**Scheme 6.2** Synthesis of an ABC miktoarm star using the DPE method.

to the formation of the living coupled product containing active sites, which in turn is used for the polymerization of *t*BuMA, resulting in the formation of the final star polymer.

### 6.3.1.4 *Iterative Based Method*

The methodologies described in Sections 3.1 and 3.2 can be used to develop a variety of miktoarm polymers with up to 4 or 5 arms. None of these methods are well-established to design and synthesize complex miktoarm stars with more than five polymer chains. Hirao and coworkers have developed a versatile concept called the 'iterative approach' to synthesize multi-armed star copolymers.[41] This strategy allows the regeneration of the reaction site after the introduction of each polymeric arm during the reaction sequence, thus giving the freedom to incorporate a limitless number of polymer chains in a single star just by repeating the reaction sequence itself. The iterative methodology basically involves two steps: (i) the reaction of a living anionic polymer segment with another polymer chain functionalized with a DPE derivative, and (ii) reintroduction of another DPE function to the growing star polymer (Scheme 6.3).[41] A large number of same or different polymer segments can be added until the steric hindrance impedes a new reaction sequence. It was also demonstrated that the newly synthesized chain-end- and in-chain-functionalized polymers containing benzyl halide (BnX) moieties react readily and quantitatively with living anionic polymers, resulting in a library of miktoarm stars. Moreover, it was found that the undesirable side reactions between a living anionic polymer and a BnX reaction site can be avoided if the living polymers are end-capped with DPE prior to the linking

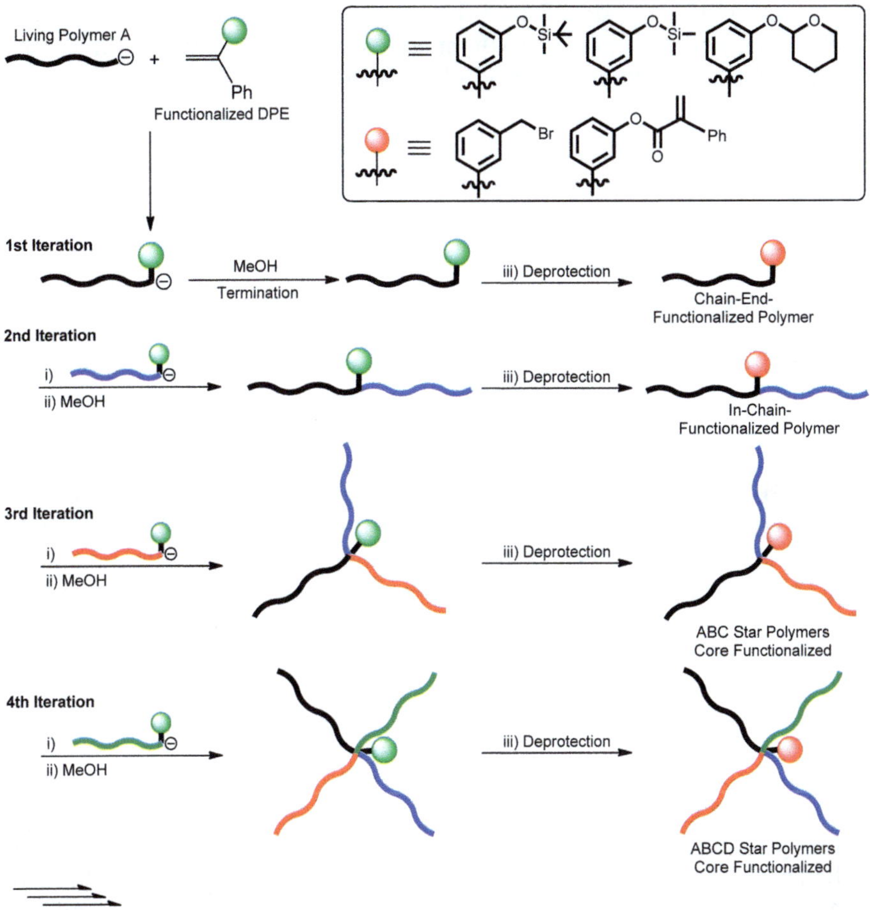


**Scheme 6.3** Iterative methodology for the synthesis of ABC, ABCD, and ABCDE miktoarm star polymers.

reaction, and by carrying out the reactions in THF at temperatures below −40 °C. As a result, a variety of highly complex and synthetically challenging multicomponent star polymers have been successfully synthesized using this methodology. The approach is based on living anionic polymerization and requires a DPE function as the linking agent. Tri-arm ABC, tetra-arm ABCD, and penta-arm ABCDE miktoarm polymers were synthesized using this 'iterative' methodology, which was further extended to develop much more complex stars including $A_2B_2$, $A_3B_3$, $A_2B_2C_2$, and up to 9-arm $A_3B_3C_3$ miktoarm stars using tri-DPE functioning as the linking core agent.[22,25,87] Using this methodology, Hirao and coworkers have successfully synthesized well-defined 3-arm $AB_2$, 7-arm $AB_2C_4$, 15-arm $AB_2C_4D_8$, and even 31-arm $AB_2C_4D_8E_{16}$ miktoarm stars, where A, B, C, D, and E correspond to polystyrene, poly(α-methylstyrene), poly(4-methylstyrene), poly(4-methoxystyrene), and poly(4-trimethylsilylstyrene) chains, respectively.[88–91] The precision and homogeneity in the structure and composition of the miktoarm stars was confirmed by $^1H$ nuclear magnetic resonance (NMR) spectroscopy, size-exclusion chromatography (SEC), and right angle laser light scatterings (RALLS) analyses.

#### 6.3.1.5 Other Methodologies

Hillmyer and others[92–95] have demonstrated the successful synthesis of a series of 3-arm ABC miktoarm-star polymers in a methodology based also on anionic polymerization involving two successive living anionic polymerization steps, and a polymer–polymer coupling reaction.[92] As illustrated in Scheme 6.4, a dual functional ethylene oxide derivative is used. In brief, the active poly(butadienyllithium) (PBLi) is prepared and end-capped with 2-methoxymethoxyoxirane yielding a di-functional poly(1,2-butadiene) PB with a hydroxyl group and a protected hydroxyl functionality at the chain end. Upon catalytic hydrogenation of PB, the corresponding poly(ethylethylene) (PEE) with a terminal hydroxyl group is obtained. It is then transformed into a potassium alkoxide, followed by the anionic ring-opening polymerization of ethylene oxide (EO), which is subsequently end-capped with ethyl bromide, affording a PEE-block-PEO with a protected hydroxyl group. After deprotection of the methoxymethyl group, followed by coupling with acyl chloride end-capped poly(perfluoropropylene oxide) (PFPO), the targeted ABC star-branched polymer is obtained. The same intermediate in-chain-OH-functionalized PEE-block-PEO serves as a macroinitiator in the anionic ring-opening polymerization of γ-methyl-ε-caprolactone (γMCL) to give an ABC star polymer of PEE-PEO and PγMCL.[95]

3-Arm ABC and 4-arm ABCD star polymers were also prepared using a di-benzyl chloride DPE derivative.[96] As shown in Scheme 6.5, a large excess of di-benzyl chloride DPE derivative is reacted with DPE-end-capped PILi anion affording a chain-end-functionalized PI with both benzyl chloride and DPE moieties. After isolation, the latter reacts selectively with DPE-end-capped poly(4-methoxystyryl)lithium (PMOSLi) at the benzyl chloride functionality, followed by reaction with PSLi with the residual DPE functionality, resulting in a 3-arm ABC star-branched polymer. The 3-arm ABC star polymer with an anion at the core can then undergo a quantitative linking reaction with chain-end-butyl bromide-functionalized poly(4-trimethylsilylstyrene) (PSiS), affording the desired 4-arm ABCD star polymer.

Wang *et al.* utilized 1,4-bis(1-phenylethenyl)benzene as a di-functional DPE derivative and successfully synthesized a 4-arm ABCD star polymer.[85] A macroinitiator was obtained through a 1 : 1 addition reaction between PSLi and the di-functional DPE derivative. Subsequently, the new microinitiator was able to initiate *t*BMA yielding an in-chain-DPE functionalized AB diblock copolymer, PS-*b*-P*t*BMA. In a similar manner, PαMSLi was inserted, and later on, 2VP was anionically polymerized affording the targeted well-controlled 4-arm ABCD stars, as illustrated in Scheme 6.6.

Hadjichristidis group developed a silyl chloride DPE derivative to be used as a dual-functionality compound, and reported the synthesis of a series of 4-arm $A_2B_2$ and $A_2BC$, and 5-arm $A_2B_2C$ stars, where A, B, C are PDMS, PS, and

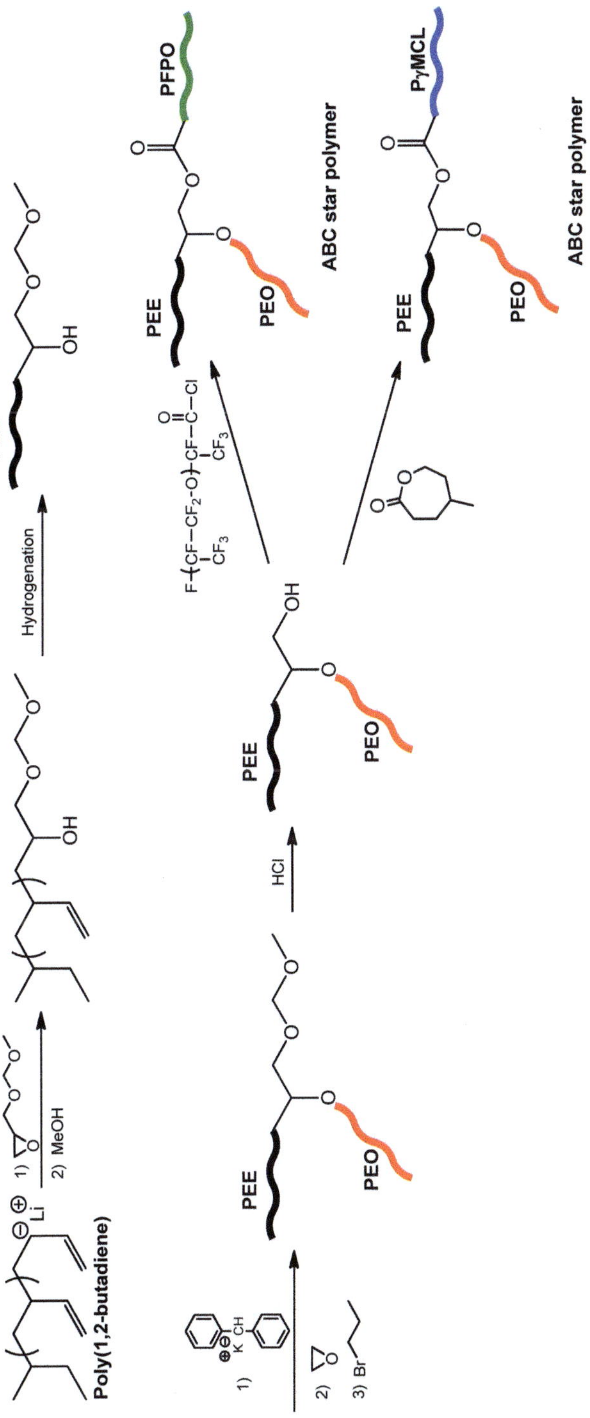


**Scheme 6.4** Synthesis of an ABC star polymer using a di-functional ethylene oxide derivative.

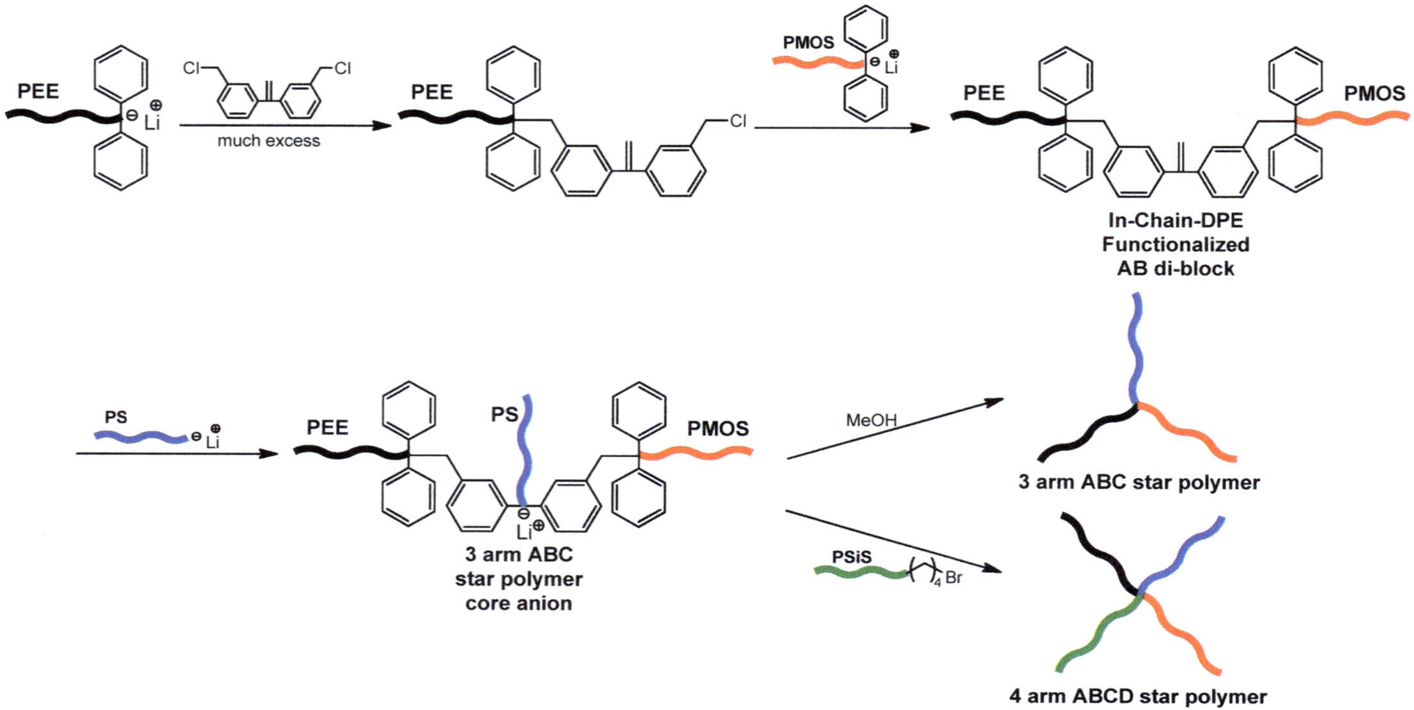


**Scheme 6.5** Synthesis of 3- and 4-arm well-defined star polymers.

PI, respectively.[66,73] Scheme 6.7 illustrates the general approach, where two titration steps are needed in order to replace the active silyl chloride groups with the corresponding living anion polymers PI and PDMS. The in-chain-DPE functional AB diblock thus prepared can be used to end-cap another living polystyryl segment, and finally the macroinitiator is used to polymerize 2-vinylpyridine (VP). All the analyses showed good agreement with the calculated values and the 4-arm star polymer showed mono-modal Gel Permeation Chromatography (GPC) analysis on separation.

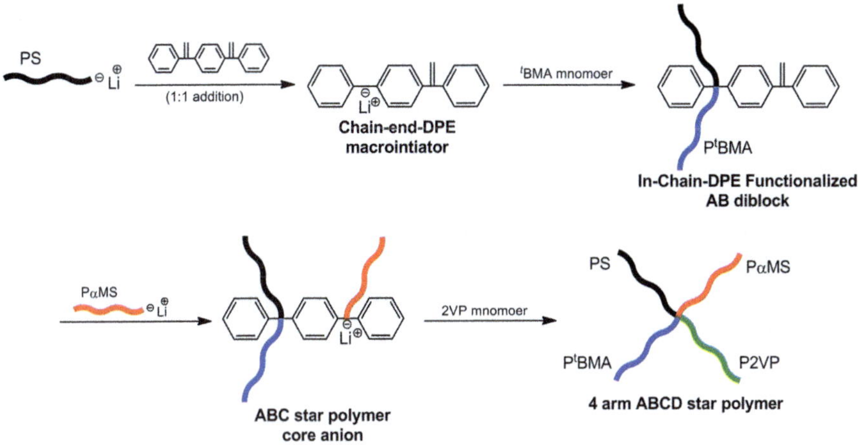


**Scheme 6.6** Utilization of a double DPE-functional derivative to synthesize a 4-arm ABCD star polymer.

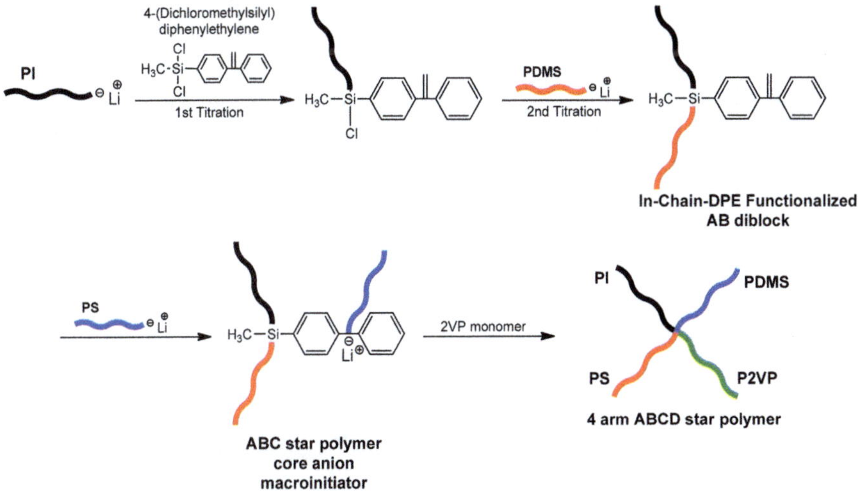


**Scheme 6.7** Synthesis of a 4-arm ABCD star polymer using a dichlorosilyl-DPE derivative.

## 6.3.2 Methods Based on Living/Controlled Radical Polymerization

### 6.3.2.1 Arm First Method[97–99]

Matyjaszewski and coworkers were the first to use the simple and general 'arm first' method for the synthesis of complex miktoarm polymers using controlled radical polymerization (CRP) techniques.[100] The 'arm-first' method involves the synthesis of individual polymeric chains first, followed by cross-linking of their ends to generate the core of the star. The linear polymeric chain macroinitiators are generated using the CRP technique and a cross linking agent with two vinyl groups, for example divinyl benzene. Compared to living anionic polymerization, which involves stringent reaction conditions, CRP techniques have been increasingly adopted by polymer chemists due to the benign experimental conditions and easy availability of the variety of monomers that can be used.[101,102] By using the 'arm first' technique in a simple one-pot synthetic reaction, heteroarm polymers with a large number of polymeric segments can be obtained (Scheme 6.8). Matyjaszewski *et al.* have reported the synthesis of a series of miktoarm stars containing two or more arm species and a number of arms ranging up to 84.[100] The polymers were synthesized in high yields with different arm numbers and ratios *via* atom transfer radical polymerization (ATRP) using the arm first technique. The species and molar ratios of the arms of miktoarm polymers were directly dependent on the species and molar ratios of the macroinitiators. The resulting miktoarm stars were well characterized using GPC, NMR spectroscopy, and liquid adsorption chromatography, demonstrating the presence of miktoarm stars rather than a mixture of various homoarm star polymers.

Although the classical 'arm first' method using a mixture of linear initiators can conveniently yield high molecular weight miktoarm stars with a large number of polymer arms, these miktoarm stars usually suffer from high polydispersities due to various side reactions such as star–star coupling reactions. The polydispersities could be reduced by decreasing the concentration of divinyl linking agent, but this in turn can lower the number of

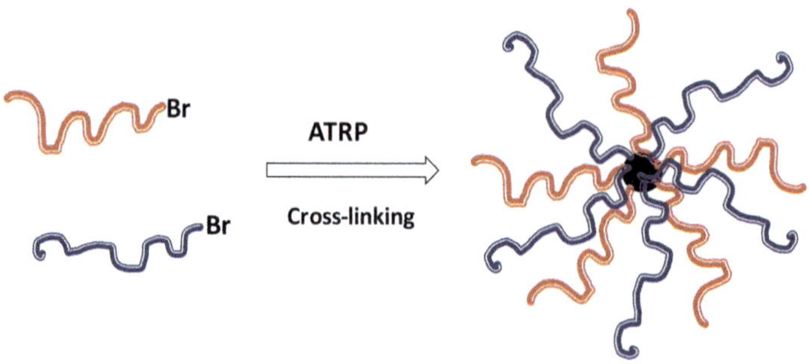


**Scheme 6.8** 'Arm first' method for the synthesis of miktoarm star polymers.

arms and overall weight of the miktoarm star. The synthesis of miktoarm stars with low polydispersities and high molecular weights can be cleverly designed by decreasing the number of initiating sites in the star while keeping the number of arms unchanged. This is highly challenging with the classical 'arm first' technique, but can be obtained by adopting a modified 'arm first' method, which involves the use of linear macromonomers (LM) in the synthesis of miktoarm stars, which increases the number of arms per star while maintaining the initiating sites unchanged. The use of low ratios of initiator-to-LM decreases the number of initiating sites, thus reducing the polydispersity resulting from side reactions.[99]

### 6.3.2.2 Core First Method[103,104]

The 'core first' method, as the name suggests, involves the use of a core containing multiple reactive sites from which grow the polymeric chains. The core first method presents better control of the growth of the polymer to introduce a predetermined number of arms, but has restrictions on the number of arms compared to the arm first technique. Huang and coworkers synthesized miktoarm stars *via* the core first method using ATRP and radical-addition fragmentation chain transfer (RAFT) polymerization techniques (Scheme 6.9).[105] The polymers were synthesized in three steps: (1) synthesis of a hyperbranched polyglycerol core, (2) addition of the first monomer (styrene, St) *via* ATRP, and (3) RAFT polymerization of the second monomer (*tert*-butyl acrylate, *t*BA) from the core, followed by hydrolysis to obtain poly(acrylic acid). The arm number and arm length have a significant effect on the structure of the miktoarm stars. The initiation efficiency of styrene for ATRP is nearly 100%, but the initiation efficiency of *tert*-butyl acrylate for RAFT can go up to 93%, when the number of polystyrene arms is less than 66 and the molecular weight is less than 1.8 kD.

Although the core first technique can allow the precise synthesis of miktoarms with full control of the number of arms, it requires multifunctional initiators with orthogonal functions, which requires additional steps for the synthesis.

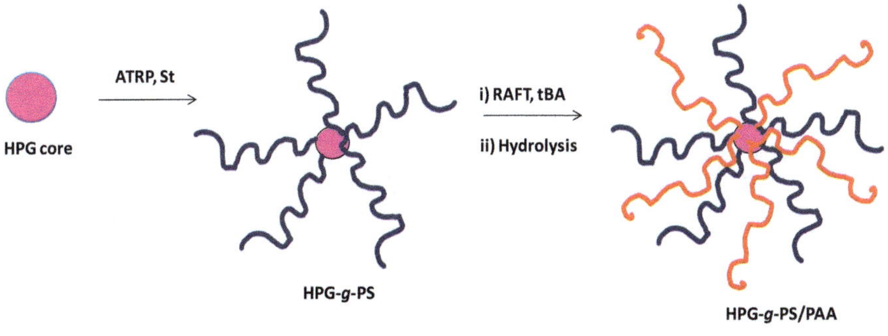


**Scheme 6.9** Schematic representation of core first method for the synthesis of miktoarm stars.

### 6.3.2.3 Click Chemistry[36,106–110]

The potential of miktoarm star polymers in designing nanostructures specific for biological applications has constantly prompted researchers to develop facile and efficient strategies for their synthesis. Significant advances in controlled radical polymerization techniques (ATRP, RAFT) with less stringent conditions have simplified the synthesis of these star polymers.[111–114] However, the advent of the 'click chemistry' concept completely changed the scenario and the way these polymers are synthesized. Click chemistry was introduced by Sharpless *et al.* in 2001, who denoted reactions that fulfill certain requirements such as high yield, mild reaction conditions, simple purification, minimum side products, and tolerance to a series of solvents and functional groups as 'click reactions'.[115] The click chemistry concept has already been adapted for the synthesis of small molecules, such as drugs, to macromolecules, including dendrimers, for applications ranging from therapeutics to nanoengineering. In addition to the requirements mentioned by Sharpless for a click reaction, Barner-Kowollik *et al.* have specified some additional traits for the reactions in polymer synthesis to fit into the click category, such as the use of polymeric building blocks in equimolar ratios, simple large-scale purification, and acceptable reaction times.[116]

The last couple of years have witnessed a plethora of reports on the use of click chemistry to synthesize a variety of miktoarm stars. Tunca *et al.* have recently published an excellent review article on the use of modular ligation strategies to construct star polymers.[38] Most commonly used click reactions are the Cu(I) catalyzed azide–alkyne addition (CuAAC), Diels–Alder, thiol-ene and thiol-yne coupling.[117] Kakkar and coworkers reported the synthesis of ABC miktoarm stars using sequential CuAAC and ring opening polymerization (ROP) reactions (Scheme 6.10).[118] A well-defined versatile core with three orthogonal functions is synthesized to carry out the controlled addition of arms *via* click and ROP reactions in sequence. A series of miktoarm stars with polyethylene glycol (PEG), polycarprolacone (PCL), and PS were successfully synthesized and well characterized using a variety of techniques. The miktoarm stars were found to self-assemble to form micelles, and the size

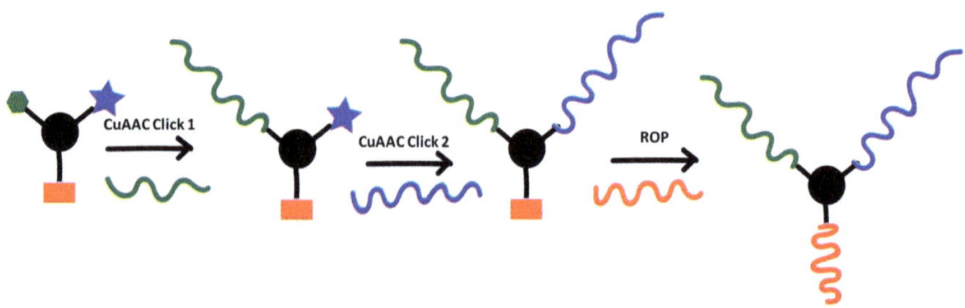


**Scheme 6.10** Synthesis of ABC miktoarm polymers by sequential CuAAC and ROP reactions.

of the micelles was inversely proportional to the length of the hydrophilic polymeric segments.

In another study, ABCD polymers with A = PCL, B = poly(*t*-butyl acrylate) (P*t*BA), C = PS, and D = poly(methyl methacrylate) (PMMA) were synthesized by Tunca *et al. via* a Diels–Alder click reaction using core-in and core-out processes.[119] PCL with an anthracene-end function was synthesized by ROP and, separately, P*t*BA with furan protected maleimide as the end-function was constructed *via* ATRP. A Diels–Alder reaction was employed to connect these homopolymers to quantitatively produce a PCL-*b*-P*t*BA block co-polymer, which was subsequently utilized as the macroinitiator to polymerize styrene and the free radical photopolymerization of MMA, to obtain an ABCD quaterpolymer.

Tunca *et al.* have also demonstrated the use of triple click reactions (CuAAC, Diels–Alder, and nitroxide radical coupling (NRC)) for the synthesis of ABC miktoarm star polymers.[120] The purpose of using click reactions was to obtain miktoarm stars with precise control of their molecular weight, polydispersity, topology, and composition. An orthogonal core with three different end functions (anthracene, alkyne and bromide) was used to selectively perform three different click reactions. Linear polymer precursors furan-protected maleimide-terminated poly(ethylene glycol) (PEG-MI), tetramethylpiperidine-1-oxyl-terminated poly(ε-caprolactone) (PCL-TEMPO), and azide terminated polystyrene (PS-$N_3$) or -poly(*N*-butyl oxanorbornene imide) (PONB-$N_3$) were separately prepared. The end-functionalized linear polymers were ligated onto a multifunctional core through sequential Diels–Alder, CuAAC, and NRC reactions in a two-pot reaction to give PEG-PCL-PS and PEG-PCLPONB miktoarm star polymers with narrow polydispersities (Scheme 6.11).

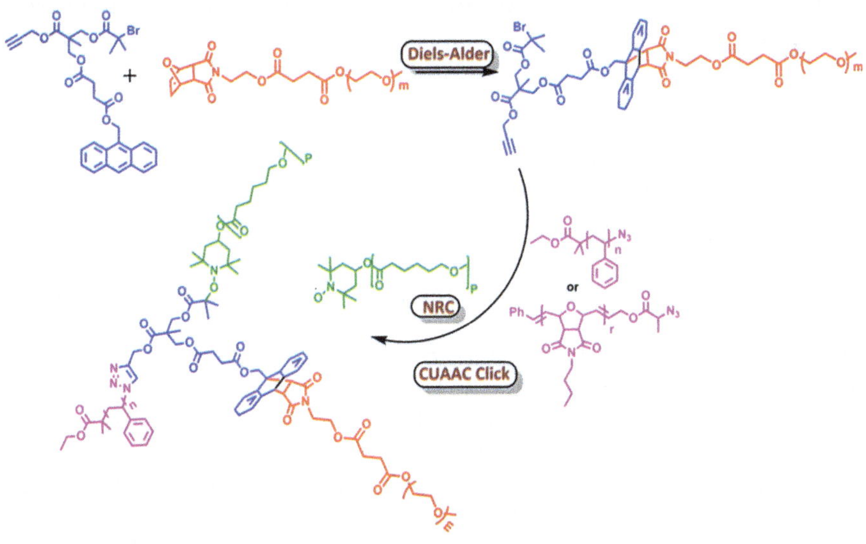


**Scheme 6.11** Synthesis of ABC miktoarm star polymers using triple click reactions (Diels–Alder, CuAAC, and NRC).

## 6.4 Miktoarm Star Polymers for Drug Delivery Applications

Tremendous efforts have been devoted to design nanocarriers to efficiently deliver drugs at targeted sites and simultaneously trace their path.[121,122] Miktoarm stars have attracted considerable attention for drug delivery applications due to their exceptionally unique solution and bulk properties, such as prolonged drug circulation, nanometer size range, and core–shell architecture. Miktoarm polymers for drug delivery applications are often composed of Food and Drug Administration (FDA)-approved biodegradable and biocompatible polymers. The drugs can be conveniently loaded into the interior of micelles of miktoarm stars or the termini of multi-arm mixed stars can be chemically conjugated to therapeutic molecules or other bioactive ligands.

### 6.4.1 Physical Encapsulation of Drugs into Miktoarm Polymer-Based Self-Assembled Nanostructures

Reports from various publications have suggested that the self-assembled nanostructures generated from amphiphilic star polymers have several remarkable properties, for example, exceptionally high drug loading, sustained drug release, introduction of multiple biological functions, *etc.*[16] In order to self-assemble into nanostructures, the miktoarm stars need to be composed of suitably balanced hydrophilic and hydrophobic polymeric segments. The most commonly reported linear polymers utilized as the arms of miktoarm star polymers for biological applications include PEG, PCL, and poly(L-lactic acid) (PLLA). Hydrophobic polymers such as PCL form the core of micelles that can incorporate hydrophobic drug molecules, increasing thus their aqueous solubility, and provide a biocompatible controlled drug release matrix. PEG is an FDA approved biocompatible polymer that is hydrophilic and forms the corona of micelles in aqueous solution. It provides stability to the micelles and also prolongs their systemic residence time by preventing the adsorption of proteins.

Amphiphilic miktoarms stars, depending on the constituents of their arms, the length of the polymer chains, and the type of polymers, can generate a variety of self-assembled structures in solution, such as polymersomes, micelles, and multiple-compartment micelles. For example, Bae and coworkers designed the formation of polymersomes through self-assembly of $AB_2$ miktoarm (mPEG-*b*-(polyHis)$_2$) stars to mimic naturally existing phospholipids for drug delivery applications.[42] The tri-arm star polymers were synthesized using a combination of 'core first' click chemistry and a ring-opening polymerization reaction to successfully synthesize mPEG2kDa-*b*-(polyHis29kDa)$_2$ with narrow polydispersity. Upon self-assembly in aqueous solution at pH 9, the resulting polymers generated nanometer-size range (70.2 nm), low cytotoxic, and pH-sensitive polymersomes. The polymersomes showed a pH dependent transition to cylindrical micelles, spherical micelles, and finally unimers below pH 7.4. Moreover, the polymersomes showed the sustained release of an encapsulated model hydrophilic dye at pH 7.4, which was considerably accelerated at endosomal (slightly acidic) pH, proving the potential of these polymersomes as drug carriers for cancer therapy.

Miktoarm polymers have immense potential to overcome several issues related to linear block-co-polymer analogues for drug delivery applications, such as burst drug release *in vivo*, low drug loading, and insufficient stability. Kakkar *et al.* conducted several studies on miktoarm polymers by exploring their potential for biological applications. For example, they developed a versatile and facile methodology to construct $A_2B$ (A = PEG; B = PCL) miktoarm polymer-based nanocarriers for the delivery of the highly hydrophobic drug nimodipine with poor aqueous solubility.[64] The polymers were synthesized using sequential CuAAC click cycloaddition and ring opening polymerization. The resulting polymers were able to self-assemble into spherical micelles and could encapsulate up to 78% nimodipine. The encapsulation of the drug into micelles remarkably improved its aqueous solubility, which was 200-fold that of the free drug. Furthermore, the encapsulated drug demonstrated a sustained release pattern *in vitro*, making these polymers excellent candidates to deliver hydrophobic drug molecules. Recently, they further explored these PEG and PCL based miktoarm polymers by introducing a third PEG chain in order to evaluate their self-assembly behaviour as well as their capability to deliver another lipophilic drug, curcumin (Figure 6.4).[123] The polymers were synthesized

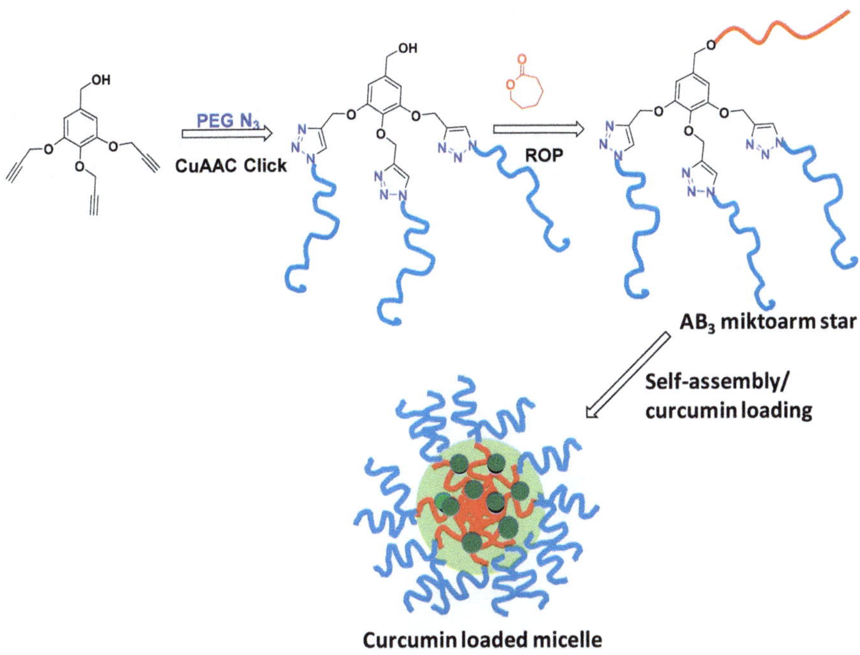


**Figure 6.4** Schematic representation of the synthesis of $AB_3$ miktoarm stars, their self-assembly, and drug loading.

from an orthogonal core with three acetylenes and a hydroxyl functional group by carrying out three simultaneous click reactions to introduce three hydrophilic PEG segments, followed by ROP of caprolactone to introduce a hydrophobic chain. They demonstrated the use of asymmetrical flow field-flow fractionation (AF4) as a highly efficient technique to precisely determine the size, size distribution, and curcumin content in the micelles. The curcumin encapsulated in the micelles was able to retain its biological activity and showed concentration and time-dependent activity toward killing glioblastoma cells in spheroids. This efficient and rapid methodology allows the facile construction of PEG and PCL based miktoarm polymers with variable chain lengths as efficient vehicles for the delivery of hydrophobic or lipophilic drugs with poor bioavailability.

Although a plethora of reports are being published every day demonstrating novel designs of drug delivery system, truly intelligent drug delivery systems need to be multifunctional in order to deliver novel therapeutics intra-cellularly at targeted sites with improved efficacy and reduced toxicity. In this regard, miktoarm polymers are highly advantageous owing to their multiple arms and tailor made architecture. While the hydrophobic interior of these polymers offer a nano-compartment to load therapeutic agents, the exterior polymeric segments can be utilized to append required targeting ligands and imaging agents through covalent linking. Kakkar *et al.* reported a multitasking mitochondria-targeting drug delivery system based on amphiphilic ABC (A = PEG, B = PCL, and C = triphenylphosphonium bromide (TPPBr)) type miktoarm star polymers for the delivery of CoQ10 (Figure 6.5).[124] TPPBr was used as the targeting ligand to target mitochondria based on its membrane potential. The polymers were conveniently synthesized using click chemistry and ring opening polymerization in sequence and formed nanosized (25–60 nm) micelles in aqueous solution. The micelles displayed extraordinary high loading (60 wt%) of CoQ10, which had not been reported before with any other nanocarriers. The Flory–Huggins interaction parameter proved to be quite low for PCL and CoQ10, which could explain the extremely high drug loading in the micelles. The biological results suggested that the miktoarm polymers described here

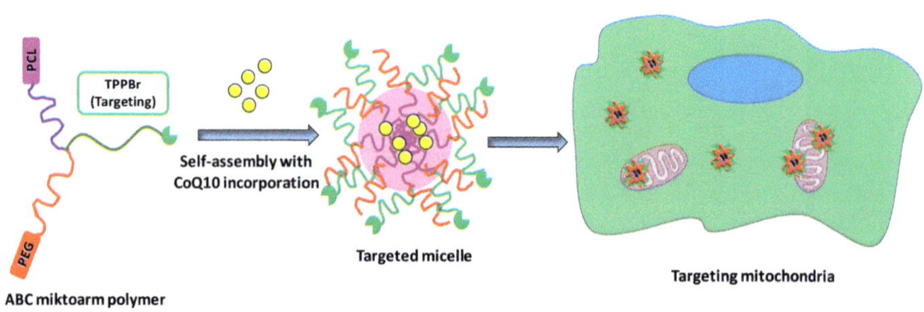


**Figure 6.5** Targeted ABC miktoarm polymer micelles for mitochondrial drug delivery.

afforded excellent nanocarriers for the delivery of CoQ10 without any loss in its efficacy. Moreover, these ABC miktoarm polymer-based site-directed drug delivery system could be efficiently used to deliver hydrophobic drugs to their specific targets.

Recently, in an effort to develop multifunctional nanocarriers with enhanced solubility and tracing capabilities, Kakkar and colleagues reported the construction of an inherently fluorescent ABC (tetraiodofluorescein (TIF) dye, PEG, and PCL) miktoarm polymer-based nanodelivery vehicle for curcumin (Figure 6.6).[50] The synthetic methodology used to construct these miktoarm stars is highly versatile, consisting of an orthogonal core that can be used to synthetically articulate the attachment of a variety of functions in a single nanosystem. The micelles resulting from these polymers could incorporate 25–69 wt% curcumin and provided a 25 000-fold increase in its solubility with a sustained release pattern for over seven days. The traceable micelles with the covalently linked imaging dye showed accumulation at discrete cellular compartments, compared to physically encapsulated dye molecules that diffuse all over the cytoplasm. These nanocarriers with integrated multiple functions offer considerable potential to address current drug delivery challenges.

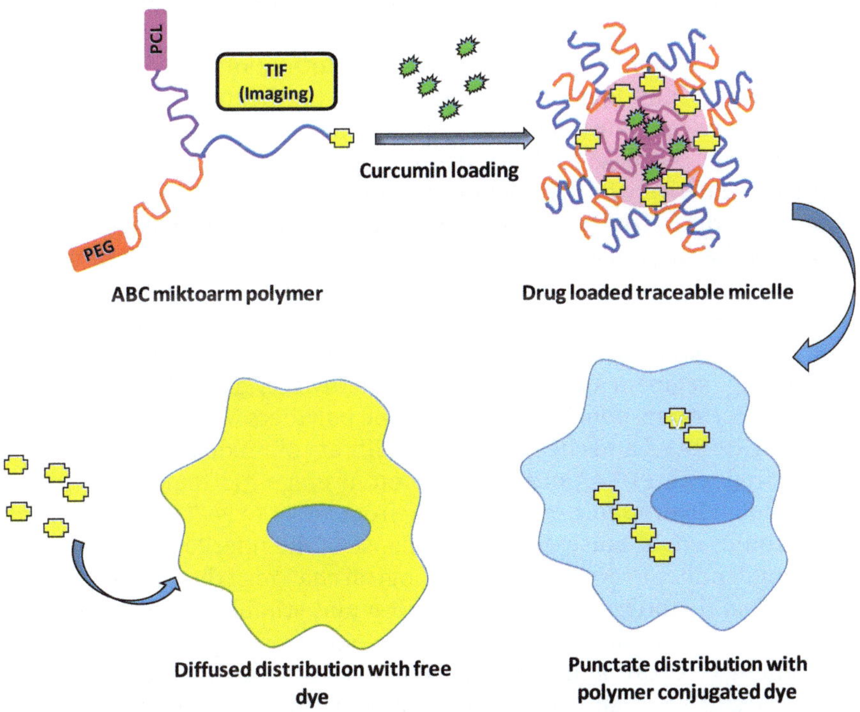


**Figure 6.6** Specific distribution of an imaging dye (TIF) upon covalent conjugation to ABC miktoarm stars in phagocytic cells compared to the diffused distribution of the free dye.

### 6.4.2 Stimulus-Responsive Miktoarm Star Polymers

Stimulus-sensitive polymers, also referred to as 'smart' polymers, are the class of polymers able to respond to particular environmental or chemical stimuli, and are considered the most promising platform for drug delivery applications.[125–130] The stimuli can be internal (*e.g.*, a change in pH in unhealthy organs or tissues, temperature, redox or different concentration of chemicals or enzymes, *etc.*) or external (*e.g.*, light or ultrasound). Stimulus-responsive polymers are continuously gaining interest in the scientific community for the design of effective delivery systems for controlled or site specific drug delivery.[131–133] The design of novel drug delivery systems for cancer therapy is one of the major growing areas, where stimulus-sensitive polymers are being widely explored.[133,134] Zhang and co-workers have reported pH-sensitive self-assembled micelles of $A_3(BC)_3$-type amphiphilic miktoarm star polymers poly(3-caprolactone)$_3$-[poly(2-(diethylamino)ethyl methacrylate)-*b*-poly(poly(ethylene glycol) methyl ether methacrylate)]$_3$ [(PCL)$_3$-(PDEAEMA-*b*-PPEGMA)$_3$] and have explored their potential to deliver hydrophobic anticancer drug doxorubicin (DOX).[49] The blank and drug loaded micelles had spherical morphology, as observed by transmission electron microscopy (TEM), with a size range of 110–240 nm. The micelles were able to effectively deliver DOX while maintaining a comparable antitumor activity to that of the free drug. By decreasing the pH, the micelles demonstrated a transition from spherical to a globular-uneven-extended conformation due to protonation of the tertiary amine groups of DEAEMA, significantly accelerating the *in vitro* drug release rate (Figure 6.7). These studies suggest that pH-responsive micelles have potential and can be prospective candidates to significantly improve the delivery of therapeutics to cancer cells, while reducing the side effects and minimizing the toxicity of the drugs on healthy tissues.

Most of the stimulus-sensitive polymers published so far are based on single stimulus. Oriol *et al.* have reported an interesting example where they developed $AB_3$ polymeric micelles responsive to dual thermal and photo stimuli.[135] The polymers were composed of an azopolymer constituting the photoresponsive segment and three poly(*N*,*N*-diethylacrylamide) (PDEAA) chains showing thermo-responsive properties. The polymers were synthesized combining ATRP, RAFT, and click chemistry. A library of miktoarm polymers were synthesized with PDEAA of three different average molecular weights with hydrophobic/hydrophilic ratios ranging from 56/44 to 26/74. While the thermal stimulus could cause the direct collapse of the micelles, exposure to UV light could induce a reversible morphological change of the micelles, causing their partial distortion. The encapsulation and stimulus-responsive release of Nile Red was studied using fluorescence spectroscopy, which showed the photo-induced controlled release of Nile Red from the distorted micelles. In another example, Qiao *et al.* designed and synthesized thermal-responsive homo- and miktoarm stars composed of either poly(*N*-isopropylacrylamide) (PNIPAM), poly(2-hydroxyethyl acrylate) (PHEA), or PEG *via* one pot copper-mediated reversible deactivation radical polymerization (RDRP) in

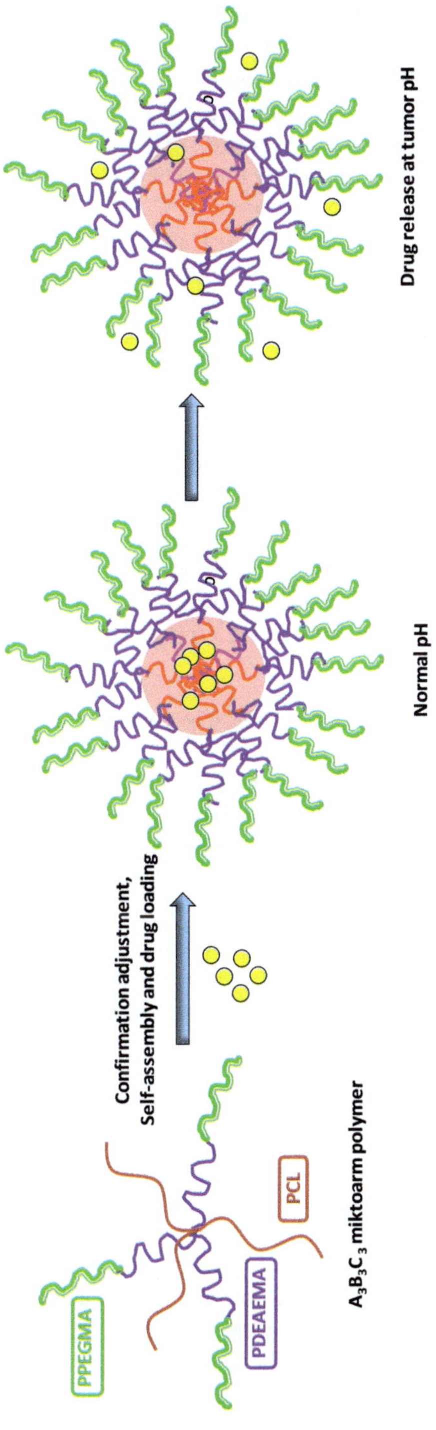


**Figure 6.7** Schematic illustration of the pH-dependent drug release from [(PCL)3-(PDEAEMA-*b*-PPEGMA)3] miktpoarm polymer micelles.

pure water using the 'arm first' technique.[60] Due to the temperature-responsive properties of PNIPAM, the polymers demonstrated reversible supramolecular self-assembly, as confirmed by dynamic light scattering (DLS) and 2D Nuclear Overhauser Effect Spectroscopy (NOESY) NMR studies. The results from biological experiments suggested the polymers to be relatively non-cytotoxic up to a concentration of 2 mg $mL^{-1}$. The use of simple chemistry and easy purification processes and their low toxicity make these polymers good candidates for biomedical applications.

Most recently, in an attempt to develop smart nanocarriers for theranostic (therapy and diagnosis) applications, Chen *et al.* introduced vesicles composed of dual pH/reduction-responsive ABC (A = PEG, B = PCL, C = PNIPAM) miktoarm star terpolymers for on-demand drug delivery.[28] The vesicles exhibited monodisperse size-distribution and were constructed by self-assembly of terpolymers with cyanine dye and the anticancer drug doxorubicin (DOX). The vesicles demonstrated enhanced cellular uptake and tumor accumulation and retention. In addition, the vesicles expressed various theranostic properties, including ultrahigh imaging contrast, fluorescence dequenching, and optimal thermo-chemotherapy-synergized tumor ablation, as 'proof of concept' smart nanocarriers for cancer treatment and diagnosis.

### 6.4.3 Chemical Conjugation of Drugs to Multi-Arm Mixed Star Polymers

Currently, great interest has been generated toward the design of drug conjugated polymeric micelles to overcome the pitfalls associated with traditional systems of physical encapsulation of drug molecules. Simple incorporation of hydrophobic drugs in the hydrophobic core of polymeric micelles results in weak hydrophobic forces, which are not strong enough and cause premature or burst release of therapeutic agents prior to reaching the targeted site of action. Covalent attachment of the drugs not only overcomes these problems, also increases the drug loading capacity as well as provides enhanced stability and site directed and sustained drug delivery. In order to develop covalently conjugated drug delivery system, nanoparticles should possess terminal handles in which perform chemical conjugation to append therapeutic agents. In this regard, the multiple arms of miktoarm stars can be conveniently employed to attach drug molecules or other bioactive ligands, making them appropriate drug delivery platforms.

Shen and co-workers have combined the concepts of both covalent conjugation and physical encapsulation in miktoarm polymers by developing self-assembled nanostructures from drug-conjugated miktoarm stars and also by incorporating free drugs in the interior of the micelles.[136] Ibuprofen-conjugated $A_2B_2$ miktoarm stars of PEG and PCL were synthesized using controlled ring opening polymerization (CROP) and click chemistry. Bromide functionalized PCL was synthesized *via* ROP and a hydroxyl terminal was utilized to attach ibuprofen through its carboxylic acid group *via* an ester linkage. The bromo group on PCL was then converted to an azide to participate in a click reaction with the alkyne-functionalized PEG to obtain

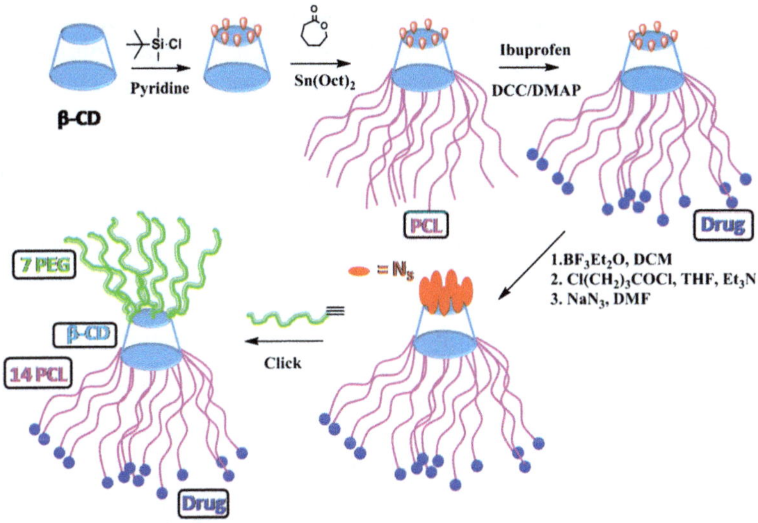


**Figure 6.8** Synthetic pathway to drug-conjugated cyclodextrin-core amphiphilic $A_{14}B_7$ miktoarm star copolymers.

$A_2B_2D$ (A = PEG, B = PCL, D = drug) miktoarm stars. These miktoarm stars self-assembled into multimorphological aggregates from aqueous solutions and showed significantly higher drug loading efficiencies compared to non-drug conjugated miktoarm polymer analogues. One could argue that this type of $A_2B_2$ drug conjugated miktoarm stars can have very low loading of covalently conjugated drug molecules. This issue was addressed by Shen's group in another report, where they developed biodegradable and biocompatible cyclodextrin cored drug-conjugated $A_{14}B_7$ miktoarm star copolymers composed of 14 PCL chains and 7 PEG arms using a combination of CROP and click chemistry (Figure 6.8).[137] These miktoarm stars were conjugated to 14 ibuprofen molecules and had the capability to self-assemble from aqueous solutions into multi-morphological aggregates that could encapsulate higher amounts of drug than similar non-drug conjugated star polymers. Once again, the drug release profile was dependent on both the physically encapsulated and covalently linked drug molecules. These types of drug-conjugated nanocarriers have the potential to show unique drug release profiles since the physically encapsulated drug can be quickly released, while the ester-linked drug would cleave only in the presence of an esterase, showing thus a sustained release pattern.

## 6.5 Summary and Future Outlook

Miktoarm star polymers have provided a platform to construct architectures with controlled and tailorable compositions, which can be conveniently modified further or adopted for a variety of applications, including biology, and especially for efficient and smart drug delivery. Amphiphilic miktoarm star polymers possess unique and beneficial solution and bulk properties compared to linear block co-polymers. Self-assembled structures obtained from miktoarm star polymers have been shown to possess better stability, high drug loading capability, enhanced stimulus responsiveness, and sustained release profiles. The presence of multiple arms in miktoarm stars provides an opportunity to use these nanocarriers for both physical encapsulation and covalent conjugation of therapeutic agents, making them useful for combination or synergistic drug delivery. In particular, the incorporation of stimulus-sensitive polymeric segments in miktoarm star is highly attractive for the design and development of intelligent and on-demand drug delivery platforms. Moreover, in addition to the delivery of drugs, the outer surface of miktoarm self-assembled nanostructures can be employed to append targeting ligands or imaging dyes for theranostic applications, making them attractive future polymeric materials for biological applications.

Despite the several advantages and demonstrated potential for biological applications, there are still plenty of challenges that need to be addressed. Significant effort has been made in the past to develop efficient synthetic strategies to prepare complex miktoarm star polymers. Several groups have reported an increasing number of different polymeric chains in miktoarm stars in good yield and low polydispersity. While there has been tremendous activity on homo-star polymers for biomedical applications, from tissue engineering and medical devices to the delivery of therapeutic molecules and genes, miktoarm stars have not been fully explored for biological applications despite their huge potential. Given the limited number of studies conducted so far regarding the biological application of miktoarm stars, it is highly unlikely that the true potential of these unique polymeric structures will be realized in the next few years. Only a few research groups have reported them for biomedical purposes. This might be attributed to the fact that construction and purification of well-defined hetero-arm polymers with precisely designed arms is slightly tricky, but modern synthetic tools and methodologies have paved the way for the preparation of well-defined miktoarm stars with a large number of arms and low polydispersity. Now, there is an urgent need to address structure–activity relationships in relation to the internal environment of the human body in order to successfully utilize these miktoarm stars for biological applications These can be achieved by conducting *in vivo* studies on these polymers to further confirm their pharmacokinetic properties. The future challenge is now to consider novel design requirements for developing smart drug delivery systems for clinical translation with enhanced stability, biocompatibility, biodegradability, and stimulus responsiveness.

## References


1. W. Park, D. Kim, H. C. Kang, Y. H. Bae and K. Na, *Biomaterials*, 2012, **33**, 8848–8857.
2. M. Stevanović, *Advanced Biomaterials and Biodevices*, John Wiley & Sons, Inc., 2014, pp. 343–366, DOI: 10.1002/9781118774052.ch10.

3. C. Peters, M. Hoop, S. Pané, B. J. Nelson and C. Hierold, *Adv. Mater.*, 2016, **28**, 533–538.
4. H. Li, T. Lee, T. Dziubla, F. Pi, S. Guo, J. Xu, C. Li, F. Haque, X.-J. Liang and P. Guo, *Nano Today*, 2015, **10**, 631–655.
5. M. Karimi, A. Ghasemi, P. Sahandi Zangabad, R. Rahighi, S. M. Moosavi Basri, H. Mirshekari, M. Amiri, Z. Shafaei Pishabad, A. Aslani, M. Bozorgomid, D. Ghosh, A. Beyzavi, A. Vaseghi, A. R. Aref, L. Haghani, S. Bahrami and M. R. Hamblin, *Chem. Soc. Rev.*, 2016, **45**, 1457–1501.
6. S. A. Papadimitriou, Y. Salinas and M. Resmini, *Chem.-Eur. J.*, 2016, **22**, 3612–3620.
7. R. Liu, S. Liang, C. Jiang, L. Zhang, T. Yuan, P. Li, Z. Xu, H. Xu and P. K. Chu, *J. Mater. Chem. B*, 2016, **4**, 1100–1107.
8. D. Bhattacharya, B. Behera, S. K. Sahu, R. Ananthakrishnan, T. K. Maiti and P. Pramanik, *New J. Chem.*, 2016, **40**, 545–557.
9. G. A. Williams, R. Ishige, O. R. Cromwell, J. Chung, A. Takahara and Z. Guan, *Adv. Mater.*, 2015, **27**, 3934–3941.
10. J. Khandare and M. Calderon, *Nanoscale*, 2015, **7**, 3806–3807.
11. M. Soleimani, J. C. Haley, D. Majonis, G. Guerin, W. Lau and M. A. Winnik, *J. Am. Chem. Soc.*, 2011, **133**, 11299–11307.
12. J. M. Ting, T. S. Navale, F. S. Bates and T. M. Reineke, *Macromolecules*, 2014, **47**, 6554–6565.
13. Y. Zhu and N. S. Hosmane, *ChemistryOpen*, 2015, **4**, 408–417.
14. Y. Deng, S. Zhang, G. Lu and X. Huang, *Polym. Chem.*, 2013, **4**, 1289–1299.
15. J. M. Ren, J. T. Wiltshire, A. Blencowe and G. G. Qiao, *Macromolecules*, 2011, **44**, 3189–3202.
16. W. Wu, W. Wang and J. Li, *Prog. Polym. Sci.*, 2015, **46**, 55–85.
17. T. K. Georgiou, *Polym. Int.*, 2014, **63**, 1130–1133.
18. D. Kuckling and A. Wycisk, *J. Polym. Sci., Part A: Polym. Chem.*, 2013, **51**, 2980–2994.
19. H. Gao, *Macromol. Rapid Commun.*, 2012, **33**, 722–734.
20. S. Pispas, N. Hadjichristidis, I. Potemkin and A. Khokhlov, *Macromolecules*, 2000, **33**, 1741–1746.
21. K. Yamauchi, K. Takahashi, H. Hasegawa, H. Iatrou, N. Hadjichristidis, T. Kaneko, Y. Nishikawa, H. Jinnai, T. Matsui, H. Nishioka, M. Shimizu and H. Furukawa, *Macromolecules*, 2003, **36**, 6962–6966.
22. A. Hirao, M. Hayashi, T. Higashihara and N. Hadjichristidis, *Complex Macromolecular Architectures*, John Wiley & Sons (Asia) Pte Ltd, 2011, DOI: 10.1002/9780470825150.ch4, pp. 97–132.
23. G. M. Soliman, A. Sharma, D. Maysinger and A. Kakkar, *Chem. Commun.*, 2011, **47**, 9572–9587.
24. G. Ungar, C. Tschierske, V. Abetz, R. Holyst, M. A. Bates, F. Liu, M. Prehm, R. Kieffer, X. Zeng, M. Walker, B. Glettner and A. Zywocinski, *Adv. Funct. Mater.*, 2011, **21**, 1296–1323.
25. K. Khanna, S. Varshney and A. Kakkar, *Polym. Chem.*, 2010, **1**, 1171–1185.
26. W. Shi, Y. Tateishi, W. Li, C. J. Hawker, G. H. Fredrickson and E. J. Kramer, *ACS Macro Lett.*, 2015, **4**, 1287–1292.

27. J. Qiu, J. Zhang, F. Yu, J. Wei and L. Ding, *J. Polym. Sci., Part A: Polym. Chem.*, 2016, **54**, 692–701.
28. A. Zhu, K. Miao, Y. Deng, H. Ke, H. He, T. Yang, M. Guo, Y. Li, Z. Guo, Y. Wang, X. Yang, Y. Zhao and H. Chen, *ACS Nano*, 2015, **9**, 7874–7885.
29. H. Gao and K. Matyjaszewski, *Macromol. Symp.*, 2010, **291–292**, 12–16.
30. A. Sharma and A. Kakkar, *Molecules*, 2015, **20**, 16987.
31. J. Gaitzsch, V. Chudasama, E. Morecroft, L. Messager and G. Battaglia, *ACS Macro Lett.*, 2016, 351–354, DOI: 10.1021/acsmacrolett.5b00913.
32. M. Asadul Hoque, H. Komiyama, H. Nishiyama, K. Nagai, T. Kawauchi and T. Iyoda, *J. Polym. Sci., Part A: Polym. Chem.*, 2016, **54**, 1175–1188.
33. J. A. Kalow and T. M. Swager, *ACS Macro Lett.*, 2015, **4**, 1229–1233.
34. S. S. Patil, B. V. Tawade and P. P. Wadgaonkar, *J. Polym. Sci., Part A: Polym. Chem.*, 2016, **54**, 844–860.
35. O.-T. Eric, Y. Chen, K. Takada, S.-i. Sato, T. Satoh and T. Kakuchi, *Polym. Chem.*, 2015, **6**, 7841–7850.
36. L. Ma, Y. Lin, H. Tan, J. Zheng, F. Liu, G. Zhang, J. Liu and T. Tang, *RSC Adv.*, 2015, **5**, 34466–34474.
37. B. Huang, M. Chen, S. Zhou and L. Wu, *Polym. Chem.*, 2015, **6**, 3913–3917.
38. O. Altintas, A. P. Vogt, C. Barner-Kowollik and U. Tunca, *Polym. Chem.*, 2012, **3**, 34–45.
39. A. Sharma, A. Khatchadourian, K. Khanna, R. Sharma, A. Kakkar and D. Maysinger, *Biomaterials*, 2011, **32**, 1419–1429.
40. S. Ito, R. Goseki, T. Ishizone, S. Senda and A. Hirao, *Macromolecules*, 2013, **46**, 819–827.
41. S. Ito, R. Goseki, T. Ishizone and A. Hirao, *Eur. Polym. J.*, 2013, **49**, 2545–2566.
42. H. Yin, H. C. Kang, K. M. Huh and Y. H. Bae, *J. Mater. Chem.*, 2012, **22**, 19168–19178.
43. H. Yin, S.-W. Kang and Y. H. Bae, *Macromolecules*, 2009, **42**, 7456–7464.
44. D. A. Christian, S. Cai, D. M. Bowen, Y. Kim, J. D. Pajerowski and D. E. Discher, *Eur. J. Pharm. Biopharm.*, 2009, **71**, 463–474.
45. D. H. Levine, P. P. Ghoroghchian, J. Freudenberg, G. Zhang, M. J. Therien, M. I. Greene, D. A. Hammer and R. Murali, *Methods*, 2008, **46**, 25–32.
46. D. E. Discher and A. Eisenberg, *Science*, 2002, **297**, 967–973.
47. L. Messager, J. Gaitzsch, L. Chierico and G. Battaglia, *Curr. Opin. Pharmacol.*, 2014, **18**, 104–111.
48. W. Lin, S. Nie, D. Xiong, X. Guo, J. Wang and L. Zhang, *Nanoscale Res. Lett.*, 2014, **9**, 1–12.
49. W. Lin, S. Nie, Q. Zhong, Y. Yang, C. Cai, J. Wang and L. Zhang, *J. Mater. Chem. B*, 2014, **2**, 4008–4020.
50. G. M. Soliman, R. Redon, A. Sharma, D. Mejía, D. Maysinger and A. Kakkar, *Macromol. Biosci.*, 2014, **14**, 1312–1324.
51. W. Lin, S. Nie, D. Xiong, X. Guo, J. Wang and L. Zhang, *Nanoscale Res. Lett.*, 2014, **9**, 243.

52. Y. Tselikas, H. Iatrou, N. Hadjichristidis, K. S. Liang, K. Mohanty and D. J. Lohse, *J. Chem. Phys.*, 1996, **105**, 2456–2462.
53. S. Sioula, N. Hadjichristidis and E. L. Thomas, *Macromolecules*, 1998, **31**, 5272–5277.
54. S. Pispas, Y. Poulos and N. Hadjichristidis, *Macromolecules*, 1998, **31**, 4177–4181.
55. Y. Zheng, S. Li, Z. Weng and C. Gao, *Chem. Soc. Rev.*, 2015, **44**, 4091–4130.
56. A.-M. Caminade, D. Yan and D. K. Smith, *Chem. Soc. Rev.*, 2015, **44**, 3870–3873.
57. A. Sharma, D. Mejia, D. Maysinger and A. Kakkar, *RSC Adv.*, 2014, **4**, 19242–19245.
58. A. Sharma, K. Neibert, R. Sharma, R. Hourani, D. Maysinger and A. Kakkar, *Macromolecules*, 2011, **44**, 521–529.
59. H. Li, D. Yang, Y. Gao, H. Li and J. Xu, *RSC Adv.*, 2015, **5**, 96377–96386.
60. T. G. McKenzie, E. H. H. Wong, Q. Fu, S. J. Lam, D. E. Dunstan and G. G. Qiao, *Macromolecules*, 2014, **47**, 7869–7877.
61. H. Hu and G. Liu, *Macromolecules*, 2014, **47**, 5096–5103.
62. J. Morsbach, A. Natalello, J. Elbert, S. Winzen, A. Kroeger, H. Frey and M. Gallei, *Organometallics*, 2013, **32**, 6033–6039.
63. W. Sun, X. He, C. Gao, X. Liao, M. Xie, S. Lin and D. Yan, *Polym. Chem.*, 2013, **4**, 1939–1949.
64. G. M. Soliman, R. Sharma, A. O. Choi, S. K. Varshney, F. M. Winnik, A. K. Kakkar and D. Maysinger, *Biomaterials*, 2010, **31**, 8382–8392.
65. R. Alizadeh and M. Ghaemy, *Polymer*, 2015, **66**, 179–191.
66. P. Fragouli, H. Iatrou, N. Hadjichristidis, T. Sakurai, Y. Matsunaga and A. Hirao, *J. Polym. Sci., Part A: Polym. Chem.*, 2006, **44**, 6587–6599.
67. D. Cho, S. Park, T. Chang, A. Avgeropoulos and N. Hadjichristidis, *Eur. Polym. J.*, 2003, **39**, 2155–2160.
68. H. Iatrou and N. Hadjichristidis, *Macromolecules*, 1993, **26**, 2479–2484.
69. J. Allgaier, K. Martin, H. J. Räder and K. Müllen, *Macromolecules*, 1999, **32**, 3190–3194.
70. H. Iatrou and N. Hadjichristidis, *Macromolecules*, 1992, **25**, 4649–4651.
71. T. Tsoukatos and N. Hadjichristidis, *J. Polym. Sci., Part A: Polym. Chem.*, 2002, **40**, 2575–2582.
72. G. Velis and N. Hadjichristidis, *Macromolecules*, 1999, **32**, 534–536.
73. A. Mavroudis, A. Avgeropoulos, N. Hadjichristidis, E. L. Thomas and D. J. Lohse, *Chem. Mater.*, 2006, **18**, 2164–2168.
74. A. Avgeropoulos, Y. Poulos, N. Hadjichristidis and J. Roovers, *Macromolecules*, 1996, **29**, 6076–6078.
75. A. Mavroudis, A. Avgeropoulos, N. Hadjichristidis, E. L. Thomas and D. J. Lohse, *Chem. Mater.*, 2003, **15**, 1976–1983.
76. O. Okay and W. Funke, *Macromolecules*, 1990, **23**, 2623–2628.
77. O. Okay and W. Funke, *Die Makromolekulare Chemie*, 1990, **191**, 1565–1573.

78. C. Tsitsilianis, D. Papanagopoulos and P. Lutz, *Polymer*, 1995, **36**, 3745–3752.
79. C. M. Fernyhough, R. N. Young and R. D. Tack, *Macromolecules*, 1999, **32**, 5760–5764.
80. R. P. Quirk and B. Lee, *Makromol. Chem., Macromol. Symp.*, 1992, **53**, 201–210.
81. H. Hückstädt, V. Abetz and R. Stadler, *Macromol. Rapid Commun.*, 1996, **17**, 599–606.
82. O. Lambert, P. Dumas, G. Hurtrez and G. Riess, *Macromol. Rapid Commun.*, 1997, **18**, 343–351.
83. M. Nasser-Eddine, S. Reutenauer, C. Delaite, G. Hurtrez and P. Dumas, *J. Polym. Sci., Part A: Polym. Chem.*, 2004, **42**, 1745–1751.
84. A. Mavroudis and N. Hadjichristidis, *Macromolecules*, 2006, **39**, 535–540.
85. X. Wang, J. He and Y. Yang, *J. Polym. Sci., Part A: Polym. Chem.*, 2007, **45**, 4818–4828.
86. T. Fujimoto, H. Zhang, T. Kazama, Y. Isono, H. Hasegawa and T. Hashimoto, *Polymer*, 1992, **33**, 2208–2213.
87. Y. Zhao, T. Higashihara, K. Sugiyama and A. Hirao, *J. Am. Chem. Soc.*, 2005, **127**, 14158–14159.
88. A. Hirao, M. Hayashi and T. Higashihara, *Macromol. Chem. Phys.*, 2001, **202**, 3165–3173.
89. A. Hirao and T. Higashihara, *Macromolecules*, 2002, **35**, 7238–7245.
90. A. Hirao, T. Higashihara, M. Nagura and T. Sakurai, *Macromolecules*, 2006, **39**, 6081–6091.
91. T. Higashihara, T. Sakurai and A. Hirao, *Macromolecules*, 2009, **42**, 6006–6014.
92. Z. Li, M. A. Hillmyer and T. P. Lodge, *Macromolecules*, 2004, **37**, 8933–8940.
93. C. Liu, M. A. Hillmyer and T. P. Lodge, *Langmuir*, 2009, **25**, 13718–13725.
94. H. Liu, C. Li, H. Liu and S. Liu, *Langmuir*, 2009, **25**, 4724–4734.
95. N. Saito, C. Liu, T. P. Lodge and M. A. Hillmyer, *Macromolecules*, 2008, **41**, 8815–8822.
96. T. Higashihara and A. Hirao, *J. Polym. Sci., Part A: Polym. Chem.*, 2004, **42**, 4535–4547.
97. X. Wei, G. Moad, B. W. Muir, E. Rizzardo, J. Rosselgong, W. Yang and S. H. Thang, *Macromol. Rapid Commun.*, 2014, **35**, 840–845.
98. Z.-M. Wu, H. Liang, J. Lu and W.-L. Deng, *J. Polym. Sci., Part A: Polym. Chem.*, 2010, **48**, 3323–3330.
99. H. Gao and K. Matyjaszewski, *Macromolecules*, 2008, **41**, 4250–4257.
100. H. Gao and K. Matyjaszewski, *J. Am. Chem. Soc.*, 2007, **129**, 11828–11834.
101. N. V. Tsarevsky and K. Matyjaszewski, *J. Polym. Sci., Part A: Polym. Chem.*, 2006, **44**, 5098–5112.
102. K. Matyjaszewski and J. Spanswick, *Mater. Today*, 2005, **8**, 26–33.
103. H. Gao and K. Matyjaszewski, *Macromolecules*, 2008, **41**, 1118–1125.

104. T. Erdogan, Z. Ozyurek, G. Hizal and U. Tunca, *J. Polym. Sci., Part A: Polym. Chem.*, 2004, **42**, 2313–2320.
105. C. Liu, Y. Zhang and J. Huang, *Macromolecules*, 2008, **41**, 325–331.
106. E. Doganci, M. A. Tasdelen and F. Yilmaz, *Macromol. Chem. Phys.*, 2015, **216**, 1823–1830.
107. B. Iskin, G. Yilmaz and Y. Yagci, *Polym. Chem.*, 2011, **2**, 2865–2871.
108. T. He, D. Li, X. Sheng and B. Zhao, *Macromolecules*, 2004, **37**, 3128–3135.
109. A. Vora, K. Singh and D. C. Webster, *Polymer*, 2009, **50**, 2768–2774.
110. E. Doganci, M. Gorur, C. Uyanik and F. Yilmaz, *J. Polym. Sci., Part A: Polym. Chem.*, 2014, **52**, 3390–3399.
111. W. A. Braunecker and K. Matyjaszewski, *Prog. Polym. Sci.*, 2007, **32**, 93–146.
112. E. Rizzardo, J. Chiefari, R. Mayadunne, G. Moad and S. Thang, *Macromol. Symp.*, 2001, **174**, 209–212.
113. C. J. Hawker, A. W. Bosman and E. Harth, *Chem. Rev.*, 2001, **101**, 3661–3688.
114. K. Matyjaszewski and J. Xia, *Chem. Rev.*, 2001, **101**, 2921–2990.
115. H. C. Kolb, M. G. Finn and K. B. Sharpless, *Angew. Chem., Int. Ed.*, 2001, **40**, 2004–2021.
116. C. Barner-Kowollik, F. E. Du Prez, P. Espeel, C. J. Hawker, T. Junkers, H. Schlaad and W. Van Camp, *Angew. Chem., Int. Ed.*, 2011, **50**, 60–62.
117. G. Franc and A. K. Kakkar, *Chem. Soc. Rev.*, 2010, **39**, 1536–1544.
118. K. Khanna, S. Varshney and A. Kakkar, *Macromolecules*, 2010, **43**, 5688–5698.
119. O. Altintas, G. Hizal and U. Tunca, *Des. Monomers Polym.*, 2009, **12**, 83–98.
120. U. S. Gunay, H. Durmaz, E. Gungor, A. Dag, G. Hizal and U. Tunca, *J. Polym. Sci., Part A: Polym. Chem.*, 2012, **50**, 729–735.
121. J. Li, Y. Wang, R. Liang, X. An, K. Wang, G. Shen, Y. Tu, J. Zhu and J. Tao, *Nanomedicine*, 2015, **11**, 769–794.
122. J. Kreuter, *Adv. Drug Delivery Rev.*, 2014, **71**, 2–14.
123. A. Moquin, A. Sharma, Y. Cui, A. Lau, D. Maysinger and A. Kakkar, *Macromol. Biosci.*, 2015, **15**, 1744–1754.
124. A. Sharma, G. M. Soliman, N. Al-Hajaj, R. Sharma, D. Maysinger and A. Kakkar, *Biomacromolecules*, 2012, **13**, 239–252.
125. M. A. C. Stuart, W. T. S. Huck, J. Genzer, M. Muller, C. Ober, M. Stamm, G. B. Sukhorukov, I. Szleifer, V. V. Tsukruk, M. Urban, F. Winnik, S. Zauscher, I. Luzinov and S. Minko, *Nat. Mater.*, 2010, **9**, 101–113.
126. M. Joglekar and B. G. Trewyn, *Biotechnol. J.*, 2013, **8**, 931–945.
127. F. Meng, Z. Zhong and J. Feijen, *Biomacromolecules*, 2009, **10**, 197–209.
128. A. Feng and J. Yuan, *Macromol. Rapid Commun.*, 2014, **35**, 767–779.
129. A. S. Hoffman, *Adv. Drug Delivery Rev.*, 2013, **65**, 10–16.
130. S.-M. Lee and S. T. Nguyen, *Macromolecules*, 2013, **46**, 9169–9180.
131. R. Cheng, F. Meng, C. Deng, H.-A. Klok and Z. Zhong, *Biomaterials*, 2013, **34**, 3647–3657.

132. W. Chen, P. Zhong, F. Meng, R. Cheng, C. Deng, J. Feijen and Z. Zhong, *J. Controlled Release*, 2013, **169**, 171–179.
133. G. H. Gao, Y. Li and D. S. Lee, *J. Controlled Release*, 2013, **169**, 180–184.
134. W. Cheng, L. Gu, W. Ren and Y. Liu, *Mater. Sci. Eng., C*, 2014, **45**, 600–608.
135. E. Blasco, B. V. K. J. Schmidt, C. Barner-Kowollik, M. Pinol and L. Oriol, *Polym. Chem.*, 2013, **4**, 4506–4514.
136. P.-F. Gou, W.-P. Zhu, N. Xu and Z.-Q. Shen, *J. Polym. Sci., Part A: Polym. Chem.*, 2009, **47**, 6962–6976.
137. P.-F. Gou, W.-P. Zhu and Z.-Q. Shen, *Biomacromolecules*, 2010, **11**, 934–943.

CHAPTER 7

# *Supramolecular (Miktoarm) Star Polymers: Self-Assembly and Applications*

BERNHARD V. K. J. SCHMIDT[a]

[a]Max Planck Institute of Colloids and Interfaces, Department of Colloid Chemistry, 14424, Potsdam, Germany
*E-mail: bernhard.schmidt@mpikg.mpg.de

## 7.1 Introduction

In recent years, supramolecular chemistry has found more and more applications in chemical research.[1] Especially, materials and polymer scientists have turned their attention towards supramolecular motifs.[2–4] Supramolecular chemistry allows the formation of chemical bonds with a variety of properties, *e.g.*, *via* hydrogen bonding, host–guest complexes or metal complexation. On the one hand, supramolecular bonds are dynamic and reversible. Nevertheless, the dynamic character is dominated by the strength of the association constant. The reversibility of the bonds gives the opportunity to break the bonds more easily than with covalent bonds – a feature that gives rise to stimulus-responsive disassembly or formation of bonds. Dynamic and reversible bonds further enable the formation of self-healing materials.[5] On the other hand, supramolecular bonds can be selective with respect to specific acceptors and donor pairs, also known as supramolecular recognition.[6]

Polymer Chemistry Series No. 25
Miktoarm Star Polymers: From Basics of Branched Architecture to Synthesis, Self-assembly and Applications
Edited by Ashok Kakkar

Published by the Royal Society of Chemistry, www.rsc.org

In this way, orthogonal recognition and bond formation can be facilitated if the recognition pairs are carefully chosen. Eventually, supramolecular chemistry has found its way into the synthesis of polymers and (miktoarm) star polymers in the last few years.

Star polymers have emerged as a frequently utilized class of polymers, mainly due to the development of reversible deactivation radical polymerization (RDRP) techniques and the introduction of the click chemistry/modular ligation concept.[7–9] In the area of RDRP, three techniques are used mainly, *i.e.* atom transfer radical polymerization (ATRP),[10] nitroxide mediated polymerization (NMP),[11] and the reversible addition fragmentation radical transfer (RAFT) process.[12,13] Each technique has its advances and merits and has to be chosen with the final application in mind. In principle, RDRP allows for the synthesis of polymers with adjustable molecular weight, narrow dispersity, and well-defined end groups; the latter being an important factor in the synthesis of supramolecular (miktoarm) star polymers, since end groups can be utilized to conjugate polymer arms into a star polymer or a multifunctional initiator can be used from the beginning. Another useful technique is click chemistry/modular ligation,[14] which allows for the connection of molecular building blocks in a convenient and efficient way, *e.g.*, the copper(I) catalyzed azide–alkyne cycloaddition (CuAAC),[15] the thiol–ene reaction,[16] the Diels–Alder reaction/hetero Diels–Alder reaction,[17,18] or the reaction between nitrile oxides and alkynes.[19] Moreover, RDRP and modular conjugation can be combined to obtain a broad variety of polymer architectures with interesting properties.

Needless to say, a combination of RDRP, modular ligation, and supramolecular chemistry has found broad interest in polymer science – and particularly in the formation of supramolecular (miktoarm) star polymers. In general, polymers can be synthesized *via* RDRP methods or other controlled polymerization mechanisms with specific supramolecular motifs as end groups (Figure 7.1(a)). Of course, modular ligation can be utilized as well to introduce supramolecular motifs (Figure 7.1(b)). In a subsequent step, polymers with complimentary supramolecular motifs are mixed together to form supramolecular connections. If a multifunctional or multi arm building block is present in this step, a supramolecular star polymer is formed (Figure 7.1(c)). Depending on the type of multifunctional blocks, supramolecular miktoarm star polymers are formed (Figure 7.1(d)). Since a variety of blocks can be combined, the supramolecular approach is highly modular and several architectures can be formed from a rather small number of building blocks or core molecules. Furthermore, the reversible nature of supramolecular bonds allows the disassembly of the formed (miktoarm) star polymers *via* external stimuli and, thus, modulation of the polymer properties is obtained. In this chapter, a variety of supramolecular motifs are presented in line with a discussion on orthogonal supramolecular building blocks, as well as the utilization of supramolecular motifs in the formation of (miktoarm) star polymers.

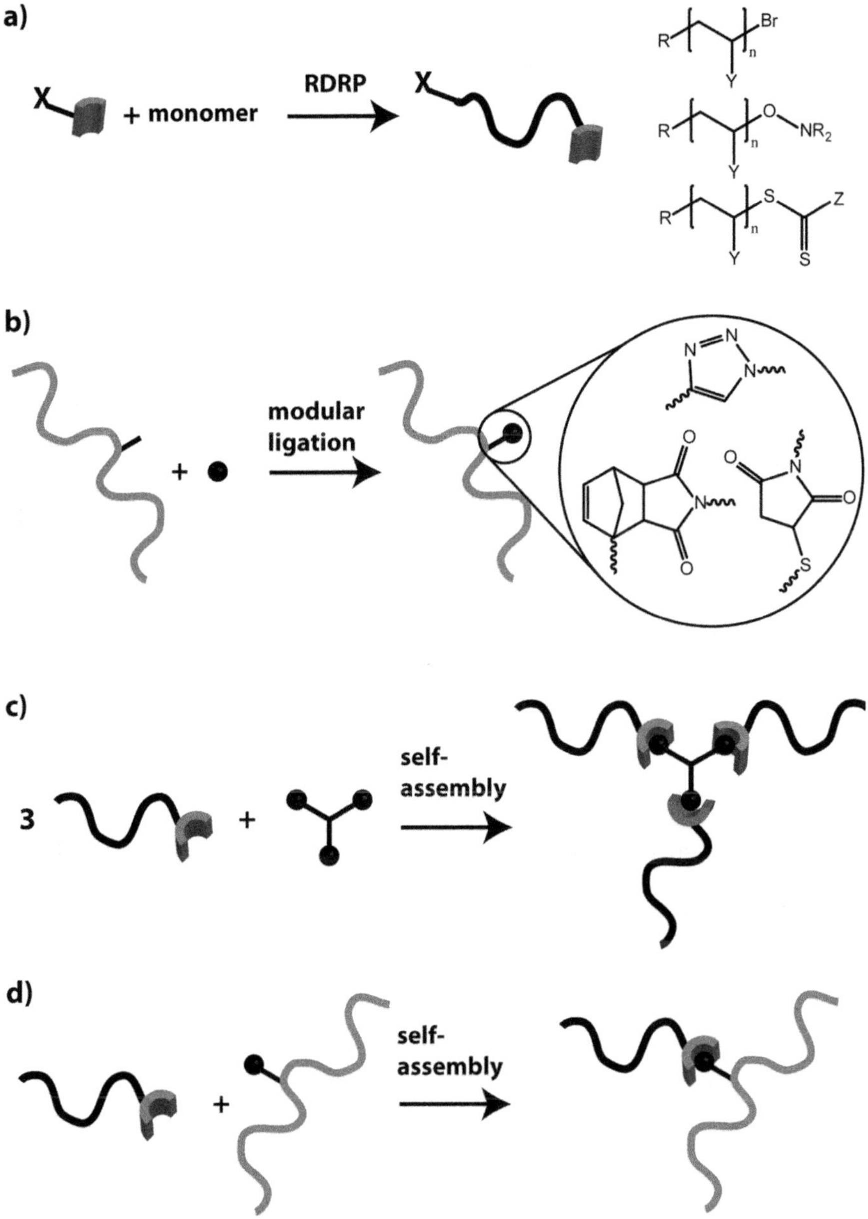


**Figure 7.1** Schematic overview of the formation of supramolecular (miktoarm) star polymers: (a) Formation of supramolecular motif (half-sphere) end functionalized polymers *via* RDRP, (b) formation of supramolecular motif (half-sphere) end-functionalized polymers *via* modular ligation, (c) formation of a supramolecular star polymer, and (d) formation of a supramolecular miktoarm star polymer.

# 7.2 Supramolecular Chemistry

## 7.2.1 Supramolecular Motifs

Hydrogen bonding is one of the most frequently utilized supramolecular motifs.[20] Usually, hydrogen bond-based supramolecular assemblies are formed in aprotic solvents. Nevertheless, protic solvents can be used as well if the association constant is large enough, which is the case for large multi receptor systems, *e.g.* the DNA. In order to obtain significant complexation, multiple donor and acceptor sites are necessary. The specificity of complexation is based on the geometrical alignment of the donor (D) and acceptor (A) units, which have to match in order to form a stable complex. An important complex is based on thymine (ADA system) and diaminopyridine (DAD system), which is based on three associating units (Figure 7.2(a)). The ureidopyrimidinone (UPy) system entails two hydrogen bond donors and two hydrogen bond acceptors units – a DDAA quadruple hydrogen bond system (Figure 7.2(b)).[21] A very efficient system is the Hamilton receptor (DADDAD system), which efficiently complexes with cyanuric acid (ADAADA system) in a 6-fold association (Figure 7.2(c)).[22]

Next to hydrogen bonding, inclusion complexes play a significant role in supramolecular polymer chemistry and star polymer formation. Mostly, cyclodextrins (CDs) are utilized in that regard.[23–25] CDs are water soluble cyclic oligosaccharides consisting of six, seven, or eight repeating units of glucopyranosides, namely α-, β- and γ-CD, respectively. Supramolecular complexes are formed with hydrophobic guest molecules in the hydrophobic cavity inside of the CD torus. The most common CD is β-CD, which forms stable complexes, for example, with adamantyl, ferrocenyl, or cholesteryl moieties in water (Figure 7.2(d)).[26] Interestingly, there is a variety of guest molecules that form stimulus-responsive complexes with CDs. In such a way, β-CD/ferrocenyl complexes can be broken *via* oxidation,[27,28] β-CD/azobenzene complexes can be disassembled by UV light irradiation[29,30] or β-CD and benzimidazole can be pH-responsively complexed.[31] Another frequently studied host–guest system is based on cucurbituril (CB), a macrocycle of glycouril units connected *via* methylene bridges that forms complexes with hydrophobic molecules in aqueous solution.[32] In this case, ternary complexes between one CB molecule and two guest molecules are formed. Charge transfer interactions between the guest molecules add significantly to the complex stability. A common combination of host and guests is CB[8], electron-poor methylviologen, and electron-rich naphthol (Figure 7.2(e)).[33] Crown ethers and cryptands are another type of host compounds well known for their complexes with cationic species, *e.g.*, secondary ammonium salts or paraquat derivatives (Figure 7.2(f)).[34,35] Another macrocyclic host is pillararene, a macrocycle composed of hydroquinone units connected *via* methylene bridges in *para* position.[36] Pillar[5]arene, consisting of five hydroquinone units, is especially applied in polymer science. Common guest groups are viologen and pyridinium derivatives, due to their electron-deficient nature that forms strong interactions

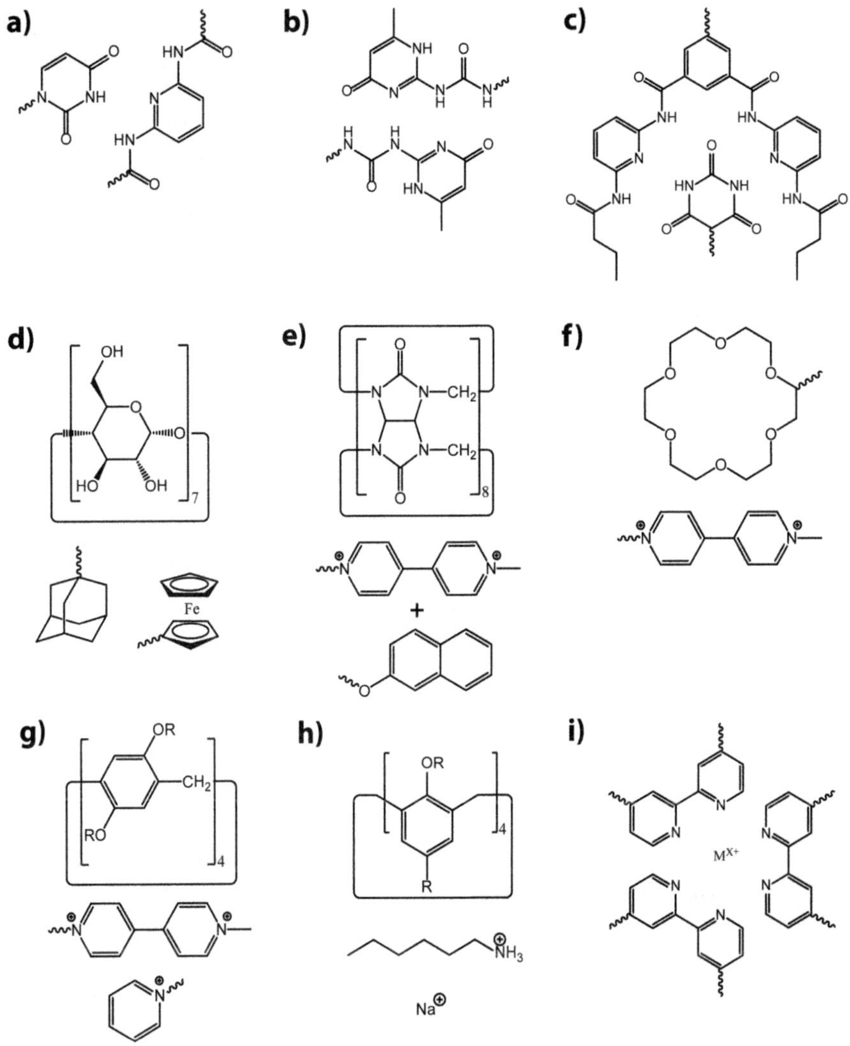


**Figure 7.2** Common supramolecular motifs: (a) thymine/diaminopyridine, (b) ureidopyrimidinone (UPy), (c) Hamilton-receptor/cyanuric acid, (d) β-CD/adamantyl or ferrocenyl, (e) cucurbit[8]uril/naphthol/viologen, (f) 18crown[6]/viologen, (g) pillar[4]arene/viologen or pyridinium, (h) calix[4]arene/ammonium or sodium ion, and (i) bipyridine/metal ion.

with the electron-rich pillararene (Figure 7.2(g)). A similar frequently used macrocycle is calixarene, which consists of phenol units connected *via* methylene groups in both *ortho* positions.[37] As guest groups, sodium ions as well as ammonium ions are qualified (Figure 7.2(h)). Resorcinarenes are closely related to calixarenes, yet based on dihydroxy benzene resorcin with methylene bridges in the *ortho* and *para* positions of one hydroxyl.

Another area of supramolecular chemistry that has found frequent utilization in polymer science is metal chelation. In that regard, several ligand/metal pairs are utilized in polymer chemistry. Metal complexation allows a variety of design opportunities, *e.g.*, the metal allows for the ligation of several molecules depending on its coordination number. In fact, the possibility to form tris complexes is an advantage over the other supramolecular motifs that are mostly only capable of forming 1:1 complexes, thus connecting only two species. Very frequently used pairs are complexes of two tridentate terpyridine (tpy) moieties with $Ru^{2+}$ or $Fe^{2+}$.[38] Bipyridine is another common ligand that is bidentate and forms tris complexes with $Ru^{2+}$ and $Fe^{2+}$ or bis complexes with $Pt^{4+}$ (Figure 7.2(i)).[38] Furthermore, $Pd^{2+}$ bis complexes with monodentate phosphines have been described.[39,40] Likewise, two 2,6-bis(1′-methylbenzimidazolyl)pyridine (MEBIP) units coordinate effectively with $Fe^{2+}$ or $Zn^{2+}$ ions.[41]

Overall, there are a lot of different supramolecular motifs that can be utilized by polymer chemists. When choosing a suitable type of supramolecular chemistry, aspects such as the solvent, stimulus-response, unwanted secondary interactions with the utilized polymers, and strength of the association are to be considered. Another useful property of some supramolecular interactions is orthogonality, which will be described in the next section.

### 7.2.2 Orthogonality of Supramolecular Motifs

In principle, supramolecular recognition motifs can be utilized in an orthogonal way.[6,42,43] Nevertheless, several factors have to be taken into account. First of all, the complexation constant should be significantly higher for the respective pairs than for the exchanged pairs. Otherwise, there will be no selective recognition. Furthermore, other factors that may disturb the complexation have to be considered, *e.g.*, solvent, temperature, concentration, or other competing donors or acceptors in the utilized molecules. As shown by Isaacs and coworkers, self-sorting of various supramolecular recognition units in chloroform solution is possible, *e.g.*, for a mixture of 10 compounds *via* hydrogen bonding and ion–dipole interactions.[44] Later, the studies were transferred to aqueous media and optimized towards ion–dipole interactions, electrostatic interactions, charge-transfer, metal complexation, and the hydrophobic effect.[45]

Multi donor/acceptor systems have proven to be extremely orthogonal, *e.g.*, in the DNA. Accordingly, orthogonal hydrogen bonding systems have found significant utilization in polymer science. For example, Altintas *et al.* synthesized a two point selective single-chain folding system based on diaminopyridine/thymine and Hamilton receptor/cyanuric acid recognition.[46] Weck and coworkers formed a supramolecular ABC triblock copolymer *via* Hamilton receptor/cyanuric acid and 2,7-diamido-1,8-naphthyridine/ureidoguanosine complexes.[47] Also, the self-recognition of UPy and benzene-1,3,5-tricarboxamide has been utilized to form complex materials *via* orthogonal hydrogen bonding recognition.[48] A combination of hydrogen bonding and host–guest chemistry has been described by Wang and coworkers. Thus, UPy and pillar[5]arene/viologen were utilized for the formation of supramolecular polymers.[49] Chen and coworkers utilized UPy and crown ether/pyromellitic diimide for the formation of supramolecular polymers.[50] Additionally, Liu and coworkers synthesized a supramolecular polymer based on CD/adamantyl and CB[8]/naphthol/viologen recognition, showing the orthogonality of CB and CD chemistry.[51] Another combination of two inclusion complex systems was introduced by Pan and coworkers.[52] A combination of pillar[5]arene/bromo alkyl and cryptand/viologen complexes was utilized in the formation of supramolecular polymers. Interestingly, cryptand and pillar[5]arene were connected *via* catenane formation. These linker catenanes were connected *via* diviologen and dibromo alkyl compounds that formed complexes with cryptand and pillar[5]arene, respectively. Huang and coworkers combined two different crown ethers (10 and 8) with a paraquat derivative and a dibenzyl ammonium salt for the formation of supramolecular polymers.[53] Tian and coworkers recently showed a supramolecular polymer based on orthogonal interactions of pillar[5]arene and crown ethers as well.[54] In particular, metal coordination has been utilized as an orthogonal recognition motif.[42] Fiore and coworkers showed the orthogonality of MEBIP/$Zn^{2+}$ or $Fe^{2+}$ metal complexation with the self-recognition of UPy in supramolecular polymer blends.[41] A supramolecular polymer was designed by Schubert and coworkers as well.[55] Orthogonal combination of UPy and tpy/$Zn^{2+}$ or $Fe^{2+}$ metal complexation was utilized to form high molecular weight polymers. Recently, a combination of metal complexation and pillar[5]arene was described for the formation of a supramolecular gel.[56] A pillar[5]arene/trimethyl ammonium based supramolecular polymer was crosslinked *via* a complex of a bidentate Schiff base group and $Cu^{2+}$. The tpy/$Fe^{2+}$ or $Ru^{2+}$ system has been further combined with the biological biotin/streptavidin system.[57]

Although a variety of orthogonal supramolecular motifs exist, the combination of different recognition units has not found a lot of application in the formation of miktoarm star polymers so far. Nevertheless, orthogonal complexation allows for the formation of complex polymers with a variety of properties, *e.g.*, stimulus response, and would be ideal for miktoarm star formation.[42]

## 7.3 (Miktoarm) Star Polymers Based on Supramolecular Recognition

In the formation of supramolecular star polymer architectures, several factors and design principles have to be considered. One factor is the utilized supramolecular motif. As presented in the previous sections, factors such as solvent, stability, and orthogonality have to be taken into account. Furthermore, the stoichiometry of the complexes is very important. Usually, 1 : 1 complexes are formed; however, in the case of higher orders, *e.g.*, for metal complexes, very large structures can be generated effectively. Another aspect is the influence of polymer chemistry – or more precisely, what structures polymer chemistry is capable of forming. RDRP and modular ligation, as well as other controlled polymerization methods, allow for the formation of specifically designed polymers. A way to form a homo 3-arm star polymer is the utilization of end-functionalized polymers and a central trifunctional unit that brings the arms together *via* supramolecular interactions to form an $A_3$ star. Of course, the central unit can have '*n*' supramolecular motifs to form an *n*-arm star polymer $A_n$. On the other hand, a miktoarm star polymer can be generated from a mid-functionalized polymer and an end-functionalized polymer endowed with the respective complimentary bonding units to form an $A_2B$ miktoarm star. The combination of two mid-chain functionalized polymers allows the formation of an X-shaped star polymer $A_4$ or miktoarm star polymer $A_2B_2$ depending on the chosen polymer arms. These examples show the possibility to generate a variety of structures from a reasonable number of precursors. The formation of supramolecular (miktoarm) star polymers has usually a modular nature and the combination of different building blocks provides the opportunity to generate a plethora of structures (Figure 7.3).

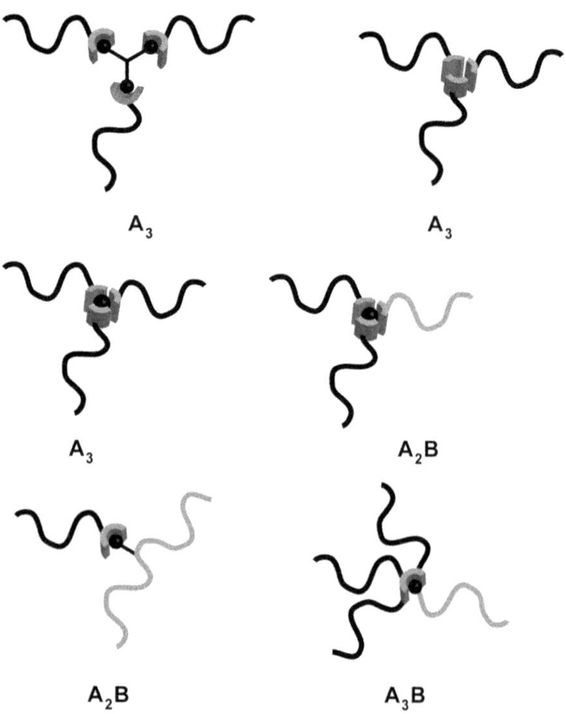


**Figure 7.3** Various supramolecular star polymer and miktoarm star polymer architectures.

### 7.3.1 Hydrogen Bonding

One of the most frequently utilized types of supramolecular recognition is hydrogen bonding. As a result, hydrogen bonding has found its way into the synthesis of (miktoarm) star polymers in many cases.[20,58] In recent years, hydrogen bonding-based homoarm star polymers have received significant research interest. Zimmerman and coworkers utilized bis-ureidodeazapterin (bis-DeAP; an AADD-DDAA receptor) in the formation of 6-arm star polymers (Figure 7.4(a)).[59] A bis-DeAP ATRP initiator was utilized to synthesize poly(styrene) (PS) or an alcohol functionalized bis-DeAP was utilized to synthesize poly(lactide) (PLA). Mixing of the end-functionalized polymers in toluene led to the formation of supramolecular star polymers. The supramolecular motif is designed in a way that no double complexes can be formed due to the arrangement of the hydrogen bond acceptors and donors in space. The formation of hexamers is mainly governed by the kinetics of complex formation and the geometry of the bis-DeAP unit. Nevertheless, pentamers and heptamers may also be formed. The formation was confirmed by size exclusion chromatography (SEC) in toluene. Later, the same group showed also poly(methyl methacrylate) (PMMA) star polymers with the same supramolecular motif and a thorough investigation on the geometry of the formed complexes and the consequences for the arm number of the resulting star polymers.[60] Bernard and coworkers synthesized a supramolecular 3-arm star polymer consisting of poly(vinyl acetate) (PVAc) arms.[61] The RAFT process was utilized to form mid-chain diaminopyridine PVAc as well as thymine end-functionalized PVAc. Both building blocks were mixed in chloroform and the self-assembly was probed by $^{1}H$ nuclear magnetic resonance (NMR) spectroscopy and SEC. Das and coworkers carried out research combining DNA with star polymers.[62] An azide end-functionalized 7-arm star poly(oligo ethylene glycol methacrylate) (POEGMA) was coupled to two complementary single strand DNAs with alkyne function. Coupling of the star polymers was possible *via* DNA hybridization with the respective complementary strands and investigated by dynamic light scattering (DLS). The assembly could be broken *via* addition of free DNA strands. Barner-Kowollik and coworkers utilized a Hamilton receptor for the formation of complex self-assemblies.[63] An alkyne end-functional copolymer of styrene and a Hamilton receptor functional repeating unit was synthesized *via* NMP and etherification to form the arms of the star polymers. Four-arm star polymers were formed *via* CuAAc. Next, cyanuric acid end-functionalized PS was added in dichloromethane to form a star polymer with arms that contained supramolecular grafts. The self-assembly was confirmed by $^{1}H$ NMR spectroscopy, DLS and static light scattering (SLS). A G-quadruplex core 8-arm star polymer was described by Tam and coworkers (Figure 7.4(b)).[64] Guanosine end-functionalized PMMA was synthesized *via* ATRP. Quadruplex formation was induced upon addition of $K^{+}$ ions, leading to self-assembly *via* hydrogen bonds and π–π stacking between the guanosine units, as well as ion–dipole interactions between the $K^{+}$ ions and guanosine. The complexes were characterized by $^{1}H$ NMR and

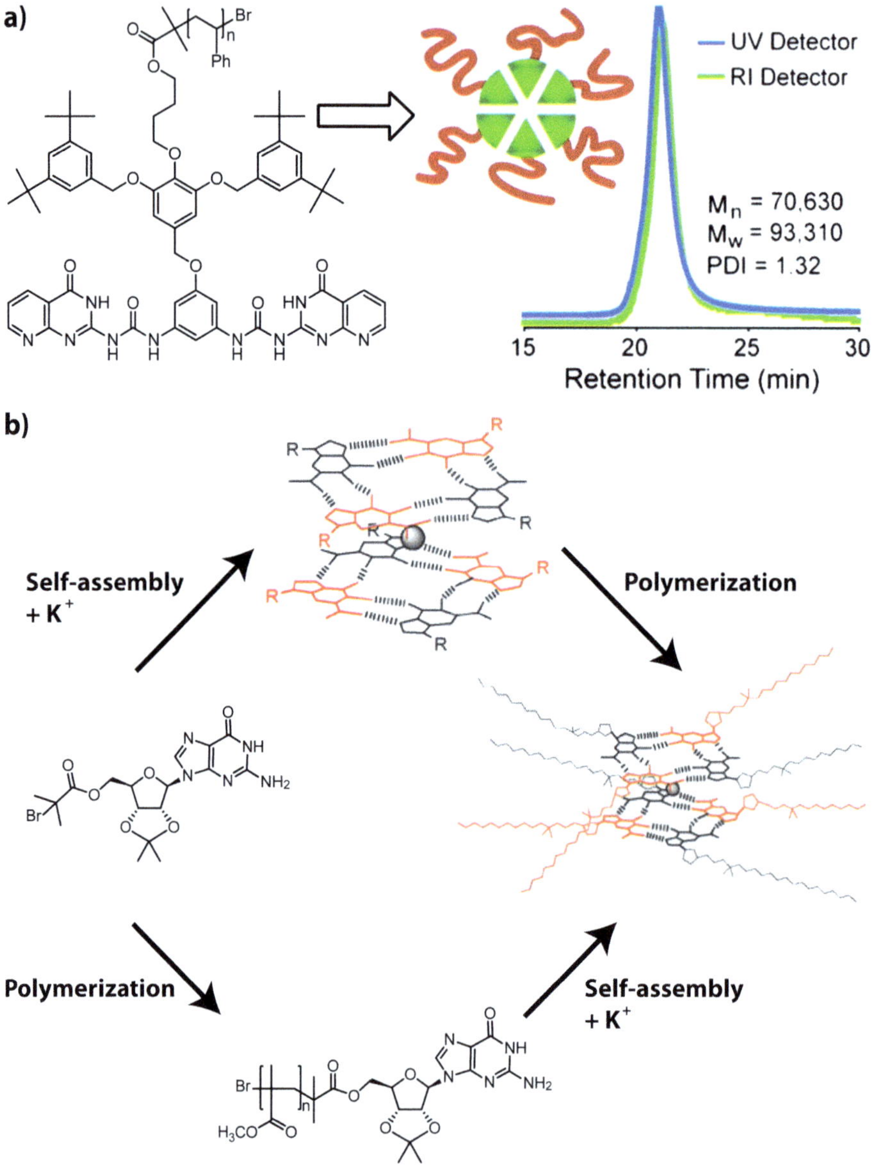


**Figure 7.4** Homoarm star polymers based on hydrogen bonding. (a) Six-arm star PS coupled *via* a bis-ureidodeazapterin (bis-DeAP) receptor. (Reprinted with permission from ref. 59. Copyright 2007 American Chemical Society.) (b) Eight-arm star polymer based on a guanosine derived $K^+$-centered G-quadruplex core. (Adapted from ref. 64 with permission from the Royal Society of Chemistry.)

circular dichroism spectroscopy. Similarly, Khan and coworkers investigated poly(butadiene) (PBD) and poly(ethylene oxide) (PEO) based G-quadruplex core star polymers and showed that non-polar polymer arms, *e.g.*, PBD, lead to star polymers with higher stability than polar polymers, *e.g.*, PEO.[65,66] Furthermore, PS-*b*-PMMA as well as PEO-*b*-poly(propylene oxide) (PEO-*b*-PPO) arms were described.[67]

In addition to homoarm star polymers, a variety of hydrogen bonding based miktoarm star polymers has been described in the literature. Bernard and coworkers presented a supramolecular $A_2B$ miktoarm star polymer based on thymine/diaminopyridine recognition (Figure 7.5(a)).[68] Thymine end-functional polymers, namely PVAc, poly(isoprene) (PI), and PMMA, and mid-chain diaminopyridine-functionalized PS were synthesized *via* RAFT polymerization and combined in chloroform to form a supramolecular complex. The star formation was monitored by $^1$H NMR spectroscopy. Furthermore, the miktoarm star polymers were analyzed by transmission electron microscopy (TEM) after solvent evaporation. The same hydrogen bonding

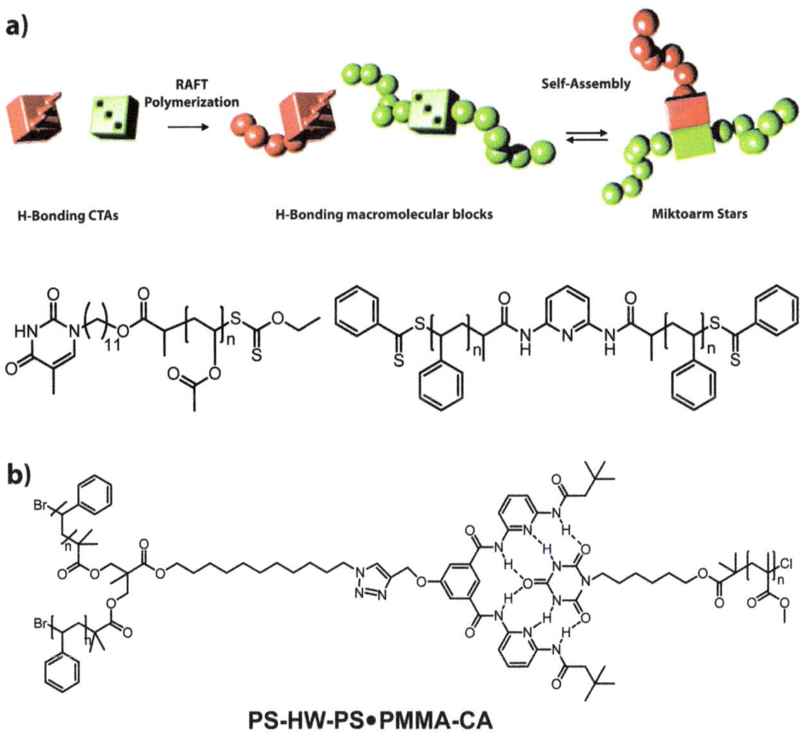


**Figure 7.5** Supramolecular miktoarm star polymers based on hydrogen bonding: (a) diaminopyridine/thymine system (PS and PVAc as examples). (Adapted with permission from ref. 68. Copyright 2010 American Chemical Society.) (b) Hamilton receptor/cyanuric acid system (PS and PMMA as examples). (Reprinted from ref. 70 with permission from the Royal Society of Chemistry.)

complex system was realized by Sawamoto and coworkers, yet the polymerization was performed *via* $Ru^{2+}$ based living radical polymerization.[69] A mid-chain thymine-functionalized PMMA and a mid-chain diaminopyridine-functionalized PS were synthesized in order to form an $A_2B_2$ miktoarm star polymer. Diffusion-ordered spectroscopy (DOSY) as well as $^1H$ NMR spectroscopy were utilized to prove the complex formation. In a similar fashion, Barner-Kowollik and coworkers synthesized Hamilton receptor/cyanuric acid functionalized polymers *via* $Cu^+$-based ATRP (Figure 7.5(b)).[70] Cyanuric acid end-functionalized PMMA was combined with mid-chain Hamilton receptor functionalized PS. Again, $^1H$ NMR spectroscopy was utilized to prove the complex formation and thus the formation of an $A_2B$ miktoarm star polymer. A hydrogen bonding based ABC miktoarm star polymer was described by Altintas *et al.*[71] A poly(*n*-butyl acrylate) (P*n*BA) with cyanuric acid end-functionalization was combined with a Hamilton receptor coupled to a PS-*b*-PEO block copolymer in between the blocks *via* CuAAc. The complex formation was thoroughly investigated *via* NMR studies, *e.g.*, DOSY and $^1H$. Furthermore, a Job plot was shown to prove the ratio of the complex components. The association constant was determined by NMR spectroscopy and the complexation was followed by DLS. In a similar fashion, H-shape star block copolymers were formed by the same group.[72] The aforementioned Hamilton receptor with PS-*b*-PEO ligation was combined with a double cyanuric acid end-functionalized P*n*BA. The complex formation was investigated by $^1H$ NMR spectroscopy, Job plot, NMR titration, DOSY, and DLS. These examples of ABC and H-shaped block copolymers nicely show the modularity of supramolecular miktoarm star polymer formation.

An interesting concept was introduced by Liu and coworkers.[73] An ABA triblock copolymer based on PMMA outer blocks and an inner copolymer block mainly composed of amine-containing monomers was synthesized. Furthermore, a diblock copolymer with a poly(*tert*-butyl acrylate) (P*t*BA) block and a block mainly based on carboxylic acid-containing monomers were synthesized as well. After association of the building blocks *via* acid–base association in dichloromethane, an effective coupling agent was added to couple the blocks together *via* amidation. Thus, a supramolecular preformation was utilized to form a covalent miktoarm star. The same group extended the concept towards nanocapsules with incorporated channels.[74] This time, an ABC triblock copolymer was utilized composed of P*t*BA and a cinnamoyl containing monomer as outer blocks and an acid containing inner block. Furthermore, a diblock copolymer consisting of PEO and an amine functional block were synthesized. Finally, both block copolymers were combined to give a supramolecular miktoarm star polymer and connected covalently *via* addition of coupling reagents. The miktoarm star polymer was capable of forming vesicles in water with the PEO block being on the solvent side and P*t*BA and the cinnamoyl blocks on the inside. The cinnamoyl blocks formed the matrix while the P*t*BA formed cylinders in the cinnamoyl phase. Next, the cinnamoyl blocks were crosslinked *via* UV irradiation and, subsequently, the *tert*-butyl groups were removed from the P*t*BA cylinders *via* trifluoroacetic acid (TFA) treatment leading to poly(acrylic acid)

(PAA) cylinders. The PAA cylinders could be swollen and deswollen depending on the pH in the fashion of a pH-responsive gate of nanocapsules used for the release of pyrene-labeled PEO. Another example focusing on supramolecular hydrogen bonding chemistry as template for the formation of miktoarm stars was described by the group of Schroeder.[75] Structurally defined miktoarm star polymers were synthesized based on single-stranded DNA (ssDNA). A dibenzocyclooctyne-functionalized nucleotide was incorporated into ssDNA *via* the polymerase chain reaction at a specific position of the chain. Subsequently, azide end-functional polymers, PEO or other DNA branches, were attached *via* a copper-free click reaction. Thus, 3-arm miktoarm stars (in the case of mid-chain functionalization), H-shape stars- and brushes were obtained. The formation of the products was investigated by Matrix-Assisted Laser Desorption/Ionization-Time of Flight mass spectrometry (MALDI-ToF MS) and gel electrophoresis. O'Reilly and coworkers described a giant surfactant based on DNA and thermoresponsive polymers.[76] Azide-functionalized poly(4-acryloylmorpholine), poly(*N*,*N*-dimethylacrylamide) (PDMA), poly(*N*-isopropylacrylamide) (PNIPAM), and PS were synthesized *via* RAFT polymerization and conjugated to a mid-chain alkyne-functionalized ssDNA *via* CuAAC. After addition of three other appropriate ssDNA strands, self-assembly into a DNA tetrahedron conjugated to a polymer strand occurred. Depending on the water solubility of the chosen polymer, *i.e.*, water insolubility of PNIPAM at elevated temperatures and water insoluble PS, micellization was studied by cryo-TEM, DLS, and atomic force microscopy (AFM).

Overall, hydrogen bonding has been used for the formation of a broad variety of (miktoarm) star polymers. The orthogonality of different hydrogen bonding motifs allows for the synthesis of well-defined supramolecular miktoarm star polymers.

### 7.3.2 Cyclodextrins

CDs have been utilized frequently for star and miktoarm star polymers, mostly due to the easy access to functionalized derivatives and, especially, mono functionalized derivatives. Furthermore, CDs have a high density of functionalities that can be utilized for star polymer formation. Therefore, CDs belong to the most utilized cores for covalent star polymers.[23,24] A very early example of β-CD-centered star polymers comes from the group of Haddleton, who synthesized a 21-arm star polymer with PS or PMMA arms, as well as block copolymer arms *via* ATRP.[77] Stenzel *et al.* utilized half-metallocene iron carbonyl mediated RDRP for an 18-arm PS star.[78] A variety of other CD-centered star polymers has been described, *e.g.*, azobenzene-functionalized polymers,[79] glycomonomers,[80,81] oligo ethyleneimine (OEI),[82] P*n*BA,[83] P*n*BA-*b*-P*t*BA,[83] poly(ε-caprolactone) (PCL),[84] PEO,[85] PEO-*b*-PLA,[86] poly(2-ethyl-2-oxazoline) (PEtOX),[87] poly(hexyl acrylate),[80] PNIPAM,[88,89] PNIPAM-*b*-PDMA,[89,90] POEGMA,[91] PS,[78,79] PS-*b*-poly(ethyl acrylate),[81] and P*t*BA.[92,93] An interesting structure, namely a polyrotaxane of α-CD-centered PMMA was described recently by Wenz and coworkers.[94] A β-CD-centered star polymer for gene transfection was described by Zhang and coworkers.[95] Poly(amidoamine) dendrons were conjugated to β-CD and polyplexes were formed with plasmid DNA. The carriers showed low cytotoxicity and efficient plasmid DNA transfection. In a similar way, ElSayed and coworkers utilized a β-CD core and amine-containing polymers for small interfering RNA (siRNA) delivery.[96] A β-CD-centered copolymer of hexyl acrylate, dimethyl aminoethyl methacrylate (DMAEMA), and methyl quaternized DMAEMA was formed *via* ATRP. Next, polyplexes with siRNA were formed and delivery of siRNA was probed. Furthermore, the polymer strands were connected to the β-CD core *via* pH-sensitive hydrazone linkages to allow more efficient siRNA decomplexation in the cell upon cleavage of the polymer arms. Another gene carrier based on γ-CD-centered OEI star polymers was described by Li and coworkers,[97] as well as a carrier based on β-CD-centered poly((2-dimethylamino) ethyl methacrylate)-*co*-poly(hydroxyl ethylacrylate) (PDMAEMA-*co*-PHEA) by Loh and coworkers.[98] A 21-arm star block copolymer for drug delivery was described by Wang and coworkers.[99] PLA was grafted from β-CD and the polymers extended by POEGMA. In water, unimolecular micelles were formed that were capable of drug encapsulation. An α-CD-centered poly(3-hydroxybutyrate) (PHB) was presented by Li and coworkers.[100] The adamantyl end groups of PHB were capable of forming complexes upon addition of β-CD. Thus, vesicles were formed in aqueous solution that could be used for drug delivery experiments.

Apart from CD-centered homo star polymers, there have also been some examples of CD-centered miktoarm star polymers. To obtain covalent β-CD-centered miktoarm star polymers, the different reactivities of the CD hydroxyls can be utilized, for example. Kakuchi and coworkers reported a β-CD-centered $AB_{20}$ miktoarm star *via* combination of NMP and ATRP.[101] A TEMPO initiator was conjugated to the acetylated mono amino β-CD and subsequently PS was synthesized *via* NMP. After removal of the acetyl protection groups, the formed hydroxyl groups were esterified in order to obtain ATRP initiators. At last, methyl methacrylate and *t*BA were polymerized *via* ATRP to obtain the miktoarm star polymer. Another example was introduced by Shen and coworkers.[102] The hydroxyl groups at the 6 position of β-CD were protected *via* silyl ethers. Subsequently, ring-opening polymerization of CL starting from the residual hydroxyl groups was conducted and endcapped with ibuprofen. In the next step, the protecting group was removed and the hydroxyls converted into azide moieties. Next, PEO was conjugated *via* CuAAc and the formed miktoarm star amphiphiles studied with regard to the formed self-assembled morphologies in solution. Furthermore, drug release of encapsulated and conjugated drugs was probed, as well as the biodegradability of the drug carriers. Shen and coworkers also synthesized a β-CD-centered P*t*$BA_7$-$PCL_{14}$ star polymer.[103] First, the primary hydroxyls were protected. Subsequently, PCL was synthesized from the remaining hydroxyls. Next the protection group was removed, an ATRP initiator conjugated, and *t*BA polymerized. Finally, the P*t*BA was deprotected to form PAA. In a similar fashion, an $A_7B_{14}$ miktoarm star polymer was described by Haddleton and coworkers.[104] At first, the more reactive hydroxyls at the 6 position of β-CD were transformed into thiol groups, which were subsequently used for thiol Michael addition with various vinyl terminated building blocks including PEO and poly((2-methoxyethoxy)ethyl methacrylate) ($PMEO_2MA$). The remaining hydroxyl groups were utilized as initiating sites for the ring opening polymerization of CL. Wu and coworkers presented a β-CD-centered 20-arm star block copolymer with one clickable extra arm.[105] A mono azide β-CD was ligated with an alkyne end-functional PS that had a TMS protected alkyne on the other end. Next, the remaining hydroxyls were esterified with bromo isobutyryl bromide in order to graft P*t*BA-*b*-PS or P*t*BA-*b*-PEO block copolymers *via* ATRP and/or CuAAc. After the deprotection of the TMS protecting group, further click chemistry could be utilized, *e.g.*, to graft the star polymers onto surfaces or on silica nanoparticles. Fan and coworkers presented a complex miktoarm star polymer based on a cationic hyperbranched β-CD containing core.[106] The shell consisted of poly(diethyl aminoethyl methacrylate) (PDEAEMA) conjugated to a second shell consisting of mixed arms of PNIPAM and PDMAEMA. Zhou and coworkers synthesized β-CD-centered with 6 PNIPAM or POEGMA and one PEO arm to study the differences in the thermoresponse.[107] As a more complicated structure, Chen and coworkers formed an H-shape polymer with an inner PEO block conjugated with a β-CD and two PNIPAM arms on each end.[108] A β-CD based amphiphile for drug delivery was synthesized by Mazzaglia and coworkers, who placed 7 hexyl units on one face of β-CD and 7 oligo ethylene glycol units on the other face.[109] Recently, Li and coworkers developed a β-CD-centered miktoarm star polymer based on PEI and hyaluronic acid for gene delivery.[110]

As shown briefly in the previous paragraphs of this section, a broad variety of CD-centered (miktoarm) star polymers formed *via* covalent linkages exist. Another way of forming star polymers based on CDs is *via* supramolecular host–guest complexes of the hydrophobic cavity in CDs and hydrophobic guest molecules in aqueous solution. Of course, CD-centered star polymers can also be used for supramolecular (miktoarm) star polymer formation *via* the remaining hydrophobic cavity being capable of supramolecular recognition with suitable guest moieties. Thus, a combination of covalent bond based miktoarm star polymers with supramolecular recognition allows for the formation of very complex miktoarm star polymer structures.

So far, CD-centered star polymers solely based on covalent bonding have been described in this section. Of course, CD also possesses the function of supramolecular host–guest complex formation. This feature allows the formation of more complex star architectures. Additionally, the stimulus response of the host–guest association can be utilized as well. In the case of supramolecular homo star polymers, Barner-Kowollik and coworkers presented a supramolecular 3-arm star polymer formed *via* a 3-fold β-CD functionalized core molecule and adamantyl end-functionalized PDMA or poly(*N*,*N*-diethylacrylamide) (PDEA).[111] The star formation was probed by rotating-frame nuclear Overhauser enhancement spectroscopy (ROESY) and DLS. Furthermore, the response of the complex towards the increasing temperature was investigated. Heating to 70 °C led to disconnected star polymers, while the star polymers formed again after cooling to ambient temperature. In a similar way, Fleischmann *et al.* utilized phenolphthalein end-functionalized PEO to form a supramolecular 6-arm star with a 6-fold β-CD core molecule.[112] The formation of the star polymer was acquainted with a visible change of the solution color from violet to colorless, as has been shown previously for brush polymers as well.[113] Jiang and coworkers synthesized a supramolecular star polymer utilizing two supramolecular interactions – namely the β-CD/adamantyl interaction and the Concanavalin A/α-mannopyranoside complex (Figure 7.6(a)).[114] An α-mannopyranoside was linked to β-CD (Man-β-CD) and an adamantyl end-functionalized PEO was synthesized. Mixing of the Man-β-CD linker with the adamantyl-functionalized PEO and concavalin A led to star polymer formation due to the tetravalent interaction of concavalin A with four α-mannopyranoside units. In order to prove the structure formation, isothermal titration calorimetry (ITC), DLS, and SEC studies were performed. Furthermore, gelation was observed after addition of free α-CD. Ritter and coworkers formed β-CD end-functionalized hyperbranched poly(glycerols) (hPG) and attached azobenzene end-functionalized PNIPAM to form a supramolecular homoarm star polymer, which was characterized by DLS.[115] Another hyperbranched approach, yet based on supramolecular connections, was presented by Xu and coworkers.[116] An α-CD-centered poly(glycidyl methacrylate) star polymer was conjugated with ethanolamine and adamantyl amine. Furthermore, a β-CD-centered poly(glycidyl methacrylate) star polymer was conjugated with ethanolamine and mixed with the α-CD based conjugate. Host–guest complexation led to the formation of hyperbranched structures that were then utilized for gene delivery.

There are also supramolecular nanoparticle-centered star polymers with a variety of spherical cores. Jiang and coworkers formed β-CD grafted quantum dots that were able to complex with azobenzene end-functionalized PDMA-*b*-PNIPAM.[117] These star polymers showed photoresponse as well as thermoresponse and were able to form hydrogels. In a similar way, Ag nanoparticles were grafted with PNIPAM-*b*-PAA,[118] $SiO_2$ nanoparticles were grafted with PEO arms,[119] Au nanoparticles were coated with α-CD, and a complex with azobenzene end-functionalized PNIPAM[120] was formed as well as PNIPAM-*b*-PDMA.[121] Redox responsive CdS-centered star polymers were described by Jiang as well.[110] Ferrocene end-functionalized PDMA-*b*-PNIPAM was synthesized and hydrogels formed *via* the thermoresponsive behavior of the PNIPAM block. Hirschbiel *et al.* prepared nanoparticles based on CD host–guest chemistry *via* photochemical means (Figure 7.6(b)).[122] A β-CD end-functionalized PDMA was complexed with an adamantyl end-functionalized PNIPAM to form a diblock copolymer. The PNIPAM was further decorated with 2-methoxy-6-benzaldehyde moieties capable of light-triggered conjugation with maleimides. Heating of the supramolecular block copolymer above the lower critical solution temperature (LCST) of the PNIPAM block led to micellization with a PNIPAM core and a PDMA shell. A tetrafunctional maleimide linker was added to enable crosslinking under light irradiation. Thus, nanoparticles were formed

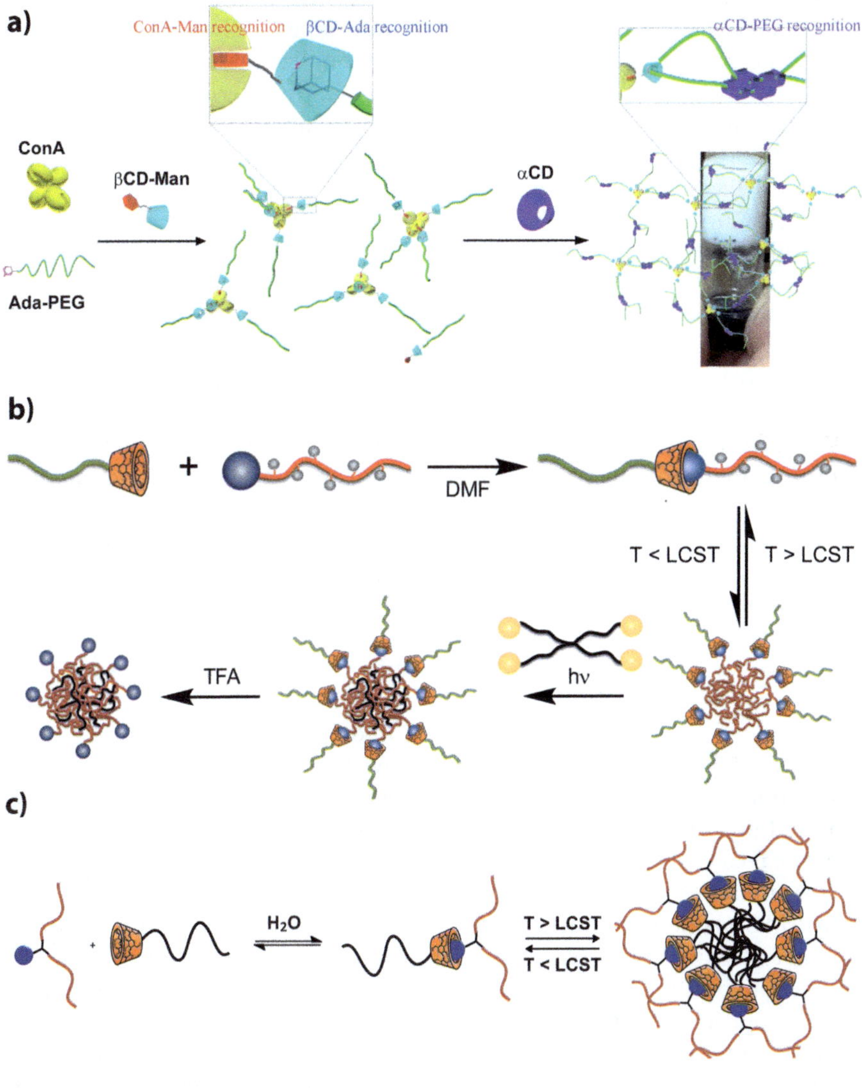


**Figure 7.6** Examples for CD based architectures. (a) Supramolecular 4-arm star based on a combination of Concanavalin A/α-mannopyranoside and β-CD/adamantyl recognition and α-CD induced gelation. (Reprinted with permission from ref. 114. Copyright 2013 American Chemical Society.) (b) Formation of a nanoparticle-centered star polymer *via* CD-based recognition and photochemical micelle core crosslinking. (Reprinted with permission from ref. 122. Copyright 2015 American Chemical Society.) (c) Formation of a supramolecular $A_2B$ miktoarm star polymer *via* β-CD/adamantyl complexation and thermo-induced micellization. (Adapted from ref. 124 with permission from the Royal Society of Chemistry.)

*via* a supramolecular connection that could be transformed into adamantyl-decorated nanoparticles after cleavage of the complex.

A triblock miktoarm star polymer was presented by Zhu and coworkers.[123] First, a mono azide and mono tosylate functionalized β-CD was synthesized, PEG was conjugated *via* CuAAC, and the remaining tosylate converted into an azide. An ATRP initiator was added in a subsequent CuAAC reaction and DMAEMA was polymerized *via* ATRP. Finally, an adamantyl end-functionalized PMMA was added in order to form an inclusion complex yielding an ABC miktoarm star polymer. Reversible assembly and disassembly was probed by TEM and DLS. Schmidt *et al.* synthesized a miktoarm star polymer based on β-CD end-functionalized PDEA and mid-chain adamantyl functionalized PDMA (Figure 7.6(c)).[124] Self-assembly of these blocks led to an $A_2B$ miktoarm star polymer that was characterized by ROESY and DLS. The thermoresponse of the PDEA block was characterized *via* turbidimetry and temperature-sequenced DLS measurements. A similar route was described for supramolecular $A_2B_2$ miktoarm star polymers, yet a mid-chain CD-functionalized polymer was utilized instead of a CD end-functionalized polymer.[125] Investigation of the temperature-induced micellization due to the PDEA blocks showed a two-step behavior for both the $A_2B$ and the $A_2B_2$ miktoarm star polymers. At temperatures below the LCST of the thermoresponsive block, unimeric self-assemblies were present. In a defined temperature range of around 5–10 °C directly above the LCST, aggregation occurred leading to small aggregates. Further heating induced agglomeration of the smaller aggregates yielding larger particles. A micelle-forming miktoarm star polymer was presented by Allcock and coworkers as well.[126] A β-CD-centered POEGMA star was complexed with an adamantyl end-functionalized poly(bis-(trifluoroethoxy)phosphazene) yielding poly(phosphazene)-centered micelles. A photoresponsive miktoarm star polymer based on azobenzene β-CD complexation was described by Wu and coworkers.[127] A 21-arm β-CD-centered PLA was synthesized *via* ring-opening polymerization of lactic anhydride. Azobenzene-functionalized PEO was connected to the homo star polymer *via* supramolecular inclusion, leading to a supramolecular amphiphile. Depending on the length of the PLA arms, various self-assembled morphologies were obtained, *e.g.*, spheres or tubes. Furthermore, the azobenzene moiety could be used to break the miktoarm star polymers *via* disassembly of the complex upon irradiation with UV light. More complicated structures were described by the same team as well, *e.g.*, H-shape star polymers.[128] Yan and coworkers presented a β-CD-centered hPG that was complexed with an adamantly-functionalized alkyl chain ($C_{12}$, $C_{18}$, and $C_{30}$), again a supramolecular amphiphile.[129] In that way, ductile vesicles were formed, which were characterized by DLS, TEM, scanning electron microscopy (SEM), and AFM. The vesicles could be broken *via* addition of free β-CD molecules. Another supramolecular miktoarm star polymer was prepared by Liu and coworkers.[130] A β-CD-centered poly(glutamic acid) was complexed with adamantyl end-functionalized PEO. Furthermore, the polypetide block was utilized for Pt complexation and cytotoxicity essays.

As is the case for CD-centered star and miktoarm star polymers, supramolecular miktoarm star polymers based on CD host–guest complexes have attracted diverse interest in DNA delivery as well as drug delivery. Jiang and coworkers combined a 4- or 7-arm β-CD-centered poly(*N*-vinylpyrrolidone) (PVP) with adamantyl end-functionalized PCL (Figure 7.7(a)).[131] The resulting miktoarm star formed micelles in water, which were characterized by DLS, TEM, and nuclear Overhauser enhancement spectroscopy (NOESY). Next, the stability in saline environment and protein adsorption was probed. Finally, drug loading with Cabazitaxel or Paclitaxel was performed and the release studied *in vitro* and *in vivo*. An enhanced efficacy in tumor inhibition and a higher amount of cargo in tumor sites was found for the CD-based micellar carrier. A β-CD-centered star shaped polycation was utilized recently for DNA delivery.[132] The polycation was synthesized *via* a Michael addition of PEI on a multi acryl functionalized α-CD. Furthermore, mono azobenzene end-functionalized PEO was added in order to form a supramolecular inclusion complex. Finally, DNA was conjugated *via* polyplex formation with the polycation to form particles with a size around 150 nm, and the formed assemblies were utilized in cellular uptake investigations. A similar approach utilizing PDMAEMA was described by Xu and coworkers (Figure 7.7(b)).[133] In this work, four PDMAEMA arms were polymerized from β-CD with initiating sites coupled to the core *via* bioreducible disulfide bonds. After polymerization, adamantyl end-functionalized poly(ethylene glycol)ethyl ether methacrylate was added to form a supramolecular complex. The PDMAEMA arms were utilized to facilitate the efficient complexation with DNA, while the PEO based arms were used for enhanced stability of the carriers in the serum and decreased toxicity of the carriers. Finally, DNA was added to form polyplexes and gene delivery was probed.

More complex structures have also been derived from β-CD-based polymer chemistry. A core crosslinked supramolecular miktoarm star polymer was described by Liu and coworkers.[134] PNIPAM arms were synthesized *via* ATRP and crosslinked *via* methylene bisacrylamide. Next, β-CD functionalized acrylamide was grafted on the core-crosslinked star polymer *via* the remaining ATRP initiation sites and, finally, adamantyl functionalized poly(4-vinylpyridine) (P4VP) was attached *via* supramolecular host–guest complex formation to yield a pH- and thermoresponsive supramolecular miktoarm star polymer. Li and coworkers presented a dumbbell-shaped star block copolymer. Two β-CD-centered PNIPAMs were connected *via* a double adamantyl end-functionalized PEO.[135] NOESY and turbidity measurements were applied to prove the complex formation. In a subsequent study, the same team described dumbbell-shaped star block copolymers based in β-CD-centered PNIPAM and a PPO mid-block.[136] Again, NOESY was performed to show inclusion complexation. Furthermore, the thermoresponsive nature of the pseudo-star block copolymers was probed. Temperature-sequenced DLS showed two-step micellization behavior. Unimers were present at temperatures below 9 °C. Between 9 and 20 °C, micelles were present and characterized *via* AFM and TEM. Above 25 °C, the micelles started to aggregate,

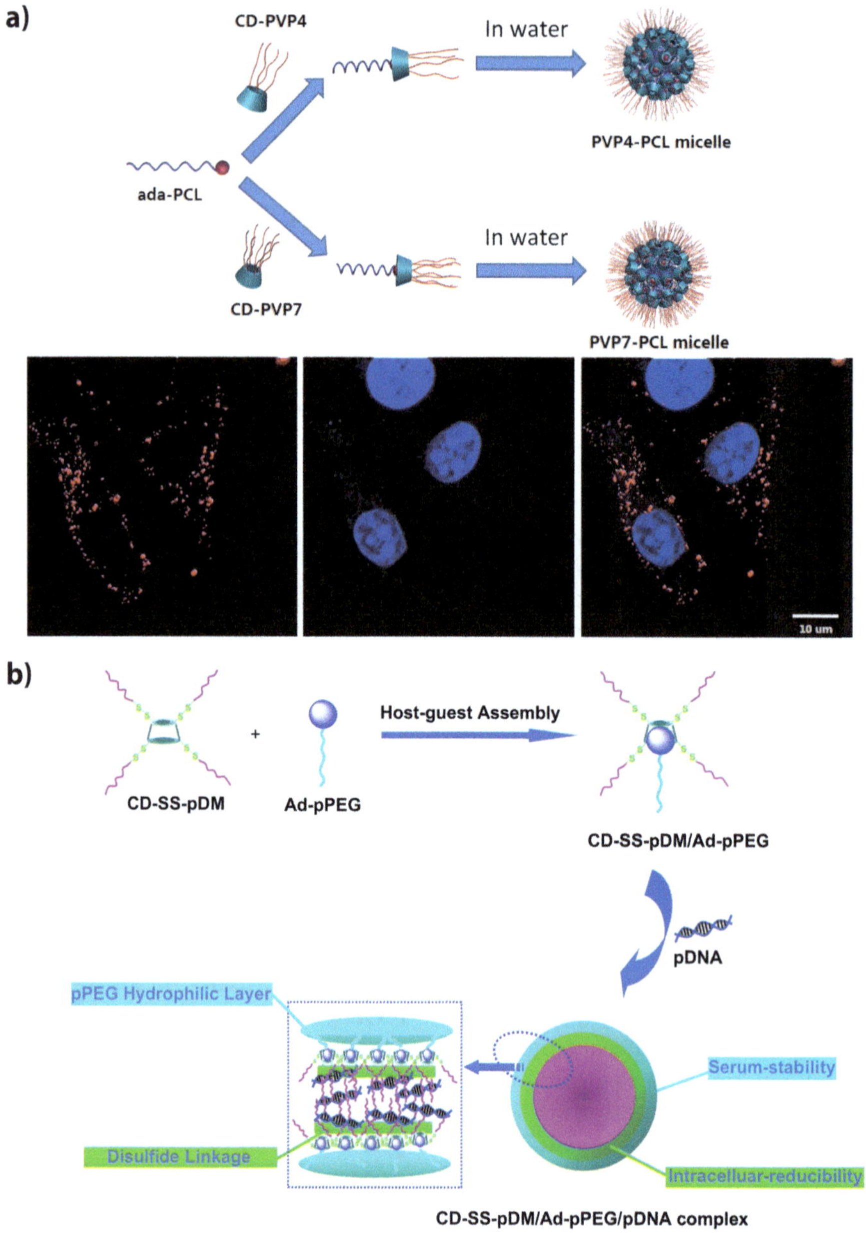


**Figure 7.7** (a) Formation of a β-CD/adamantyl based $PVP_4$-PCL or $PVP_7$-PCL miktoarm star polymers for vesicle formation, drug encapsulation and delivery; confocal laser scanning microscopy images showing vesicle internalization *in vitro*. (Adapted from ref. 131 with permission from the Royal Society of Chemistry.) (b) Formation of a bioreducible supramolecular ($PDMAEMA_4$-poly(ethylene glycol)ethyl ether methacrylate) miktoarm star polymer *via* β-CD/adamantyl interaction and polyplex formation for gene delivery. (Reprinted from ref. 133, *Biomaterials*, **34**, Yang Hu, Wei Yuan, Na–Na Zhang, Jie Ma, Wan-Tai Yang, Fu-Jian Xu, Supramolecular pseudo-block gene carriers based on bioreducible star polycations, 5411–5422, 2013, with permission from Elsevier.)

forming large particles. The inverse structure was recently described by Winnik and coworkers.[137] A double adamantyl end-functionalized PNIPAM was brought together with β-CD-centered PEO star polymers. NOESY, DOSY, and DLS studies were utilized to prove the complex formation, as well as isothermal titration calorimetry. Depending on the concentration of the building blocks, different architectures were observed in solution by DLS. At low concentrations, no complex formation was observed; at higher concentrations, dumbbell shaped complexes were formed. Heating above the LCST of PNIPAM led to aggregates of PNIPAM in the case of low polymer concentrations, while the β-CD-centered stars remained in solution. At higher concentrations, mesoglobules were formed upon heating above the LCST of the PNIPAM block. In a similar way, Wei and coworkers connected β-CD-centered PEO star polymers with double adamantyl-functionalized PS.[138] Schmidt *et al.* showed a supramolecular H-shape block copolymer in addition to the supramolecular $A_2B_2$ miktoarm star block copolymer described above.[125] The supramolecular H-shape block copolymer showed similar thermoresponsive behavior to that of the $A_2B_2$ miktoarm star block copolymer and the dumbbell shaped block copolymers reported by Li's group.[136] Two-step aggregation behavior with the formation of micellar aggregates in a defined temperature range and further agglomeration of the aggregates upon heating was found. In addition to their hPG/alkyl miktoarm star polymer,[129] Yan and coworkers presented a dumbbell shaped supramolecular block copolymer based on β-CD-centered hPG and double azobenzene end-functionalized PS.[139] The complexation was followed by UV-Vis titration measurements. Furthermore, the self-assembly in water was probed by DLS, SEM, and TEM. Finally, the formed micelles were disassembled under UV irradiation, as probed by DLS.

### 7.3.3 Other Inclusion Complexes

Besides CDs, other hosts have been utilized for star polymer formation. Su and coworkers utilized cucurbit[6]uril as the core molecule for covalent poly(acrylamide) stars *via* a free radical polymerization pathway.[140] Thus, no controlled structure was obtained and branched architectures were obtained that contained several core molecules. Nevertheless, vesicles or nanocapsules were formed depending on the polymerization conditions and were characterized by TEM and SEM. Next, butylamine hydrochloride was added to form inclusion complexes and change the properties of the obtained nanostructures. Furthermore, the synthesis was extended to another monomer yielding pH responsive behavior. A crown ether-based 3-fold homoarm star polymer was described by Gibson and coworkers.[141] A paraquat end-functionalized PS was synthesized *via* NMP as well as a 3-fold crown ether functional core molecule. The respective molecules were mixed in chloroform yielding the supramolecular 3-arm star polymers that showed yellow color due to charge transfer interactions between the electron-poor paraquat units and electron-rich crown ethers. The complex formation was monitored by viscosity studies and $^1$H NMR spectroscopy. Dendrimer functionalized pillar[5]arenes were described by Li and coworkers.[142] Pillar[5]arene was mono functionalized with a generation 2 or 3 benzyl alcohol dendrimer. Next, the dendrimers were mixed with double or triple adiponitrile-functionalized core molecules to form complex structures that were characterized by $^1$H NMR spectroscopy and DOSY. Zhou and coworkers presented a calix[4]arene-centered star polymer for photodynamic therapy.[143] PEO chains were conjugated to calix[4]arene *via* ether bridges to form a 4-arm star polymer. Next, an inclusion complex with the drug Chlorin e6 was formed and characterized by $^1$H NMR spectroscopy and a Job plot. The guest drug molecule was not completely embedded into the host. Therefore, hydrophobic groups remained on the outside leading to micelle formation, which was investigated by DLS and TEM. The micelles were applied for cytotoxicity tests *in vivo* upon photoirradiation. A decrease of cell viability was observed for the micelle system compared to the free Chlorin e6 drug. Calixarene- and resorcinarene-centered star polymers were described by Corbin and coworkers.[144] Hydroxyl groups were utilized as ROP initiators for LA. In the case of calix[4]arene, 4-arm stars were obtained, and 8-arm stars were obtained for resorcin[4]arene. Furthermore, star polymers with cleavable arms were designed. The formed star polymers were characterized by SEC, $^1$H NMR spectroscopy, MALDI-ToF, and dynamic scanning calorimetry (DSC). Rangelov and coworkers studied the biocompatibility and aggregation behavior of calix[4]arene-based star polymers.[145] EO polymerization was started from the supramolecular host leading to 4-arm star polymers. In water, aggregation was observed and studied by AFM, DLS, and SLS. Finally, cell viability in the presence of the star polymers was investigated *in vitro*. A calix[8]arene was utilized by Ten'kovtsev and coworkers and functionalized with an initiator for cationic ROP of 2-isopropyl-2-oazoline.[146] Thus, thermoresponsive calix[8]arene 8-arm stars with poly(2-isopropyl-2-oazoline) arms were formed. Later, an anthracene unit was conjugated to the end groups of the poly(oxazoline) arms in order to investigate micelle formation and dynamics by luminescence studies.[147] Furthermore, poly(methacrylic acid) (PMAA) was added to study the supramolecular interactions between the acid functions and the oxazoline units. Bisht and coworkers synthesized 4-arm (utilizing protected hydroxyls) and 8-arm PCL stars centered by a resorcin[4]arene.[148] The products were characterized by SEC and $^1$H NMR spectroscopy, and the crystallization behavior was studied in detail. Shen and coworkers also utilized resorcinarene as the core, yet for amphiphilic star block copolymers.[149] An 8-arm PCL star with a resorcin[4]arene core was formed *via* ROP, subsequently PEO was conjugated and characterized by SEC and $^1$H NMR spectroscopy. Furthermore, micelles were formed, and the drug loading and release of Indomethacin was studied. A covalently linked miktoarm star polymer based on calixarenes was synthesized by Shen and coworkers.[150] As such, a double hydroxyl/double alkyne functional calixarene was synthesized. The hydroxyl functions were utilized for ROP of CL and the alkynes were utilized for CuAAc with azide end-functionalized PEO. Finally, the self-assembly of the amphiphilic miktoarm star polymers was investigated by TEM and DLS.

Additionally, there are some examples of miktoarm star polymer formation based on inclusion complexes. Takahata and coworkers presented a sliding miktoarm star polymer (Figure 7.8).[151] A pseudo-rotaxane initiator based on a crown ether/quaternary amine was synthesized with a bulky headgroup and an initiating site for the ROP of CL. The polymerization was terminated with another bulky group in order to fix the macrocycle on the chain and obtain a rotaxane. Actually, the crown ether was functionalized before with either an alkyne, a RAFT initiator, or an ATRP initiator. Next, PEO was attached *via* CuAAc, styrene was polymerized from the polyrotaxane *via* RAFT, or MMA was polymerized *via* ATRP to obtain a miktoarm star polymer formed *via* a mechanical bond. In fact, a miktoarm star polymer was only obtained

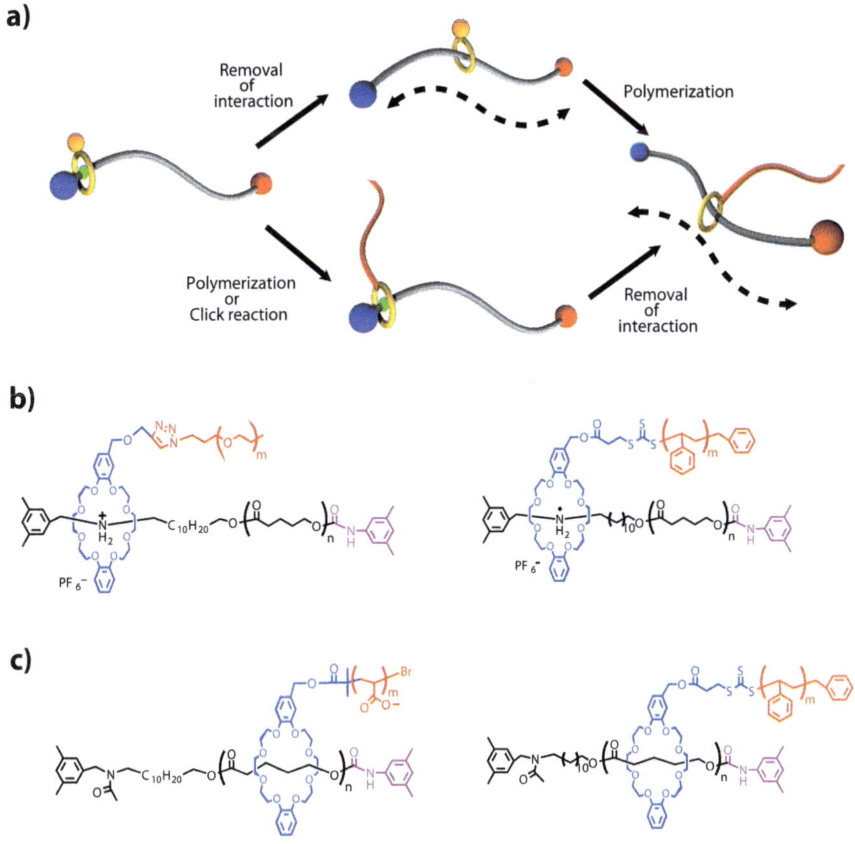


**Figure 7.8** Formation of a mechanically locked $A_2B$ miktoarm star polymer: (a) schematic overview; (b) mechanically locked diblock copolymers (PEG-*b*-PCL and PS-*b*-PCL) due to crown ether/cation interaction; (c) mechanically locked sliding $A_2B$ miktoarm star polymer ($PCL_2$-PS and $PCL_2$-PMMA) due to removal of the crown ether/cation interaction. (Adapted with permission from ref. 151. Copyright 2014 American Chemical Society.)

when the amine was protected and no interaction between the crown ether and the quaternary amine was possible. Since the quaternary amine caught the sliding crown ether on one end of the chain without protection, the formed product was rather a diblock copolymer. With a truly sliding crown ether, and thus a sliding arm over the full length, a miktoarm star was obtained. In a similar fashion, the same group synthesized a 3-arm PCL star where one arm was connected to a linear PCL *via* a mechanical bond.[152] The 3-arm star was formed due to the mid-chain quaternary ammonium functionality of the longer PCL chain and a crown ether end-functionalized PCL. Of course, all chain-ends were capped with bulky benzyl end-stoppers to keep the chains from dissociation. Interestingly, the 3-arm star could be converted to a linear polymer *via* acetylation of the mid-chain amine and removal of the quaternary amine/crown ether interaction. Therefore, sliding of the crown ether towards the end of the polymer, where a benzyl stopper was attached *via* an amide due to the preferred interaction of the crown ether with the amide, was possible. Thus, a linear polymer with mechanical links was obtained.

### 7.3.4 Metal Complexes

Metal complexation allows the formation of supramolecular star polymers in a facile fashion.[38] Since metals usually bind multiple ligands, metals can easily serve as a core for star polymers. In particular, homoarm star polymers have been formed in that regard. Very frequently, bipyridine ligands have been utilized to form 3-arm star polymers and block copolymers, *e.g.*, *via* the formation of bipyridine end-functionalized polymers in the first step and then mixing with appropriate metal ions. Several types of metal-centered 3-arm star polymers *via* bipyridine ligand end functionalized polymers have been described in such a way, *e.g.* PNIPAM[153] and PS[153,154] with $Ru^{2+}$ cores. Four-arm star polymers based on bipyridine have been synthesized as well *via* two mid-chain functionalized polymers and one non-functionalized ligand, *e.g.*, PAA,[155] PEtOx[156] or PS[157,158] with $Ru^{2+}$ cores. Furthermore, 6-arm stars were synthesized *via* bipyridine mid-chain functionalization, *e.g.*, PCL-*b*-PMMA,[159] PCL-*b*-P*t*BA,[159] PCL-*b*-PLA,[160] PEtOx,[161] PEO,[162] PEO-*b*-PCL-*b*-PLA,[163] PEtOx-*b*-poly(undecyl oxazoline),[164,165] PLA-*b*-PEO,[166] PLA-*b*-PCL,[160] PLA-*b*-PMMA,[159] PLA-*b*-P*t*BA,[159] PMMA,[167,168] PS-*b*-PMMA,[159] PS-*b*-P*t*BA,[159] with $Fe^{2+}$ cores or PMMA[167,168] with $Zn^{2+}$ cores, or PAA,[155] PCL-*b*-PLA,[160] PEtOx,[156] PLA,[155] PMMA,[168] PS[154,157,158,169] or PLA-*b*-PCL[160] with $Ru^{2+}$ cores. A thermochromic 6-arm star polymer based on oxazolines was described by Fraser and coworkers.[170] The star center was formed *via* bipyridine $Fe^{2+}$ complexes with a red-violet color. Upon heating, the complexes turned colorless. A $Ru^{2+}$-centered bipyridine PEI 6-arm star polymer was described by the same team.[171] First, an oxazoline star polymer was synthesized that was converted to PEI *via* acid hydrolysis. Furthermore, the formed star polymers were used for gene delivery studies. A 3-arm supramolecular star polymer based on $Fe^{2+}$ and dibenzoylmethane ligands was described by Fraser and coworkers as well.[172,173] This way, dibenzoyl methane PLA was synthesized and a tris complex formed with $Fe^{2+}$. Schubert and coworkers presented a 4-arm star polymer based on $Cu^+$ and 3,6-di(2-pyridyl)-pyridazines (Figure 7.9(a)).[84,174] The employed ligand was capable of forming a complex with two $Cu^+$ ions. Due to steric hindrance, four ligands formed a complex with four $Cu^+$ ions in a 2 × 2 grid. PEO chains were conjugated to the ligands *via* amidation before complex formation or the final formed complexes, leading to a 4-arm star polymer. Very recently, Gibson and coworkers presented a 3-arm poly(vinyl

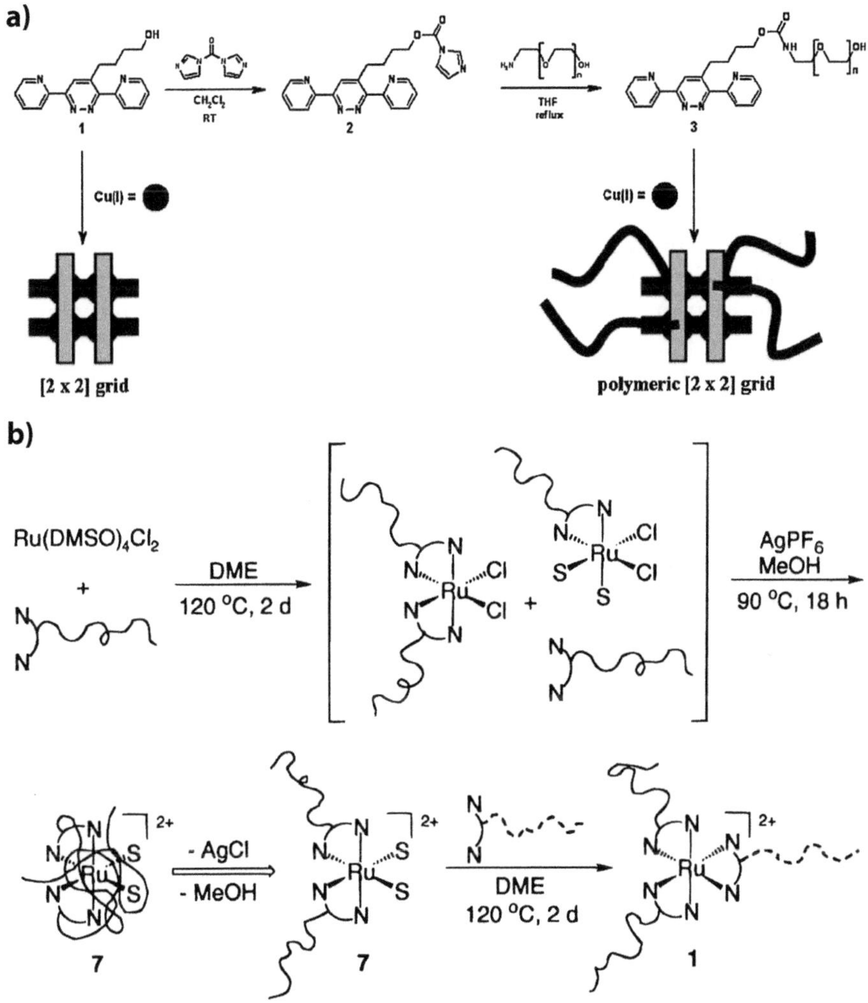


**Figure 7.9** (a) Formation of a 4-arm PEG star with a $Cu^+$/3,6-di(2-pyridyl)-pyridazine based grid-like center. (Reproduced from ref. 174 with permission of John Wiley and Sons.) (b) Formation of a $Ru^{2+}$ centered $A_2B$ miktoarm star polymer *via* sequential exchange of solvent ligands (solid line: PS; dashed line: PMMA; S: solvent, *i.e.*, DME (dimethoxyethane)). (Reprinted with permission from ref. 176. Copyright 2000 American Chemical Society.)

alcohol) (PVA) star polymer based on a $Fe^{3+}$/dopamine complex that showed protein-like antifreezing behavior in water.[175]

More complicated is the synthesis of metal complex-based miktoarm star polymers. Nevertheless, a metal centered miktoarm star polymer was described by Fraser and coworkers (Figure 7.9(b)).[176] Bipyridine end-functionalized PS and PMMA, as well as mid-chain bipyridine functionalized PS and PMMA, were synthesized *via* ATRP. The miktoarm stars were formed in a stepwise manner *via* $Ru^{2+}$ complexes. An $A_2$ polymer was formed first *via* solvent controlled reactivity at the metal center, *i.e.*, complex formation with two solvent molecules at the remaining sites. The third arm was later added *via* exchange of the ligand-functionalized polymer with the attached solvent molecules. Thus, an $A_2B$ supramolecular miktoarm star was accessible. The formation was investigated *via* SEC. Harruna and coworkers synthesized an $A_2B$ miktoarm star polymer based on bipyridine $Ru^{2+}$ complexes as well.[177] In this case, one PS arm was connected to two PNIPAM arms in aqueous solution *via* the metal complex. In addition, Fraser and coworkers synthesized an $A_3B_3$ miktoarm star polymer based on bipyridine/iron complexes.[178] Therefore, a bipyridine centered diblock copolymer was synthesized *via* a combination of ROP of CL and ATRP of styrene. Three of these ligand-centered polymers could complex with $Fe^{2+}$ to form a supramolecular miktoarm star polymer and the complexation was investigated by UV-Vis. In a similar way, the same team described miktoarm star polymers based on $Ru^{2+}$.[179] A variety of bipyridine macroligands was synthesized, *e.g.*, $PS_2$, PEG/PCL, PS/PCL, $PCL_2$, or (PCL-*b*-PLA)$_2$, and utilized for complex formation. The complexes were characterized *via* luminescence lifetime measurements and SEC. Since lanthanide ions are capable of forming complexes with high coordination numbers, they are very useful for star polymer synthesis, as shown by Bender *et al.*, who utilized dibenzoyl ligands or bipyridine ligands together with $Eu^{3+}$.[180] Therefore, dibenzoyl end-functionalized PLA and bipyridine mid-chain functionalized PCL were synthesized *via* ROP. In order to prove the formation of a miktoarm star, luminescence lifetime and AFM measurements were undertaken. Another approach based on ruthenium ions was described by Satoh and coworkers.[181] The miktoarm stars were formed *via* 2-(1*H*-1,2,3-triazol-4-yl) pyridine or 2,6-bis(1*H*-1,2,3-triazol-4-yl)pyridine ligands in a two-step procedure. First, ligand end-functionalized polymers were complexed with $Ru^{2+}$ to form a mono complex. Next, other ligand-functionalized polymers were added to form a bis complex. The number of attached chains could be controlled by the type of ligand, *i.e.*, double or triple coordinating. In this way, $A_3$ star polymers and $A_2B$ and $A_2B_2$ miktoarm star polymers were formed, *e.g.*, with PS, P*n*BA, poly(styrene oxide), poly(hexyl isocyanate), or PCL arms. The synthesis of the macro ligands was performed utilizing azide end-functionalized polymers conjugated with 2-ethynylpyridine for divalent ligands or 2,6-diethynylpyridine for trivalent ligands in a CuAAc reaction. In a similar fashion, homoarm star polymers based on poly(styrene oxide) up to an arm count of 12 were synthesized by the same team.[182]

### 7.3.5 Other Supramolecular Motifs

In addition to the already presented motifs, some other interactions have been utilized as well. Pan and coworkers worked on the π–π interactions of triphenylene with carbon nanotubes.[183] A triphenylene 6-arm star PLA was synthesized *via* ROP and the products were added to multiwalled carbon nanotubes in THF. After sonication, supramolecular complexes were formed and analyzed by a variety of methods, *e.g.*, $^1$H NMR spectroscopy, SEM, or TEM. In a similar way, Hizal and coworkers utilized pyrene-functionalized miktoarm star polymers.[184] An ABC miktoarm star polymer was formed *via* a combination of ATRP and CuAAc, namely PMMA, PEG, and PS blocks were connected in the apex. The PS block was functionalized with a pyrene unit capable of π–π interactions with carbon nanotubes. The complexes were formed in THF under sonication and, again, several characterization methods were utilized, including TEM and fluorescence spectroscopy. Lee and coworkers prepared π–π complexes from pyrene end-functionalized polymers, yielding rather undefined assemblies.[185] A supramolecular star polymer based on acid–base interactions was described by An and coworkers.[186] A 6-fold tertiary amine-functionalized core molecule was associated with an acid-functionalized RAFT agent. This supramolecular 6-fold RAFT agent was utilized in the polymerization of PS, P*t*BMA, and P*t*BMA-*b*-PS. Consecutive, the star polymers could be cleaved *via* addition of an acid. A molybdenum cluster – namely a Keplerate cluster – based star polymer was presented by Bu and coworkers.[187] Electrostatic self-assembly of the negatively charged core with ammonium end-functionalized PS led to the formation of supramolecular star polymers studied by TEM, SEM, AFM, and DLS. An europium based polyoxometalate ($EuWo_{10}O_{36}{}^{9-}$) core was utilized to form multiarm star polymers *via* interaction with ammonium end-functionalized RAFT agents that were utilized in the polymerization of styrene.[188] The association of amylose and poly(tetrahydrofuran) (PTHF) was utilized by Loos and coworkers.[189] A 3-arm star PTHF was synthesized *via* cationic polymerization and conjugated with amylose. Finally, the amylose blocks were utilized to ligate the star polymer with further PTHF homopolymers in a supramolecular fashion.

## 7.4 Conclusions and Outlook

The development of supramolecular chemistry has had a significant influence on polymer science and, in particular, on the formation of (miktoarm) star polymers. The advances in supramolecular bonding, such as reversibility, stimulus-response, orthogonality, and dynamics, are a welcome addition to the properties for star polymers. Since a broad variety of supramolecular motifs are available, a huge amount of possibilities exist for polymer chemists to form intended star polymers with desired properties. In that regard, supramolecular chemistry functions as a toolbox with hydrogen bonding, inclusion complexes, and metal complexes as the tools, which allow for the choice of the right building block for the envisaged application. Together with RDRP methods and some other controlled polymerizations, supramolecular chemistry has been used for a plethora of star architectures and star compositions. Certainly, multi homoarm star polymers have been formed *via* supramolecular interactions, mainly *via* multifunctional supramolecular motifs. There are also a lot of covalent-bound star polymers with supramolecular recognition motifs as core molecules. In the realm of miktoarm star polymers, supramolecular bound assemblies have been formed *via* a clever combination of supramolecular motifs with defined polymers, in particular *via* formation of well-defined end-functionalized polymers. The formed architectures cover $A_3$ star polymers over $A_3B_3$ miktoarm star polymers to ABC miktoarm star polymers and even more complicated star polymers with a different scope. The presented studies comprise the fundamental work to prove the formation of the respective supramolecular star architectures and the examination/utilization of the physical properties as well as the formation of nano- or microstructures, *e.g.*, micelles, vesicles or nanoparticles. On the other hand, research with supramolecular star polymers has developed in the direction of applications, mostly in the direction of drug and gene delivery.

In the future, supramolecular (miktoarm) star polymers will be designed more and more for specific applications. Certainly, drug delivery and biomedical applications will have a significant role in that regard. Nevertheless, other directions are most likely possible, *e.g.*, in the field of organic electronics and surface coatings. A rather weakly investigated field is the self-assembly of supramolecular star polymers in bulk/thin films, as it has been sparsely done with supramolecular diblock copolymers so far.[190,191] There are significant new and interesting properties to expect with covalent bound miktoarm star polymers.[192,193] Another underrepresented field is the combination of different supramolecular motifs. Although a variety of orthogonal supramolecular motifs exists, the combination of different recognition units has not found a lot of application in the formation of miktoarm star polymers so far. In particular, the utilization of orthogonal motifs would allow the formation of complex miktoarm star polymers. In summary, the field of supramolecular (miktoarm) star polymers shows a broad variety of different polymer compositions and structures. Certainly, more complex structures will be prepared in the near future and new applications will be targeted, as well as further advances on the existing structures and applications.

## Acknowledgements

Support from the Max-Planck society is gratefully acknowledged.

## References

1. J. M. Lehn, *Angew. Chem., Int. Ed.*, 1990, **29**, 1304–1319.
2. J.-M. Lehn, *Proc. Natl. Acad. Sci.*, 2002, **99**, 4763–4768.
3. O. Ikkala and G. ten Brinke, *Science*, 2002, **295**, 2407–2409.

4. L. Brunsveld, B. Folmer, E. Meijer and R. Sijbesma, *Chem. Rev.*, 2001, **101**, 4071–4098.
5. P. Cordier, F. Tournilhac, C. Soulié-Ziakovic and L. Leibler, *Nature*, 2008, **451**, 977–980.
6. S.-L. Li, T. Xiao, C. Lin and L. Wang, *Chem. Soc. Rev.*, 2012, **41**, 5950–5968.
7. H. C. Kolb, M. Finn and K. B. Sharpless, *Angew. Chem., Int. Ed.*, 2001, **40**, 2004–2021.
8. K. Kempe, A. Krieg, C. R. Becer and U. S. Schubert, *Chem. Soc. Rev.*, 2012, **41**, 176–191.
9. O. Altintas, A. P. Vogt, C. Barner-Kowollik and U. Tunca, *Polym. Chem.*, 2012, **3**, 34–45.
10. K. Matyjaszewski, *Macromolecules*, 2012, **45**, 4015–4039.
11. C. J. Hawker, A. W. Bosman and E. Harth, *Chem. Rev.*, 2001, **101**, 3661–3688.
12. J. Chiefari, Y. Chong, F. Ercole, J. Krstina, J. Jeffery, T. P. Le, R. T. Mayadunne, G. F. Meijs, C. L. Moad and G. Moad, *Macromolecules*, 1998, **31**, 5559–5562.
13. C. Barner-Kowollik, *Handbook of RAFT Polymerization*, John Wiley & Sons, 2008.
14. C. Barner-Kowollik, F. E. Du Prez, P. Espeel, C. J. Hawker, T. Junkers, H. Schlaad and W. Van Camp, *Angew. Chem., Int. Ed.*, 2011, **50**, 60–62.
15. J. F. Lutz, *Angew. Chem., Int. Ed.*, 2007, **46**, 1018–1025.
16. A. B. Lowe, *Polym. Chem.*, 2010, **1**, 17–36.
17. M. A. Tasdelen, *Polym. Chem.*, 2011, **2**, 2133–2145.
18. M. Glassner, G. Delaittre, M. Kaupp, J. P. Blinco and C. Barner-Kowollik, *J. Am. Chem. Soc.*, 2012, **134**, 7274–7277.
19. I. Singh, Z. Zarafshani, J.-F. Lutz and F. Heaney, *Macromolecules*, 2009, **42**, 5411–5413.
20. D. C. Sherrington and K. A. Taskinen, *Chem. Soc. Rev.*, 2001, **30**, 83–93.
21. F. H. Beijer, R. P. Sijbesma, H. Kooijman, A. L. Spek and E. W. Meijer, *J. Am. Chem. Soc.*, 1998, **120**, 6761–6769.
22. S. K. Chang and A. D. Hamilton, *J. Am. Chem. Soc.*, 1988, **110**, 1318–1319.
23. B. V. K. J. Schmidt, M. Hetzer, H. Ritter and C. Barner-Kowollik, *Prog. Polym. Sci.*, 2014, **39**, 235–249.
24. S. Tan, K. Ladewig, Q. Fu, A. Blencowe and G. G. Qiao, *Macromol. Rapid Commun.*, 2014, **35**, 1166–1184.
25. G. Chen and M. Jiang, *Chem. Soc. Rev.*, 2011, **40**, 2254–2266.
26. M. V. Rekharsky and Y. Inoue, *Chem. Rev.*, 1998, **98**, 1875–1918.
27. F. Szillat, B. V. K. J. Schmidt, A. Hubert, C. Barner-Kowollik and H. Ritter, *Macromol. Rapid Commun.*, 2014, **35**, 1293–1300.
28. Q. Yan, J. Yuan, Z. Cai, Y. Xin, Y. Kang and Y. Yin, *J. Am. Chem. Soc.*, 2010, **132**, 9268–9270.
29. H. Zhang, X. Fan, R. Suo, H. Li, Z. Yang, W. Zhang, Y. Bai, H. Yao and W. Tian, *Chem. Commun.*, 2015, **51**, 15366–15369.
30. B. V. K. J. Schmidt, M. Hetzer, H. Ritter and C. Barner-Kowollik, *Macromolecules*, 2013, **46**, 1054–1065.

31. Z. Zhang, J. Ding, X. Chen, C. Xiao, C. He, X. Zhuang, L. Chen and X. Chen, *Polym. Chem.*, 2013, **4**, 3265–3271.
32. S. J. Barrow, S. Kasera, M. J. Rowland, J. del Barrio and O. A. Scherman, *Chem. Rev.*, 2015, **115**, 12320–12406.
33. F. Biedermann and O. A. Scherman, *J. Phys. Chem. B*, 2012, **116**, 2842–2849.
34. B. Zheng, F. Wang, S. Dong and F. Huang, *Chem. Soc. Rev.*, 2012, **41**, 1621–1636.
35. Z. Niu, F. Huang and H. W. Gibson, *J. Am. Chem. Soc.*, 2011, **133**, 2836–2839.
36. M. Xue, Y. Yang, X. Chi, Z. Zhang and F. Huang, *Acc. Chem. Res.*, 2012, **45**, 1294–1308.
37. D.-S. Guo and Y. Liu, *Chem. Soc. Rev.*, 2012, **41**, 5907–5921.
38. U. S. Schubert and C. Eschbaumer, *Angew. Chem., Int. Ed.*, 2002, **41**, 2892–2926.
39. J. Willenbacher, O. Altintas, V. Trouillet, N. Knöfel, M. J. Monteiro, P. W. Roesky and C. Barner-Kowollik, *Polym. Chem.*, 2015, **6**, 4358–4365.
40. J. Willenbacher, O. Altintas, P. W. Roesky and C. Barner-Kowollik, *Macromol. Rapid Commun.*, 2014, **35**, 45–51.
41. S. Coulibaly, C. Heinzmann, F. L. Beyer, S. Balog, C. Weder and G. L. Fiore, *Macromolecules*, 2014, **47**, 8487–8496.
42. P. Wei, X. Yan and F. Huang, *Chem. Soc. Rev.*, 2015, **44**, 815–832.
43. M. M. Safont-Sempere, G. Fernández and F. Würthner, *Chem. Rev.*, 2011, **111**, 5784–5814.
44. A. Wu and L. Isaacs, *J. Am. Chem. Soc.*, 2003, **125**, 4831–4835.
45. P. Mukhopadhyay, A. Wu and L. Isaacs, *J. Org. Chem.*, 2004, **69**, 6157–6164.
46. O. Altintas, E. Lejeune, P. Gerstel and C. Barner-Kowollik, *Polym. Chem.*, 2012, **3**, 640–651.
47. S. K. Yang, A. V. Ambade and M. Weck, *J. Am. Chem. Soc.*, 2010, **132**, 1637–1645.
48. T. Mes, M. M. E. Koenigs, V. F. Scalfani, T. S. Bailey, E. W. Meijer and A. R. A. Palmans, *ACS Macro Lett.*, 2012, **1**, 105–109.
49. Y. Guan, M. Ni, X. Hu, T. Xiao, S. Xiong, C. Lin and L. Wang, *Chem. Commun.*, 2012, **48**, 8529–8531.
50. X.-Z. Wang, X.-Q. Li, X.-B. Shao, X. Zhao, P. Deng, X.-K. Jiang, Z.-T. Li and Y.-Q. Chen, *Chem.–Eur. J.*, 2003, **9**, 2904–2913.
51. Q. Wang, Y. Chen and Y. Liu, *Polym. Chem.*, 2013, **4**, 4192–4198.
52. Q. Wang, M. Cheng, Y. Zhao, L. Wu, J. Jiang, L. Wang and Y. Pan, *Chem. Commun.*, 2015, **51**, 3623–3626.
53. F. Wang, C. Han, C. He, Q. Zhou, J. Zhang, C. Wang, N. Li and F. Huang, *J. Am. Chem. Soc.*, 2008, **130**, 11254–11255.
54. H. Li, X. Fan, M. Qi, Z. Yang, H. Zhang and W. Tian, *Chem.–Eur. J.*, 2015, **22**, 101–105.
55. H. Hofmeier, R. Hoogenboom, M. E. L. Wouters and U. S. Schubert, *J. Am. Chem. Soc.*, 2005, **127**, 2913–2921.

56. P. Wang, H. Xing, D. Xia and X. Ji, *Chem. Commun.*, 2015, 17431–17434.
57. H. Hofmeier, J. Pahnke, C. H. Weidl and U. S. Schubert, *Biomacromolecules*, 2004, **5**, 2055–2064.
58. A. J. Wilson, *Soft Matter*, 2007, **3**, 409–425.
59. E. M. Todd and S. C. Zimmerman, *J. Am. Chem. Soc.*, 2007, **129**, 14534–14535.
60. E. M. Todd and S. C. Zimmerman, *Tetrahedron*, 2008, **64**, 8558–8570.
61. J. Bernard, F. Lortie and B. Fenet, *Macromol. Rapid Commun.*, 2009, **30**, 83–88.
62. S. Averick, E. Paredes, W. Li, K. Matyjaszewski and S. R. Das, *Bioconjugate Chem.*, 2011, **22**, 2030–2037.
63. O. Altintas, T. Muller, E. Lejeune, O. Plietzsch, S. Bräse and C. Barner-Kowollik, *Macromol. Rapid Commun.*, 2012, **33**, 977–983.
64. A. Likhitsup, S. Yu, Y.-H. Ng, C. L. L. Chai and E. K. W. Tam, *Chem. Commun.*, 2009, 4070–4072.
65. I. Gadwal, S. De, M. C. Stuparu, S. G. Jang, R. J. Amir and A. Khan, *J. Polym. Sci., Part A: Polym. Chem.*, 2012, **50**, 2415–2420.
66. I. Gadwal, S. De, M. C. Stuparu and A. Khan, *Polym. Chem.*, 2012, **3**, 2615–2618.
67. I. Gadwal, S. De, M. C. Stuparu, R. J. Amir, S. G. Jang and A. Khan, *J. Polym. Sci., Part A: Polym. Chem.*, 2012, **50**, 1844–1850.
68. S. Chen, A. Bertrand, X. Chang, P. Alcouffe, C. Ladavière, J.-F. Gérard, F. Lortie and J. Bernard, *Macromolecules*, 2010, **43**, 5981–5988.
69. S.-H. Lee, M. Ouchi and M. Sawamoto, *Macromolecules*, 2012, **45**, 3702–3710.
70. O. Altintas, U. Tunca and C. Barner-Kowollik, *Polym. Chem.*, 2011, **2**, 1146–1155.
71. O. Altintas, D. Schulze-Suenninghausen, B. Luy and C. Barner-Kowollik, *Eur. Polym. J.*, 2015, **62**, 409–417.
72. O. Altintas, D. Schulze-Suenninghausen, B. Luy and C. Barner-Kowollik, *ACS Macro Lett.*, 2013, **2**, 211–216.
73. H. Dou, L. Hong and G. Liu, *Macromolecules*, 2010, **43**, 4629–4637.
74. H. Hu and G. Liu, *Macromolecules*, 2014, **47**, 5096–5103.
75. A. B. Marciel, D. J. Mai and C. M. Schroeder, *Macromolecules*, 2015, **48**, 1296–1303.
76. T. R. Wilks, J. Bath, J. W. de Vries, J. E. Raymond, A. Herrmann, A. J. Turberfield and R. K. O'Reilly, *ACS Nano*, 2013, **7**, 8561–8572.
77. K. Ohno, B. Wong and D. M. Haddleton, *J. Polym. Sci., Part A: Polym. Chem.*, 2001, **39**, 2206–2214.
78. M. H. Stenzel-Rosenbaum, T. P. Davis, V. Chen and A. G. Fane, *Macromolecules*, 2001, **34**, 5433–5438.
79. T. He, T. Hu, X. Zhang, G. Zhong and H. Zhang, *J. Appl. Polym. Sci.*, 2009, **112**, 2120–2126.
80. L. Zhang and M. H. Stenzel, *Aust. J. Chem.*, 2009, **62**, 813–822.
81. M. H. Stenzel and T. P. Davis, *J. Polym. Sci., Part A: Polym. Chem.*, 2002, **40**, 4498–4512.

82. C. Yang, H. Li, S. H. Goh and J. Li, *Biomaterials*, 2007, **28**, 3245–3254.
83. P. Chmielarz, S. Park, A. Sobkowiak and K. Matyjaszewski, *Polymer*, **88**, 36–42.
84. R. Hoogenboom, B. C. Moore and U. S. Schubert, *Chem. Commun.*, 2006, 4010–4012.
85. C. Huin, Z. Eskandani, N. Badi, A. Farcas, V. Bennevault-Celton and P. Guégan, *Carbohydr. Polym.*, 2013, **94**, 323–331.
86. M. Hu, Y. Shen, L. Zhang and L. Qiu, *Biomacromolecules*, 2016, **17**, 1026–1039.
87. M. W. M. Fijten, C. Haensch, B. M. van Lankvelt, R. Hoogenboom and U. S. Schubert, *Macromol. Chem. Phys.*, 2008, **209**, 1887–1895.
88. J. Xu and S. Liu, *J. Polym. Sci., Part A: Polym. Chem.*, 2009, **47**, 404–419.
89. A. Wycisk, A. Döring, M. Schneider, M. Schönhoff and D. Kuckling, *Polymers*, 2015, **7**, 921.
90. H. Zhang, Q. Yan, Y. Kang, L. Zhou, H. Zhou, J. Yuan and S. Wu, *Polymer*, 2012, **53**, 3719–3725.
91. Y. Li, H. Guo, J. Zheng, J. Gan, K. Wu and M. Lu, *Colloids Surf., A*, 2015, **471**, 178–189.
92. K. Karaky, S. Reynaud, L. Billon, J. François and Y. Chreim, *J. Polym. Sci., Part A: Polym. Chem.*, 2005, **43**, 5186–5194.
93. H. Yang, J. Liu, D. Hu and Z. Lei, *Biotechnol. Appl. Biochem.*, 2016, DOI: 10.1002/bab.1474.
94. C. Teuchert, C. Michel, F. Hausen, D.-Y. Park, H. W. Beckham and G. Wenz, *Macromolecules*, 2013, **46**, 2–7.
95. B. Liang, J. J. Deng, F. Yuan, N. Yang, W. Li, J. R. Yin, S. X. Pu, L. C. Xie, C. Gao and L. M. Zhang, *Carbohydr. Polym.*, 2013, **94**, 185–192.
96. Y. Y. Durmaz, Y.-L. Lin and M. E. H. ElSayed, *Adv. Funct. Mater.*, 2013, **23**, 3885–3895.
97. F. Zhao, H. Yin, Z. Zhang and J. Li, *Biomacromolecules*, 2013, **14**, 476–484.
98. X. J. Loh and Y.-L. Wu, *Chem. Commun.*, 2015, **51**, 10815–10818.
99. Z. Xu, S. Liu, H. Liu, C. Yang, Y. Kang and M. Wang, *Chem. Commun.*, 2015, **51**, 15768–15771.
100. J. Zhu, K. L. Liu, Y. Wen, X. Song and J. Li, *Nanoscale*, 2016, **8**, 1332–1337.
101. Y. Miura, A. Narumi, S. Matsuya, T. Satoh, Q. Duan, H. Kaga and T. Kakuchi, *J. Polym. Sci., Part A: Polym. Chem.*, 2005, **43**, 4271–4279.
102. P.-F. Gou, W.-P. Zhu and Z.-Q. Shen, *Biomacromolecules*, 2010, **11**, 934–943.
103. P.-F. Gou, W.-P. Zhu, N. Xu and Z.-Q. Shen, *J. Polym. Sci., Part A: Polym. Chem.*, 2010, **48**, 2961–2974.
104. Q. Zhang, G.-Z. Li, C. R. Becer and D. M. Haddleton, *Chem. Commun.*, 2012, **48**, 8063–8065.
105. B. Huang, M. Chen, S. Zhou and L. Wu, *Polym. Chem.*, 2015, **6**, 3913–3917.
106. W. Tian, X. Lv, C. Mu, W. Zhang, J. Kong, Y. Liu and X. Fan, *J. Polym. Sci., Part A: Polym. Chem.*, 2012, **50**, 759–771.
107. C.-G. Mu, X.-d. Fan, W. Tian, Y. Bai and X. Zhou, *Polym. Chem.*, 2012, **3**, 1137–1149.

108. C.-G. Mu, X.-D. Fan, W. Tian, Y. Bai, Z. Yang, H. Yao and H. Chen, *J. Polym. Sci., Part A: Polym. Chem.*, 2013, **51**, 1405–1416.
109. M. L. Bondì, A. Scala, G. Sortino, E. Amore, C. Botto, A. Azzolina, D. Balasus, M. Cervello and A. Mazzaglia, *Biomacromolecules*, 2015, **16**, 3784–3791.
110. H. Yin, F. Zhao, D. Zhang and J. Li, *Int. J. Pharm.*, 2015, **483**, 169–179.
111. B. V. K. J. Schmidt, T. Rudolph, M. Hetzer, H. Ritter, F. H. Schacher and C. Barner-Kowollik, *Polym. Chem.*, 2012, **3**, 3139–3145.
112. C. Fleischmann, H. Wöhlk and H. Ritter, *Beilstein J. Org. Chem.*, 2014, **10**, 2263–2269.
113. M. Hetzer, C. Fleischmann, B. V. K. J. Schmidt, C. Barner-Kowollik and H. Ritter, *Polymer*, 2013, **54**, 5141–5147.
114. K. Wei, J. Li, G. Chen and M. Jiang, *ACS Macro Lett.*, 2013, **2**, 278–283.
115. G. Maatz, A. Maciollek and H. Ritter, *Beilstein J. Org. Chem.*, 2012, **8**, 1929–1935.
116. H. Hu, H.-Q. Song, B.-R. Yu, Q. Cai, Y. Zhu and F.-J. Xu, *Polym. Chem.*, 2015, **6**, 2466–2477.
117. J. Liu, G. Chen, M. Guo and M. Jiang, *Macromolecules*, 2010, **43**, 8086–8093.
118. J. Leng, X. Wen, J. Rao, G. Zou and Q. Zhang, *Polymer*, 2012, **53**, 1543–1550.
119. M. Guo, M. Jiang, S. Pispas, W. Yu and C. Zhou, *Macromolecules*, 2008, **41**, 9744–9749.
120. C. Luo, F. Zuo, Z. Zheng, X. Cheng, X. Ding and Y. Peng, *Macromol. Rapid Commun.*, 2008, **29**, 149–154.
121. K. Wei, J. Li, J. Liu, G. Chen and M. Jiang, *Soft Matter*, 2012, **8**, 3300–3303.
122. A. F. Hirschbiel, B. V. K. J. Schmidt, P. Krolla-Sidenstein, J. P. Blinco and C. Barner-Kowollik, *Macromolecules*, 2015, **48**, 4410–4420.
123. X. Huan, D. Wang, R. Dong, C. Tu, B. Zhu, D. Yan and X. Zhu, *Macromolecules*, 2012, **45**, 5941–5947.
124. B. V. K. J. Schmidt, M. Hetzer, H. Ritter and C. Barner-Kowollik, *Polym. Chem.*, 2012, **3**, 3064–3067.
125. B. V. K. J. Schmidt and C. Barner-Kowollik, *Polym. Chem.*, 2014, **5**, 2461–2472.
126. Z. Tian, C. Chen and H. R. Allcock, *Macromolecules*, 2014, **47**, 1065–1072.
127. J. Wang, X. Wang, F. Yang, H. Shen, Y. You and D. Wu, *Langmuir*, 2014, **30**, 13014–13020.
128. J. Wang, X. Wang, F. Yang, H. Shen, Y. You and D. Wu, *Langmuir*, 2015, **31**, 13834–13841.
129. W. Tao, Y. Liu, B. Jiang, S. Yu, W. Huang, Y. Zhou and D. Yan, *J. Am. Chem. Soc.*, 2012, **134**, 762–764.
130. D. Yong, Y. Luo, F. Du, J. Huang, W. Lu, Z. Dai, J. Yu and S. Liu, *Colloids Surf., B*, 2013, **105**, 31–36.
131. C. Xie, P. Zhang, Z. Zhang, C. Yang, J. Zhang, W. Wu and X. Jiang, *Nanoscale*, 2015, **7**, 12572–12580.

132. Y. Kong, W. Li, Q. Mao and Y. Wang, *Macromol. Chem. Phys.*, 2015, **216**, 1507–1515.
133. Y. Hu, W. Yuan, N.-N. Zhao, J. Ma, W.-T. Yang and F.-J. Xu, *Biomaterials*, 2013, **34**, 5411–5422.
134. Y.-Y. Liu, S. Lan and L.-Q. Xiao, *Macromol. Chem. Phys.*, 2015, **216**, 749–760.
135. Z.-X. Zhang, X. Liu, F. J. Xu, X. J. Loh, E.-T. Kang, K.-G. Neoh and J. Li, *Macromolecules*, 2008, **41**, 5967–5970.
136. Z.-X. Zhang, K. L. Liu and J. Li, *Macromolecules*, 2011, **44**, 1182–1193.
137. V. Bennevault, C. Huin, P. Guegan, K. Evgeniya, X. P. Qiu and F. M. Winnik, *Soft Matter*, 2015, **11**, 6432–6443.
138. L. Xu, X. Liang, L. Zhang, J. Wu, Z. Li, M. Yu and L. Wei, *Colloid Polym. Sci.*, 2015, **294**, 145–155.
139. D. Zhang, Y. Fan, H. Li, K. Li, Y. Yao, Y. Zhou and D. Yan, *RSC Adv.*, 2015, **5**, 47762–47765.
140. X. Huang, F. Hu and H. Su, *Macromolecules*, 2013, **46**, 1274–1282.
141. F. Huang, D. S. Nagvekar, C. Slebodnick and H. W. Gibson, *J. Am. Chem. Soc.*, 2005, **127**, 484–485.
142. Q. Li, K. Han, J. Li, X. Jia and C. Li, *Tetrahedron Lett.*, 2015, **56**, 3826–3829.
143. C. Tu, L. Zhu, P. Li, Y. Chen, Y. Su, D. Yan, X. Zhu and G. Zhou, *Chem. Commun.*, 2011, **47**, 6063–6065.
144. R. D. Dria, B. A. Goudy, K. A. Moga and P. S. Corbin, *Polym. Chem.*, 2012, **3**, 2070–2081.
145. D. Momekova, D. Budurova, E. Drakalska, S. Shenkov, G. Momekov, B. Trzebicka, N. Lambov, E. Tashev and S. Rangelov, *Int. J. Pharm.*, 2012, **436**, 410–417.
146. A. V. Ten'kovtsev, A. E. Trofimov and L. I. Shcherbinskaya, *Polym. Sci., Ser. B*, 2012, **54**, 142–148.
147. T. N. Nekrasova, V. D. Pautov, T. D. Anan'eva, R. T. Imanbaev, R. Y. Smyslov and A. V. Ten'kovtsev, *Polym. Sci., Ser. A*, 2015, **57**, 6–12.
148. R. Wu, T. F. Al-Azemi and K. S. Bisht, *RSC Adv.*, 2014, **4**, 16864–16870.
149. C. Gao, Y. Wang, W.-p. Zhu and Z.-q. Shen, *Chin. J. Polym. Sci.*, 2014, **32**, 1431–1441.
150. P.-F. Gou, W.-P. Zhu and Z.-Q. Shen, *J. Polym. Sci., Part A: Polym. Chem.*, 2010, **48**, 5643–5651.
151. D. Aoki, S. Uchida and T. Takata, *ACS Macro Lett.*, 2014, **3**, 324–328.
152. D. Aoki, S. Uchida and T. Takata, *Angew. Chem., Int. Ed.*, 2015, **54**, 6770–6774.
153. G. Zhou, J. He and I. I. Harruna, *J. Polym. Sci., Part A: Polym. Chem.*, 2007, **45**, 4225–4239.
154. X. Wu and C. L. Fraser, *Macromolecules*, 2000, **33**, 4053–4060.
155. R. M. Johnson and C. L. Fraser, *Biomacromolecules*, 2004, **5**, 580–588.
156. J. E. McAlvin and C. L. Fraser, *Macromolecules*, 1999, **32**, 6925–6932.
157. X. Wu, J. E. Collins, J. E. McAlvin, R. W. Cutts and C. L. Fraser, *Macromolecules*, 2001, **34**, 2812–2821.

158. X. Wu and C. L. Fraser, *Macromolecules*, 2000, **33**, 7776–7785.
159. R. M. Johnson and C. L. Fraser, *Macromolecules*, 2004, **37**, 2718–2727.
160. P. S. Corbin, M. P. Webb, J. E. McAlvin and C. L. Fraser, *Biomacromolecules*, 2001, **2**, 223–232.
161. J. J. S. Lamba and C. L. Fraser, *J. Am. Chem. Soc.*, 1997, **119**, 1801–1802.
162. A. Pfister and C. L. Fraser, *Biomacromolecules*, 2006, **7**, 459–468.
163. G. L. Fiore and C. L. Fraser, *Macromolecules*, 2008, **41**, 7892–7897.
164. C. Park, J. E. McAlvin, C. L. Fraser and E. L. Thomas, *Chem. Mater.*, 2002, **14**, 1225–1230.
165. J. E. McAlvin and C. L. Fraser, *Macromolecules*, 1999, **32**, 1341–1347.
166. G. L. Fiore, J. L. Klinkenberg and C. L. Fraser, *Macromolecules*, 2008, **41**, 9397–9405.
167. L. Viau, M. Even, O. Maury, D. M. Haddleton and H. Le Bozec, *Macromol. Rapid Commun.*, 2003, **24**, 630–635.
168. L. Viau, M. Even, O. Maury, D. M. Haddleton and H. Le Bozec, *C. R. Chim.*, 2005, **8**, 1298–1307.
169. S. J. Payne, G. L. Fiore, C. L. Fraser and J. N. Demas, *Anal. Chem.*, 2010, **82**, 917–921.
170. J. E. McAlvin, S. B. Scott and C. L. Fraser, *Macromolecules*, 2000, **33**, 6953–6964.
171. G. L. Fiore, J. M. Edwards, S. J. Payne, J. L. Klinkenberg, D. G. Gioeli, J. N. Demas and C. L. Fraser, *Biomacromolecules*, 2007, **8**, 2829–2835.
172. J. L. Gorczynski, J. Chen and C. L. Fraser, *J. Am. Chem. Soc.*, 2005, **127**, 14956–14957.
173. J. Chen, J. L. Gorczynski, G. Zhang and C. L. Fraser, *Macromolecules*, 2010, **43**, 4909–4920.
174. R. Hoogenboom, B. C. Moore and U. S. Schubert, *Macromol. Rapid Commun.*, 2010, **31**, 840–845.
175. D. J. Phillips, T. R. Congdon and M. I. Gibson, *Polym. Chem.*, 2016, **7**, 1701–1704.
176. C. L. Fraser, A. P. Smith and X. Wu, *J. Am. Chem. Soc.*, 2000, **122**, 9026–9027.
177. G. Zhou, J. He and I. I. Harruna, *J. Polym. Sci., Part A: Polym. Chem.*, 2007, **45**, 4204–4210.
178. A. P. Smith and C. L. Fraser, *Macromolecules*, 2002, **35**, 594–596.
179. A. P. Smith and C. L. Fraser, *Macromolecules*, 2003, **36**, 5520–5525.
180. J. L. Bender, P. S. Corbin, C. L. Fraser, D. H. Metcalf, F. S. Richardson, E. L. Thomas and A. M. Urbas, *J. Am. Chem. Soc.*, 2002, **124**, 8526–8527.
181. N. Xiao, Y. Chen, X. Shen, C. Zhang, S. Yano, M. Gottschaldt, U. S. Schubert, T. Kakuchi and T. Satoh, *Polym. J.*, 2013, **45**, 216–225.
182. Y. Chen, N. Xiao, T. Satoh and T. Kakuchi, *Polym. Chem.*, 2014, **5**, 4993–5001.
183. L.-P. Yang and C.-Y. Pan, *Macromol. Chem. Phys.*, 2008, **209**, 783–793.
184. H. Durmaz, A. Dag, U. Tunca and G. Hizal, *J. Polym. Sci., Part A: Polym. Chem.*, 2012, **50**, 2406–2414.

185. M.-C. Li, R.-M. Ho and Y.-D. Lee, *J. Mater. Chem. C*, 2013, **1**, 1601–1606.
186. K. Tao, D. Lu, R. Bai, H. Li and L. An, *Macromol. Rapid Commun.*, 2008, **29**, 1477–1483.
187. Q. Zhang, L. He, H. Wang, C. Zhang, W. Liu and W. Bu, *Chem. Commun.*, 2012, **48**, 7067–7069.
188. D. Li, H. Li and L. Wu, *Polym. Chem.*, 2014, **5**, 1930–1937.
189. R. Rachmawati, H. D. de Gier, A. J. J. Woortman and K. Loos, *Macromol. Chem. Phys.*, 2015, **216**, 1091–1102.
190. D. Montarnal, N. Delbosc, C. Chamignon, M.-A. Virolleaud, Y. Luo, C. J. Hawker, E. Drockenmuller and J. Bernard, *Angew. Chem., Int. Ed.*, 2015, **54**, 11117–11121.
191. L. M. Pitet, A. H. M. van Loon, E. J. Kramer, C. J. Hawker and E. W. Meijer, *ACS Macro Lett.*, 2013, **2**, 1006–1010.
192. Z. Li, M. A. Hillmyer and T. P. Lodge, *Langmuir*, 2006, **22**, 9409–9417.
193. K. Yamauchi, K. Takahashi, H. Hasegawa, H. Iatrou, N. Hadjichristidis, T. Kaneko, Y. Nishikawa, H. Jinnai, T. Matsui, H. Nishioka, M. Shimizu and H. Furukawa, *Macromolecules*, 2003, **36**, 6962–6966.

# Subject Index